# CAMBRIDGE LIBRARY COLLECTION

*Books of enduring scholarly value*

## Earth Sciences

In the nineteenth century, geology emerged as a distinct academic discipline. It pointed the way towards the theory of evolution, as scientists including Gideon Mantell, Adam Sedgwick, Charles Lyell and Roderick Murchison began to use the evidence of minerals, rock formations and fossils to demonstrate that the earth was older by millions of years than the conventional, Bible-based wisdom had supposed. They argued convincingly that the climate, flora and fauna of the distant past could be deduced from geological evidence. Volcanic activity, the formation of mountains, and the action of glaciers and rivers, tides and ocean currents also became better understood. This series includes landmark publications by pioneers of the modern earth sciences, who advanced the scientific understanding of our planet and the processes by which it is constantly re-shaped.

## A Long Life's Work

Despite never graduating from university, Sir Archibald Geikie (1835–1924) forged an exceptionally successful scientific career. In 1855 he was appointed to the Scottish branch of the Geological Survey, and by 1882 was Director General of the Survey. In keeping with his Edinburgh beginnings, most of his career was spent studying igneous rocks. He was a prolific and gifted writer, producing textbooks, popular science books and biographical and historical works, including the influential *Founders of Geology* (1897), as well as numerous technical publications. The only geologist to hold the post of President of The Royal Society (1908–12), he also served as President of the Geological Society of London and the British Association, and received an array of honorary degrees and medals. This autobiography, published in the year of his death, provides a readable, personal account of the life of one of the great scientific figures of the nineteenth century.

Cambridge University Press has long been a pioneer in the reissuing of out-of-print titles from its own backlist, producing digital reprints of books that are still sought after by scholars and students but could not be reprinted economically using traditional technology. The Cambridge Library Collection extends this activity to a wider range of books which are still of importance to researchers and professionals, either for the source material they contain, or as landmarks in the history of their academic discipline.

Drawing from the world-renowned collections in the Cambridge University Library and other partner libraries, and guided by the advice of experts in each subject area, Cambridge University Press is using state-of-the-art scanning machines in its own Printing House to capture the content of each book selected for inclusion. The files are processed to give a consistently clear, crisp image, and the books finished to the high quality standard for which the Press is recognised around the world. The latest print-on-demand technology ensures that the books will remain available indefinitely, and that orders for single or multiple copies can quickly be supplied.

The Cambridge Library Collection brings back to life books of enduring scholarly value (including out-of-copyright works originally issued by other publishers) across a wide range of disciplines in the humanities and social sciences and in science and technology.

# A Long Life's Work

## *An Autobiography*

Archibald Geikie

Cambridge University Press

CAMBRIDGE UNIVERSITY PRESS

Cambridge, New York, Melbourne, Madrid, Cape Town,
Singapore, São Paolo, Delhi, Mexico City

Published in the United States of America by Cambridge University Press, New York

www.cambridge.org
Information on this title: www.cambridge.org/9781108048392



This edition first published 1924
This digitally printed version 2012

ISBN 978-1-108-04839-2 Paperback

# A LONG LIFE'S WORK

## AN AUTOBIOGRAPHY

MACMILLAN AND CO., LIMITED
LONDON · BOMBAY · CALCUTTA · MADRAS
MELBOURNE

THE MACMILLAN COMPANY
NEW YORK · BOSTON · CHICAGO
DALLAS · SAN FRANCISCO

THE MACMILLAN CO. OF CANADA, LTD.
TORONTO

*Lafayette photographer* *Emery Walker ph. sc.*

Arch. Geikie

# A LONG LIFE'S WORK

## *An Autobiography*

BY

SIR ARCHIBALD GEIKIE, O.M., K.C.B.

D.C.L. OXON.; SC.D. CAMB., DUBL.; LL.D. ST. AND., GLASG., ABERD., EDIN.;
PH.D. LEIPZIG, UPSALA, PRAGUE

ASSOCIÉ ÉTRANGER DE L'INSTITUT DE FRANCE; OFFICIER DE LA LÉGION D'HONNEUR
FOR. MEMB. R. ACCAD. LINCEI, ROME; NAT. ACAD. SCI., WASHINGTON
COR. MEMB. ACAD. BERLIN, MUNICH, VIENNA, STOCKHOLM, CHRISTIANIA
FORMERLY PROFESSOR OF GEOLOGY, UNIVERSITY OF EDINBURGH; THEREAFTER DIRECTOR-GENERAL, GEOL. SURVEY, UNITED KINGDOM
PRESIDENT, GEOLOGICAL SOCIETY, 1890, 1906; PRESIDENT, ROYAL SOCIETY, 1908-1913

MACMILLAN AND CO., LIMITED
ST. MARTIN'S STREET, LONDON
1924

PRINTED IN GREAT BRITAIN

# CONTENTS

## CHAPTER I

1835-1850



## CHAPTER II

1850–1855

## CHAPTER III

1855–1860



## CHAPTER IV

1860–1867

## CHAPTER VIII

1892–1896



## CHAPTER IX

1897

CHAPTER X

1898–1903



CHAPTER XI

1903–1908

## CHAPTER XIII



## ILLUSTRATIONS

# CHAPTER I

## BOYHOOD AND SCHOOL-LIFE
(1835–1850)

For many years my family had pressed upon me their wish that I would jot down for them the recollections of my past life; but I had persistently evaded compliance with this desire. My story might be a little longer and more varied than that of most scientific men, who are quite content to have done their work and to be remembered by it, without themselves attempting to write an account of their career or collecting material for posthumous biography. But at last when I had entered into my eighty-third year, and believed myself to have done with official life, and to be finally domesticated in my pleasant home in the country, my daughters, with more insistence than ever, renewed their request that at last I would commit to writing the story of my life. Their appeal was supported by the discovery that my innate love of work demanded some steady occupation for the ample leisure which I now enjoyed. I was thus induced to look up old letters, journals and fragmentary diaries, in order to see how far they could stimulate a memory which is itself somewhat tenacious. This search among the records of the past was a task of mingled regret and pleasure; but it resulted in a definite decision to narrate the successive stages of my career, as if I were telling the tale at my own fireside. In describing the progress of geological science, wherein I have taken part, I shall generally avoid the use of technical language, and try to make that progress intelligible, perhaps even interesting, to the non-scientific reader.

My father was a respected citizen of Edinburgh, to whose sedulous and intelligent help in my boyhood and youth I owed much. He was widely known as a lover of music, which he studied as a science, as well as an art. He wrote some Scots songs that were popular in their day, but his chief labour was given to church music. His own good taste, and his familiarity with the best work of the older musicians enabled him to render considerable service to the progress of Scottish psalmody. He was for many years Music Critic to the *Scotsman* newspaper, in which capacity he attended all the concerts for which Edinburgh was noted. As he often took me with him on these occasions, my boyhood was enriched by listening to many great singers and instrumentalists.

During school-life my father took a keen interest in my daily work, and though, as he used to confess, his Latin had grown somewhat rusty, he often helped me over difficulties. A lover of good literature, he had formed a small but well-selected library, in which I browsed with pleasure and profit. He identified himself with my early progress, and from the width of his circle of acquaintance, was often able to do me timely and essential service. The pleasure and sympathy with which he watched my progress formed one of the greatest joys of my youth.

My mother, Isabella Thom, came of a seafaring race. Her father, as captain of a vessel in the mercantile marine, had seen a good deal of the world. Her only brother followed with success the same profession, and eventually settled in New York, where he formed a nautical school. She herself was an indefatigable house-keeper, looked after her family with unfailing care and affection, and entirely won their hearts. She was a true Scot, and when in the mood, could speak the vernacular language with great purity, often using words that had long ceased to be heard in the still current vocabulary.

I was born in Edinburgh on 28th December, 1835. Of one phase of my childhood I have myself no distinct recollection, but it has often been related to me. My father, discovering that I had a quick ear for musical sounds, taught me some simple nursery songs which I learnt to sing correctly, though the words were childishly pronounced. My mother would sometimes take me with her when shopping, and occasionally, not having noticed that I had strayed from her side, she would hear my voice trolling one of the familiar ditties from the back premises, where perched on a chair, I gave the audience "Little Bo-peep" or some equally affecting lay.

The visit of Queen Victoria and Prince Albert to Scotland in the autumn of 1842 is still to me a vivid memory. I remember that she arrived by sea at Granton, early in the morning, and that without delaying to land until the various official bodies had time to muster their deputations, she came ashore and drove through Edinburgh, before Lord Provost Forrest and the Town Council, it was alleged, were out of bed. The words of a humorous ballad that was sung and sold everywhere on the streets come back to one's recollection :

> Hey ! Jamie Forrest are ye wauken yet ?
> Or are your Baillies sleepin' yet ?

Afterwards I saw the Queen and the Prince as they drove through the streets, with their glittering escort of cavalry.

But the most memorable experience of this same year was being placed at the beginning of October in Mr. Black's School, 10 George Street. It was a kind of preparatory day-school, of some note in its day, where only English was taught, together with a slight infusion of science, chiefly in the form of a few chemical and physical experiments, performed from time to time by the master. The pupils included both sexes, the girls being ranged along one side of the large room, the boys on the other. Mr. Black himself was a worthy man, of no great ability, who posed as a teacher of elocution, and gave public recitations, at which some of his more promising pupils assisted him in his exhibition of the art of declamation.

During the three years spent at this school no attempt was made to teach us writing, and no written exercises were prescribed. One or two of my letters, written in pencil during the last year at the seminary, have been preserved. They are laboriously printed rather than written, capitals and small letters intermingled, each line beginning with a capital, and the spelling not altogether creditable to the school. One of the most needful requisites on advancing to a public school was to speed up this cacography. I passed through a succession of writing masters, one of whom, a patient painstaking teacher, had a strong antipathy to steel pens, and trained his pupils on quills. As he passed up and down the desks, a good part of his time was taken up in mending the quills, which he did with infinite good nature and most enviable exactness. He brought me on rapidly.

The most important public incident that occurred during my time at the George Street school was witnessed by the pupils; for St. Andrew's Church, where it took place, stood on the other side of the street, exactly opposite to our class-rooms. On the 18th May, 1843, the town was crowded with black coats—ministers and elders from every parish in Scotland, all in a ferment of excitement, outside and inside the church, like a swarm of hiveing bees. I was too young to understand the cause of this trepidation. But no schoolboy could fail to be impressed by the sight of the long procession of black coats, filing out of the building and marching off to turn down the hill to the large hall which had been engaged at Canonmills. We were witnessing the actual Disruption of the Kirk of Scotland. The processionists whom we watched, in their sombre attire, looked like a gigantic funeral, but they were on their way to inaugurate a new ecclesiastical assembly, and to found the Free Kirk of Scotland. I can remember one name that was on everybody's lips—Thomas Chalmers; and though I had never seen him, his greatness and goodness were deeply impressed on my memory by the universal tribute of respect from opponents as well as from friends. While I am writing these lines (May 1918), the newspapers announce that the General Assemblies of the Established and the Free Churches are holding their annual meetings in Edinburgh, where the main subject of their deliberations is the reunion of the two communities into one Church. For seventy-five years they have lived apart, but the original antagonism has long since abated, and there now seems to be every probability that the schism will at last be healed.

The last term at the preparatory school was followed by my first excursion to a distance from home. I had gained my teacher's medal, and had won his good-will, as well as his approbation of the way in which I had mastered his elocution lessons. He asked to be allowed to take me with him to Glasgow, that I might appear there as his pupil, at one of his public recitations. Consent having been obtained from my parents, he first took me for a few days in August 1845 to a farm on the moors between Croy and Kilsyth. For the first time in my life I found myself there in the midst of open and untouched Nature. I can recollect the delight of solitary rambles in the trackless heather, then in full bloom, with clumps of luxuriant bracken, bramble-bushes, wild roses and—what an inexperienced boy of nine could not fail to appreciate—acres of ripening blaeberries. Familiar till then only with gardens and cultivated fields, I was deeply impressed by this wild type of landscape, and the impression then made abides with me still. Having been in the habit of using the pencil, I made an attempt to sketch a part of the scene visible from a rise on the moor, whence the eye caught the silver surface of the Forth and Clyde Canal, winding along the valley like a placid river, till lost behind the distant woods.

Recollections of the visit to Glasgow are vague and confused. But it was here that I made my first appearance in public, and I well remember standing on the platform and facing the audience, without, so far as I can recall, the least feeling of trepidation. I can even recollect the dress I wore on the occasion. My memory does not retain the name of the piece or pieces which I recited, but they would no doubt be some of the stock poems which we learnt by heart at school.

By far the most vivid and delightful memory of that

August was my introduction to the Scottish Highlands. Though only a transient peep into scenery which was afterwards to become so familiar, it has left on my mind a clear-cut memory which has never been effaced, even by the larger and more detailed experience of later years. Travelling in Scotland was in those days by no means so easy as it has now become. I remember the opening of the first railway in Scotland, between Edinburgh and Glasgow, in the spring of 1842. Before that time the two cities were connected by a coach-road and a canal. I recall the pleasure of an excursion on that canal, with the novel and delightful sensation of seeing fields, trees and cottages flit past, as in one continuous panorama that unfolded itself before one's eyes, with no feeling of movement in the horse-dragged barge wherein one was slowly borne along.

From the famous Broomielaw of Glasgow we sailed down the Clyde in one of the many steamboats which even in those days threaded their way along that crowded thoroughfare. At the foot of Loch Lomond we found a tiny passenger steamer ready to start up the lake. The fairy-like beauty of the islands in the lower part of that noble sheet of water, and the huge rugged mass of Ben Lomond towering above them, were then to me the opening of a new world. Landing from the steamer at Inversnaid, we walked up the hill road to Loch Katrine, where a still more wonderful landscape unfolded itself. The sun had nearly set before we reached Stronachlacher, where we procured a boat and four stalwart Highlanders to row us down to the lower end of the lake, for at that time no steamboat had yet appeared on this lovely water. Never shall I forget the varied delights of that evening. The rays of the sunset behind us were lighting up the great circuit of mountain-crests and peaks in front, and gave an added glory to the autumn tints of the heath-clad slopes and bare crags. Every cleft and corry above, and every tree on the declivities below were accurately reflected on the mirror-like surface of the loch. The shades of evening began to gather all round us, while the rugged summits still shone in the afterglow. And to fill up my enjoyment, as we slowly slid along the placid mere, the full harvest moon appeared from behind the crest of Ben Venue. As the daylight faded, the woods on either side grew darker and more solemn, until the silvery sheen of the moon fell on the still water. No house was visible, nor any obvious trace of man. Our boatmen, however, made the echoes resound with their boat-songs, which in true Celtic fashion they sang most lustily, moving their oars to keep time with the cadence of the music. By the time we reached the landing-place at the further end of the lake the last traces of daylight had vanished, and we plunged on foot into what seemed to be a mysterious and impenetrable forest. By the light of the moon, however, glinting through the thick-set pines, we soon found a track or rough road that brought us sometimes near the verge of an impetuous river which filled the silence with its roar, as it rushed and chafed among the rocks and boulders of its channel. Late in the evening we reached the inn at the Trossachs, then a much less pretentious hostelry than its modern successor.

Sir Walter Scott had been dead a little more than three years when I was born. During these and the immediately subsequent years there was much stir in Scotland to raise in Edinburgh a worthy monument to his memory. The design by George M. Kemp having been chosen, the work of erecting his fanciful and appropriate structure was begun in 1840. I can recall the public interest in watching the gradual rise of the graceful stonework, as it mounted upwards, buttress above buttress and pinnacle above pinnacle, from the level of Princes Street, and the great excitement when the last finial was placed on the summit. In 1846 I saw the inauguration of this memorial from a window near at hand, and looked over the greatest concourse of human beings I had ever yet beheld. I remember also the shock which ran through the whole community when the news spread of the accidental drowning of the gifted architect, before the completion of his masterpiece. Some of Scott's poetry was already familiar to me, and the recent excursion into the Highlands had brought the landscape and incidents of the *Lady of the Lake* into vivid realisation.

At the beginning of October 1845 I entered the Edinburgh High School—the famous *Schola Regia Edinensis*, at which for more than three hundred years so large a proportion of eminent Scotsmen had been educated. It was still, as it had been from its earliest record, in 1517, under the management of the Town Council of Edinburgh. It had four classical masters and a Rector. Each master taught the same boys from the period of their entry until, at the end of four years, they might pass into the Rector's hands. The master who took the first or most juvenile class for the year in which I joined was Dr. James Boyd. He had since his appointment conducted three classes through the four years' course of study, and was now taking charge of his fourth battalion. He was a short and decidedly portly personage, with a large head, well covered with curly grizzled brown hair, and a full ruddy face, which displayed a great range of expression from the most playful humour to the sternest gravity. Sitting at his raised desk, where his want of height was concealed, his massive head reminded us of the plaster bust of Jupiter. No master could have a more complete command of his pupils. He could put into his features such a look of severe displeasure as made the boldest boy shrink into himself. And this mere look was enough to quell any little noise or disorder. But when occasion arose he could administer corporal punishment with the most remarkable energy. He loved a joke, and much to the gratification of his pupils, never failed to catch the humorous side of any mistake or mistranslation. Thus, while we stood in awe of him, this feeling was mingled with respect and real liking. It was outside in the yards that he appeared to least advantage, for there his short stature, notable corpulence, and wide steps, as he strode along with a sort of jerky or springy gait, made him a conspicuous figure.

I soon grew fond of Dr. Boyd, and as I did my best to master the work to which he set us, he kindly took notice of my efforts, and thus began a mutual esteem which lasted until his death, a few years after I left school. Being the most popular master in the High School, he had always the largest class. He was a sound scholar, a good teacher in a humanistic way, bringing out the meaning, allusions and beauties of an author, rather than dwelling on the niceties of grammatical construction and the quantities of syllables. He had edited several classics for educational purposes, also an edition of Adam's *Roman Antiquities*, of which large impressions were sold. To him I owe the beginning of an appreciation of the Latin classics which has been a joy to me all my life.

At first the change was rather overpowering from the quietude of the George Street school to the buzz and stir of the playground, filled with nearly five hundred boys, and to the presence of some seventy class-fellows in a specially capacious room, with windows facing towards the Bay of Prestonpans. At that time athletics formed no part of the ordinary curriculum in the High School. The only instruction provided in physical training was a non-obligatory course of fencing and gymnastics. Cricket had not yet been generally established in Scotland. A good many years had to pass before this game was properly organised, with open ground on which to play it. The one great game then in vogue on the tolerably extensive gravelly yards of the school was that of the *Cleckenbrod* or *Clecken*—an interesting survival of an old Scottish game which seems to be now extinct. In my boyhood it could still be said that " nothing was to be seen in the hands of the boys but cleckenbrods." Each member of the school, even before he had obtained all the necessary class-books, took care to provide himself with one of these implements. It was a battledore, made out of one piece of stout elm, possessing such elasticity that a ball could be driven with it through the air to a considerable distance. The game was played by two sides between two goals; the object of the side which led off being to get the ball to the opposite goal, while that of the other side was to drive it back to its starting-point. Some skill and practice were required to lift the ball from the ground with the clecken, never with the hand, before an opponent could reach it, and then to send it away as far as possible. Many a fierce tussle arose, the ball being hit along the ground until it reached some open space, where one of the players could deftly raise it into the air with his clecken, and drive it off in the right direction. The cleckenbrod was carried with the class-books back and forward to school, and usually hung from the wrist by a string fastened to the end of the shaft.

There were of course the usual games of wrestling, leap-frog and such like, together with not infrequent warlike sieges and stubborn defences of class-room entrances, as well as other escapades to keep the body active, while in the proper season there were quieter amusements in sheltered parts of the yards, such as marbles, and in the weeks when cherries were ripe, the variously-played game of *Paips*. This was a Scottish form of the ancient games of *Nuces*, played by the Greeks and Romans with nuts or acorns. One of the forms of the game, as played at the High School, was to place on the ground, at the base of the high wall on the north side of the yards, three cherry-stones, called "paips," with a fourth on the top of them, so as to form a little pyramid. The player who, from a certain measured distance could throw a paip so as to dislodge the pyramid took the top of the pile, or the whole of it, as his prize, while if he missed his aim, his missile, or another cherry-stone, was forfeited to his opponent. In this form of the game it was always the object of the player to select the largest and heaviest cherry-stone he could find. A High School boy accustomed to the game, when he came to enter on the serious business of life might literally quote the Latin phrase *Nuces reliqui* (I have given up my Paips).

Three months after my entry, namely, on 3rd January, 1846, the High School was thrown into excitement by the arrival and induction of a new Rector. Dr. Carson had resigned the rectorship in the previous autumn, and the Town Council had elected Dr. Leonard Schmitz in his place. The whole school, as on prize-giving day, was mustered in the Great Hall, each class headed by its master. Probably none of our class had ever seen anything of the kind before. There in the centre of the audience sat the Lord Provost, magistrates and town councillors in their official red robes, with their attendant halbert-bearing lictors in uniform. The sight of this dignified company brought before the boys, in its most imposing guise, our body of governors, and we naturally felt prouder than ever to be looked after by such a resplendent company.

In the centre, too, our attention was arrested by the appearance of a tall thin man, with a pale earnest student-like face, long sleek hair, wanting an arm, and wearing a black robe. This proved to be our new Rector. The galleries were filled with a crowd of gaily-dressed ladies, mostly mothers, aunts, and elder sisters of the boys. After the Principal of the University had opened the proceedings with prayer, the Lord Provost, who at that time was Adam Black, the well-known publisher, rose and made a speech wherein, after an eulogium of the retiring Rector, he gave a laudation of his successor whom he formally introduced. He had some words too for us boys. It was a pleasure to the juveniles of the first class to see that the next man who spoke was no other than their own master, Dr. Boyd. He, as the senior master, said how pleased his colleagues and he were with the choice which the Town Council had made. We knew that one of those masters had been a candidate for the rectorship, and we hoped that he was pleased too. After a few words of praise for the newcomer, whose works were already well known in this country, our master gave place to the new Rector. Of course, we all strained our ears to catch every word he said. Although a German by birth, Dr. Schmitz spoke English fluently, but with a perceptible foreign accent. He finished his speech by addressing a few words to the boys, hoping that he and we would get on well together, and assuring us that he would do all he could for our benefit.

I soon became keenly interested in Latin. The elementary grammar which we used was essentially based on *Ruddiman's Rudiments*—a book which had been in use in Scotland for nearly a century and a half. Ruddiman's metrical rules, in Latin hexameters, for genders of nouns and for the quantities of syllables we learnt off by heart, and I can still repeat some of them. The pronunciation taught was the old Scots form, which was nearly the same as the Continental, dating as it did from the time when the teaching in the Universities of Scotland was in Latin, and when Scots professors filled chairs in foreign seminaries as effectively as they did in those of their own country. The English pronunciation was as difficult to understand in Scotland as it was abroad. I well remember that a boy from the grammar-school of Skipton, in Yorkshire, who had joined our class, was called up to read and translate the familiar lines in the *Metamorphoses* of Ovid, beginning

Ante mare et terras, et quod tegit omnia, caelum.

It was the first time we boys had ever heard the English way of reading Latin, but the lad was stopped at the first two words by Dr. Boyd, who called out "What? what? Aunty Mary!: we shall be having Uncle Tom next!" In later life, the Scottish pronunciation of Latin which one had learnt at school served me in good stead in Arctic Norway, as will be told in a subsequent chapter.

It was an amusing task to translate into English verse such Latin as that of the fables of Phaedrus. By the time we reached the *Metamorphoses* of Ovid I had acquired some facility in this kind of versification, and boldly adventured to translate Ovid's account of the Four Ages into English heroic metre. Our master handed this juvenile attempt to Dr. Steven, who was at that time engaged in writing his *History of the High School of Edinburgh*, and it was inserted in his volume which appeared in 1849. This was my first production in print.

Our worthy master, in order to encourage drawing and map-making among his pupils, used to offer exemption from attendance at school on Saturday to those who handed in creditable drawings or maps, Saturday being always a half holiday. I often availed myself of the privilege, which was, however, of doubtful advantage. It is true that I gained the whole of the last day of the week for a country ramble, and the prosecution of the pursuits in natural history which had begun greatly to interest me. But on the other hand, I lost the revision of the work of the week,—in some respects the most useful lesson of all.

In July, at the end of the session, the annual prize-giving took place in the Great Hall, when a large concourse of visitors filled the galleries. I remember that on one of these occasions Thomas Babington Macaulay was present, brought thither by his trusty political supporter, Lord Provost Black, who officiated in the chair. It was worthy of lifelong remembrance that I received my prize under the eyes of the illustrious author of the *Lays of Ancient Rome*. The perusal of these poems, of which I could then repeat long passages by heart, had not a little to do in developing my boyish love of Roman history and Latin poetry.

It was during these school years that I had the unforgettable happiness of seeing Helen Faucit on the stage of the Edinburgh Theatre Royal, of which Scott's friend, William Murray, had long been manager. In February 1849 she appeared there in several of her favourite Shakespeare characters, such as Juliet, Portia and Beatrice. The charm of her voice at once captivated me, and her acting revealed to my boyish imagination the nobility of womanhood. In long subsequent years, when I became intimate with her husband, Sir Theodore Martin, these early reminiscences of her greatness were among the links that cemented our friendship. Through my father's appreciation of good acting, I was often taken, in those schoolboy years, to the Theatre Royal, where some of the best actors of the day appeared. "Will Murray" himself was excellent in comedy. He would take quite subordinate parts. Thus he played one of the witches in *Macbeth*, and made the part one of the most powerful in the drama. As long as he was manager he kept together a good company, and introduced from time to time the leading members of the profession from the south. One of my earliest dramatic recollections is to have seen the famous Charles Mackay in one of his latest appearances as Bailie Nicol Jarvie in the dramatised version of *Rob Roy*. It must have been a wonderful performance when he was in his prime, as Scott's letters bear witness.

Among the early reminiscences which the narrative of these school-days recalls is my father's intimacy with James Ballantine, a minor Scots poet who was well known in literary society from the publication of his *Gaberlunzie's Wallet* and *Life of David Roberts*, and for his revival of the art of painting on glass. Some of his songs, set to music by my father, were, as I have already said, popular in their day. He was a favourite in our family, and I can still recall his broad smiling face, his hearty laugh, his merry humour, and his broad Scots speech. His work in glass-painting having given him fame beyond his own country, he was commissioned to execute some of the stained windows in the House of Lords.

Before passing on from the subject of the life of the Schola Regia Edinensis, I may allude to my last interview with good old Dr. Boyd. It took place some years after I had quitted school and not long before his death. Having to make a call in one of the quiet terraces beyond the school, I saw him walking slowly towards me and intently reading. As I drew near he lifted his eyes from his book, and recognising me, stood still and began a friendly talk, in the course of which he alluded to the volume in his hand, which, if I remember, was the work of some old divine. "You see," he said, "all we can carry with us into the next world is the knowledge we have gained in this. And so, I fill up my spare moments in storing my mind with thoughts which I shall then be glad to remember."

During school-life, another kind of education was going on in the country rambles to which Saturdays were devoted, and in which the boys in great measure taught each other. It was thus that we learnt to recognise trees and wild flowers. In our incursions into the woodlands we now and then came upon the little black wild cherries which we knew by the Scots name of *geans*. We afterwards learnt that this is the Scots spelling of *guignes*, the French name of the fruit, and one of the verbal relics of the ancient alliance between Scotland and France.

For some years I gave myself up to collecting diurnal lepidoptera, and succeeded in forming a collection of the commoner butterflies and day-moths of Midlothian. It was while these desultory natural history pursuits were in full progress that the incident occurred which turned me into the hitherto untrodden geological field, and of which I have elsewhere given an account.[1] My comrades had come upon a group of limestone quarries a few miles to the south of Edinburgh of which they gave me so highly coloured a description that I resolved to go with them next Saturday to their "cave of wonders." When we reached the quarries, I bent over a mound of broken fragments of the limestone, and the first piece which I picked up contained my first-found fossil. It was thought by the leader of the previous excursion to be a fish, but it afterwards proved to be the seed-cone or catkin of an ancient form of club-moss.

Here was a momentous experience, which called forth the sense of thoughtful wonder more vividly than any incident had done before in the course of my boyish pursuits. It brought to my recollection the opening of the Wizard's tomb at Melrose, described in *The Lay of the Last Minstrel*, and the sudden issue of the glamorous light from the grave. As I broke up the blocks of stone and

[1] *Geological Sketches at Home and Abroad*, p. 1.

laid open fragment after fragment of plants, delicately preserved, a light seemed to stream into my mind from these buried relics. To discover that beneath the present surface of the earth with all its life and beauty, in which our young souls delighted, there lay entombed in the subterranean rocks the remains of plants, unlike, but perhaps once not less beautiful, than those with which we were familiar, yet vastly more ancient; to realise that every organism which we could lay bare to the light of day had never before been seen by human eyes and to feel that there might perhaps be no end to the variety of once living things which we might unearth—all this was new and startling. It came as a revelation that filled my mind with an indescribable pleasure, not unmingled with awe and an earnest longing to know more.[1] Thenceforward the rocks and their fossils became increasingly the subject of my everyday thoughts. Other branches of natural history were gradually relinquished, and the rambles round Edinburgh, or to a distance from that centre, became more and more geological.

In those wanderings into the Midlothian coal-field, one was brought into contact with a phase of life which seems now to be almost incredible. For many generations the colliers and salters of Scotland were *adscripti glebae*, literally slaves, who by law were bound to the places where they worked, and could be legally reclaimed by their owners if they dared to shift elsewhere. The last Act of Parliament which broke their chains was only passed so late as 1799. The older men and women whom I used to see in the wretched colliery villages had actually been born in servitude, for in spite of Acts of Parliament, their masters continued to keep hold on them by letting them run into debt. In those days I never ventured to descend into a coal pit, but I have many a time watched the salters at their task in its different stages of obtaining salt by boiling down sea water at the "pans" along the coast to the east of Edinburgh.

[1] His me rebus quaedam divina voluptas
Percipit atque horror. LUCRETIUS, III. 28.

The quarries of Burdiehouse proved to be the means of my first introduction into the company of scientific men. I had found a fossil form in the limestone which I could determine neither from Dr. Hibbert's paper on the Burdiehouse limestone in the *Transactions of the Royal Society of Edinburgh*, nor from Lindley and Hutton's treatise on *Fossil Plants*. My father having mentioned my difficulty to Robert Chambers, received a few days afterwards from him the following note: "Happening to meet Professor Fleming, I mentioned to him that your son had found a fossil vegetable in Burdiehouse quarry which he believed to be new, or supposed might prove to be so. At the same time I told the Professor how the youth had entered with such zeal upon the study of geology. The worthy Doctor was interested, and very readily agreed to my proposal that he should see and pronounce upon the fossil. The young geologist may therefore, if he pleases, call on Dr. Fleming any morning, and show him the object, when I have no doubt he will get the best decision upon it which Edinburgh can afford. I would suppose it to be worth his while thus to make acquaintance with so eminent a naturalist."

Robert Chambers, one of the two partners in the famous publishing house known by their name, was a conspicuous citizen of Edinburgh. Over and above his active devotion to the business of spreading literature through the country, he was much interested in science, and was shrewdly suspected to be the author of the *Vestiges of the Natural History of Creation*, which had been published in 1844, and had made a great stir, not only among men of science, but in the community at large. My acquaintance with him, now begun, lasted as long as he lived, and allusions to it will appear in later chapters of this narrative.

Professor John Fleming, the most distinguished naturalist of his day in Scotland, was at this time sixty-five years of age and filled the chair of Natural Science in the New College, Edinburgh. Besides his zoological works, which have become classics in science, he had published papers on geological subjects, in which he did not always agree with the general opinion of his day. When I called upon him I found him engaged with Dr. Robert Kaye Greville, the eminent botanist. This was for my errand a fortunate conjuncture, seeing that Dr. Greville was the resident authority whom one would have selected to pronounce on the fossil. The specimen was carefully looked at by both of the learned men, and was decided to be a macerated example of the fern known as *Sphenopteris bifida*. It had lain long, soaking and rotting in the soft mud which was now hard limestone, till all its softer parts had disappeared leaving only a skeleton of its structure. Dr. Greville soon after this time went into politics, and I do not remember to have ever met him again ; but from time to time I saw Professor Fleming.

During these years every book relating to geology on which I could lay hands was eagerly read. The text-books, notably those of Lyell, De La Beche, and the small school-book of David Page were my chief authorities. But the volume from which, of all others, most stimulus came was the immortal *Old Red Sandstone* of Hugh Miller. Though it supplied little of the information of which one stood most in need, it communicated some of the inspiration and enthusiasm of a true lover of nature, and revealed the spirit and methods of an ardent investigator in what was to me a hitherto unexplored field. I certainly owe more to the impulse given by that book than to any other piece of writing in scientific literature. At the same time I was captivated by the author's style. My earliest attempts in geological essay-writing showed how powerful this mastery was. But Miller's English was pure and idiomatic, based on a wide range of reading in English literature, and was therefore by no means a bad model to follow until a style of one's own could be worked out.

But above and beyond all books, the fresh open face of nature was my best source of inspiration and instruction. Few parts of the British Isles are better fitted to awaken and sustain an interest in the past history of our globe than the region in which the city of Edinburgh stands.

The stratified rocks with their inclosed fossil plants and animals are laid bare to the sky in many picturesque ravines and water-courses, and they are well displayed along the shores of the Firth of Forth. But besides the ordinary sources of interest which these rocks present all over the British Isles, they are here invested with an added attraction, in the abundant evidence which they contain of groups of ancient volcanoes that were active when the stratified rocks were in the course of deposition. Though it was the organic remains which first awoke my geological ardour, before long my attention was strongly diverted to the volcanic history of the region. Comparatively little had been done in the investigation of this subject. Some of the men, indeed, such as Robert Jameson, were still living who had scorned the idea that any of the rocks of the district had a volcanic origin. But Charles Maclaren had demonstrated the volcanic nature of Arthur's Seat and the Pentland Hills, and his excellent descriptions were of great value to me.

From this geological digression I return to my school-life and its holidays. Of these the furthest afield was a visit to London in the autumn of 1848. As in those days the continuous railway connexion between Edinburgh and London had not been fully established, the journey to and from the south of England continued to be usually made by steamboat. It was my first sea-voyage, and it has left in my mind a pleasant memory of islands, sea-cliffs and sea-gulls, sailing ships of divers sizes, distant fields, and occasional towns, and in the evening a succession of lighthouses, twinkling and flashing one after another as we passed along the coast. Looking back on the impressions left by the stay in London, I am inclined to doubt whether a boy, hardly in his teens, should be rushed through the manifold sights of the great metropolis. He does not yet possess the amount of general knowledge required to enable him to understand and enjoy a tithe of what is shown to him. I carried away a confused jumble of impressions, amidst which only a few incidents still stand out clear and sharp. One of the most prominent of these is a visit to the House of Lords, where I saw the great Duke of Wellington. I kept my eye steadily fixed on him as he sat, and occasionally moved about. The stooping figure with the white hair and high stock, fastened with a large silver buckle behind, remains indelibly fixed in my memory.

One of the pleasantest of the excursions during my school holidays was made with a school-fellow to a family of his acquaintance in the vale of the Gala Water. We were fortunate enough to take part in the great sport of "burning the water," so graphically described in *Guy Mannering*. It was indeed a stirring sight, when, as dusk set in, the company assembled on the banks of the river to spear salmon by torchlight. Two or three stalwart yeomen with lights and "liesters" (or long pronged spears), entered the water and began to move to the places which the fish frequented. The flare of light on the surface of the river, throwing the figures of the sportsmen into strong dark relief, as holding their uplifted weapons, they peered into the depths, and the shouts of the crowd on the banks when, after a sudden splash, the liester emerged with a fish, made a scene of memorable picturesqueness. This form of sport being now illegal, I think myself lucky to have seen it in full operation.

# CHAPTER II

## A WRONG START. HOW THINGS WERE RIGHTED (1850–1855)

The time duly arrived when it was necessary to settle the problem of what calling should be chosen for a lad whose whole bent lay towards natural history pursuits. It never occurred to his parents that a love of geology could possibly lead to any situation with the prospect of a desirable future, although it might always be a pleasant occupation for leisure hours and holidays. We had an old family friend, manager of one of the large banks in Edinburgh, who urged that I should join his staff, where he would be glad to look after my interests. This vocation was ultimately selected, and as some preliminary legal training was considered desirable in a banker, it was arranged that I should spend two years in the office of a Writer to the Signet. The firm selected was that of William Fraser, a kindly and picturesque old gentleman, son of Luke Fraser, the master under whom Sir Walter Scott sat for three years, at the High School. I well remember my first interview with him in his house in Castle Street, which was not far from opposite to that where Scott lived and wrote. I was a little fellow, rather delicate in appearance, while he, before he began to stoop, must have been six feet in height. He was almost completely bald, such scanty locks as remained were white, and his face beamed with benevolence. He took me by both hands, placed me on a chair and sat down beside me. After asking a number of questions about my studies, as to which and my standing in school he had previously enquired of my old master, Dr. Boyd, he told me that, as he was becoming an old man, he did not take as much business, or keep as large a staff as heretofore; and that there might now and then come times when there would not be much for me to do. "But never be idle," said he. At any time when work had slackened, I was to come into his library, which contained a good collection of books in law and literature, and make choice of a volume. Some time afterwards, on the occasion of one of these slack intervals, I had picked out of his shelves a copy of Hume's *Philosophical Essays*, which at the end of the day was duly placed inside my desk. By next morning the volume had been restored to its place in the library, the worthy lawyer probably considering it his duty not to allow such dangerous mental food to be taken under his roof, by so young a reader.

I soon found, as might have been expected, that for me the work of a lawyer's office was unspeakably dull. Notwithstanding the sympathy and friendly assistance of the head-clerk, I was not learning much law, though I certainly did acquire some little knowledge of business habits, which has stood me in good stead since then. A further gain arose from the care, neatness and legibility required in the preparation of legal documents. Having thus to cultivate caligraphy, I formed a handwriting which has always been at least legible.

During this time my heart was often far away from Castle Street, among the hills and water-courses of Midlothian, where I had found such manifold sources of interest and enjoyment. The brief summer holidays were consequently eagerly welcomed when every day from morning till night could be spent in the field. In the year 1851 the choice was offered to me of another visit to London, or a few weeks of geologising in the island of Arran. Without the least hesitation the latter was chosen. Delightful and instructive as the Great Exhibition of that year would doubtless have been, I felt sure that it would possess less real value for me, in what had become the most absorbing pursuit of my life, than the opportunity of seeing with my own eyes, the geological structure of an island which had the reputation of containing within its borders an epitome of the geology of both the Highlands and Lowlands of Scotland.

Accordingly on 3rd September I sailed from the Broomielaw of Glasgow in one of the steamboats which plyed direct between Glasgow and Arran, and took five hours on the voyage. The scenery of the river below Dumbarton and of the Firth of Clyde, with the fine group of Arran mountains, now seen for the first time, gave me a thrill of pleasure. There was then no steamboat pier at Brodick, passengers being landed in small boats. This holiday proved to be one prolonged enjoyment. Every day brought some new wonders of scenery, some fresh details of geological interest, and some unexpected sights in natural history. My early love of butterflies was revived in the midst of such a number and variety of them as I had never before seen. Even as high as the top of Goatfell, they were flying about in the warm sunshine. The day being fine and clear, the view from the summit was seen at its best. To one who had never before been so high as 2866 feet above the sea, nor seen at one time so wide an area of the earth's surface, it was an experience never to be forgotten, to gaze over an expanse of land and sea that stretched from Ben Lomond and the Highland mountains on the north, across the blue Firth and far over the Lowlands to the heights of Galloway on the south, and thence to the blue cliffs of Ireland. Nor less impressive was it to look westward from this elevation down into the glens and corries of Arran itself, and to enjoy for the first time a close view of the details of rugged mountain scenery.

The time spent in the island was not only delightful but singularly instructive. With the aid of the excellent little pamphlet by A. C. Ramsay on the geology of Arran, I was able to go straight to the sections I specially wished to visit, and to gain a clear general view of the structure of the island. Returning to duty from this memorable excursion, I wrote two articles under the title of "Three Weeks in Arran, by a young geologist," which the editor of one of the Edinburgh newspapers, who was a family friend, inserted in his columns in the month of December. They were most juvenile productions, but their publication proved unexpectedly fortunate for their author. Mr. Andrew Coventry, an advocate at the Scottish bar, asked my father whether the writer of the articles would like an introduction to Hugh Miller. Of course I was overjoyed at the prospect of coming into personal touch with one whom I already reverenced as a kind of demigod. His appearance as he walked the streets of Edinburgh was familiar to me—a man above the middle height, strongly-built and broad-shouldered, clad in a suit of rough tweed, with a shepherd's plaid across his chest, his shock of sandy hair escaping from under a soft felt hat to join his large bushy whiskers.[1] At this time he lived in a detached house between Edinburgh and Portobello. When, after the letter of introduction had reached him, I presented myself in person, he received me with great kindness, remarking that he had read my articles and had wished to find out their author, whom he had not supposed to be quite so young a geologist as he proved to be. He ushered me into his sanctum, where the treasures of his long years of research were stored, many of the specimens being familiar to me from the engravings of them published in his books.

We were soon on the most friendly footing. I remember with what surprise and pride I received his first letter written within a fortnight of our meeting, which I have carefully preserved. It ran as follows:

*Witness* Newspaper Office
Edinburgh 15th January 1852.

My dear Sir

I trust to be quite at leisure on the evening of Saturday and shall expect to see you at six o'clock to

[1] "Hugh Miller, his Work and Influence," in my *Landscape in History*, p. 257.

take a quiet cup of tea with me, and discuss a few geological facts. A return omnibus passes my house at nine in the evening for Edinburgh. Yours truly,

HUGH MILLER.

From that time onwards till he died I often spent an evening with him at his home in Portobello. We met also in Edinburgh ; for when he had a paper to read at the Royal Physical Society, he would send me a note of invitation to attend. These papers of his were always interesting, the observations in them being generally new to science and, together with the specimens that illustrated them, having been not infrequently explained to me already under his own roof. He read his communications himself with a strong northern accent and pronunciation. After his paper on a " Raised sea-bottom at Fillieside " was read, he asked me to accompany him to the locality itself, that I might judge of the reality of the evidence which some critics had called in question. He there completely convinced me of the accuracy of his observations. It was a valuable lesson to see in what a masterly way he had gathered all the evidence on which his conclusions were based.

About the same time that the much valued friendship of Hugh Miller was gained, another acquaintance was made which proved to be of hardly less importance. Through Mr. Edmond Logan, Writer to the Signet, I was introduced to his brother William, the eminent Director of the Geological Survey of Canada, who about this time had occasion to come to England, in connexion with the important collections of rocks and minerals which Canada contributed to the Great Exhibition of 1851. When on this side of the Atlantic, he usually paid a visit to Edinburgh, where he had been educated at the High School. His brother would kindly send me word of his arrival, and I enjoyed long interviews with the veteran geologist, who spread his maps on the floor where, both on our knees, he gave detailed accounts of his work, showing at the same time the diagrams and sketches in his note-books. Besides his geological qualifications, he was a true artist, gifted with the faculty of presenting, in a few rapid lines, vivid glimpses of the landscapes he had beheld as he worked his way along the great rivers, in forest-clad, unexplored regions, known only to the Indians. In the midst of his scientific expositions he would now and then pause to intercalate the narrative of some adventure or whimsical episode that had taken place on the ground to which he was referring. These hours, so freely given to me, were not only interesting as personal records; they at the same time opened up broad views of geological structure and of methods of investigation, all entirely new to me. Years afterward I was able in some slight measure to repay Sir William's kindness. When some active, but parsimonious and unsympathetic politicians, in Canada were opposing the continuance of the Canadian Geological Survey out of public funds, and were causing him a good deal of anxiety, I wrote some unsigned articles pointing out the great importance of the work done by him and his Survey staff, both from a scientific and economic point of view. These articles, appearing in an English journal, were, it was said, not without effect in his favour in the colony.

There was yet another friend of my youth who was led by the papers on Arran to interest himself in my career—Dr. George Wilson, well known as the biographer of Henry Cavendish, and as a writer and speaker of singular charm. He was likewise a chemist of repute, author of some original papers on fluorine, and lecturer on chemistry in the extra-mural school in Edinburgh. He became my warmest and most active benefactor, treating me as a son, and ever on the outlook for some further way of encouraging and advancing me. It was in his class-room that I learnt the principles of chemistry, in his laboratory that I acquired the art of analysis, and in his house that I made the acquaintance of his friend, Alexander Macmillan, the publisher—an acquaintance which afterwards ripened into the most intimate friendship. Besides all the science he taught me, there was the memorable lesson of his cheerful patience and endurance, in feeble health and often in great bodily suffering, under which he clung to his work till death released him.

With grateful recollections I name another eminent man of science who in these early years gave me encouragement. One day I received a note from James David Forbes, the distinguished Professor of Natural Philosophy in the University of Edinburgh and Alpine explorer, asking me to breakfast with him. His tall spare figure was daily to be seen in the streets, as he passed to the College or to the rooms of the Royal Society of Edinburgh—his sunken cheeks, the pallor of his well-chiselled features which had been preyed on by serious illness; his long neck, round which he always wore one of the large neckcloths then in vogue, and above this, when out of doors, the thick muffler, from under which, as he passed, one might now and then hear the cough, that told of the malady from which he was suffering.

After breakfast he took me into his study and showed some of the beautiful water-colour drawings which he had made in the course of his wanderings among the Alps and in Norway. The memory abides with me of the thin, white, slender, and almost transparent hands that turned over the sketches, as he described them and the circumstances in which they were taken. Our interview concluded with a piece of practical advice. "You will have to climb mountains," he said, "and feel thirst. Let me urge you never to carry spirituous liquor of any kind with you. I have always found by far the most effective resource to be a flask of cold tea." In later years I saw much of Forbes at intervals, and my regard for him deepened as time went on. He was one of the most beautiful and interesting personalities whom I have ever known.

Forbes early in life had shown keen interest in geology. There can be no doubt that had he thrown himself into the active prosecution of geological research with the same ardour which distinguished his study of the physics of glaciers, he would have been one of the leading geologists of his day. In his memory the remembrance still remained fresh of the warfare of the Plutonists and Neptunists in Edinburgh, and in his youth some of the combatants were still alive, including the two great Huttonian protagonists John Playfair and James Hall, and the stout partisan of Werner, Robert Jameson. Forbes used often to express to his friends the sadness with which he had watched the slow fading of the halo that shone round Scottish geology in his youth.

One of the few citizens of Edinburgh who interested themselves in geological matters in my youth was Alexander Bryson, partner in a well-known firm of watchmakers. So far as I know, he had never done any original scientific work of importance; his chief claim to notice lay in his having acquired the valuable collection of apparatus and thin slices of fossil woods and minerals which had belonged to the ingenious William Nicol, inventor of the prism that bears his name. Having made Bryson's acquaintance I had opportunities of examining the Nicol specimens, and while so engaged, met for the first time Henry Clifton Sorby of Sheffield, who came to Edinburgh for the express purpose of seeing Nicol's collection, and with whom I then began a friendship which lasted as long as he lived. It was deeply interesting to watch the dexterity and acuteness with which Sorby worked over the thin sections of minerals under the microscope. He divined the important use to which Nicol's method of investigation could be turned in the study of the minute structures of minerals and rocks, and the light which could thus be thrown on their history. In a year or two after this time he had carried his researches so far as to enable him to prepare for the Geological Society the celebrated essay which revolutionised the science of Petrology. Up to this time the ultimate chemical constitution of rocks could be determined by chemical analysis, but the component minerals could often only be more or less vaguely inferred, and no light was thrown on the inner structure of these masses.

Nicol's method, based upon the preparation of thin translucent sections of even dense rocks, opened up a new vista for research into the history of the earth's crust.

> " The invention all admired, and each how he
> To be the inventor missed ; so easy it seemed
> Once found, which yet unfound, most would have thought
> Impossible."

Before the end of my two years in the lawyer's office, it had become evident to my parents, as it had long been to myself, that the career which they had carefully planned for me was hopelessly impossible, and that the sooner I turned my face in another direction, the better it would probably be for my future. The final rupture of my mercantile life was immediately brought about by an attack of scarlet fever, which greatly reduced my strength. After several attempts to return to Castle Street, I was compelled to take a long holiday, part of which was spent in London, where I passed many happy and profitable hours among the galleries of the newly-established Museum of Practical Geology in Jermyn Street. On returning home, I finally took leave of my benevolent Writer to the Signet.

The outlook was now somewhat blank. What I would fain obtain was some post where the passion for geology could find scope. Such posts must confessedly be few, if indeed any existed at all. But obviously the most urgent course in the meantime was to go through such a training as would fit one for any post of the kind which might present itself, or indeed for any other calling. This course of action was approved by my father, cheerfully willing, out of his slender resources, to defray my expenses at the University. To ease his burden, as much as was in my power, I undertook the tuition of private pupils.

In regard to the teaching of geology there was at that time no inducement to attend the University of Edinburgh. The professor of Natural History (which included geology and mineralogy) was Robert Jameson, now a feeble old man, whose lectures were read for him by an assistant, but with no course of practical instruction which forms so essential a part of proper training. There was then in Edinburgh a private teacher, Alexander Rose by name, who, possessing a good collection of minerals, taught mineralogy to the few pupils who wished to study that branch of science. I attended his lectures and demonstrations, and met there for the first time Dr. M. Foster Heddle, with whom acquaintance ripened afterwards into intimate friendship. Eventually he became Professor of Chemistry in the University of St. Andrews, and was acknowledged to be the best mineralogist of his day in Scotland. As already mentioned, I now attended the chemistry lectures of Dr. George Wilson, and went through a course of practical analysis in his laboratory. Such training would be helpful in the further pursuit of geological studies.

At the same time I was anxious that my training should not be only in science, but should embrace also the humanistic side of culture. From the High School I had carried away a strong love of the Latin and Greek classics, and had kept up the reading of them in leisure hours. Horace in particular was a constant companion. But I now began with renewed energy to read both the languages with an excellent, but somewhat eccentric scholar, William Skae, well known by sight on the streets of Edinburgh, where he always trotted along, looking neither to right nor left, and flourishing in his hand a bright red cotton handkerchief. I had already acquired a good knowledge of French, which has enabled me ever since to read the language with ease. I had also made some progress in the acquisition of German.

While these studies were in active cultivation, I heard, for the first time, of the existence of the Geological Survey. My ever kind friend, Dr. George Wilson, seeing my strong propensity towards geology, had made some enquiries about this branch of the public service, which had been at work for some fifteen years in the south-west of England and in Wales. From the information obtained by him, this Survey seemed to be exactly what would fulfil my wishes. A little later Robert Chambers added some further particulars about the Geological Survey, telling me that he had been recently in Wales, where he met a party of the Survey officers at work. Being the senior member of the company, he was voted into the chair at the dinner in the little inn where they were lodging. He graphically described to me how he had to carve a leg of Welsh mutton, and how, long before those last served had made much way, the earlier applicants were one after another pushing in their plates for a further slice or two, so that at the end of the repast there was little left of the joint but a bare bone. Their prodigious appetites seemed to be the characteristic of the geologists that had most impressed him. " So, my young friend," said he, " if you join the Survey you will have to add the qualities of 'a valiant trencher-man' to your other accomplishments."

Not long after this conversation a more detailed and satisfactory account of the Survey and its work was given to me by the leader of the party whose healthy appetites had astonished the author of the *Vestiges of the Natural History of Creation*. Reference has already been made to A. C. Ramsay (p. 23). This able geologist had in the spring of 1841, joined the staff of the Geological Survey under De La Beche, and was now Local Director. In the month of August 1853 he was deputed by the Director-General to proceed to Scotland for the purpose of ascertaining whether the Ordnance maps, on the scale of six inches to a mile, were sufficiently advanced to be available for use by the Geological Survey, if that Survey were now extended into Scotland. He came to Edinburgh, and through Dr. Wilson's kindness I was introduced to him. After a short conversation he asked me to take him over Arthur Seat, and explain its geological structure. We had a few hours on that interesting hill, so well elucidated by Charles Maclaren. During our walk he gave a full account of the work of the Geological

Survey, and held out the hope that, should the work be extended to Scotland, a post in the service might probably be found for me. Meanwhile he counselled the continuation of my training. Thus at last a ray of daylight seemed to be let in on my future.

It was therefore with a light heart and a determination to work harder than ever, that a few weeks afterwards I set out for the island of Skye, on the warm invitation of the Rev. John Mackinnon, minister of the parish of Strath, who thought that the geology of his district would be found to be full of interest and not much known.[1] Macculloch's *Western Isles* and such other published material as could be discovered were obtained. The perusal of these authorities opened up a field of geological investigation, in which it was obvious that much fresh detail remained to be discovered by the next comer; and I started full of hope.

The travelling arrangements throughout the West of Scotland in the middle of last century would now be considered primitive and inconvenient. In no part of the country has the march of improvement been more marked during the last sixty-five years. At the beginning of September 1853 I left Edinburgh in a train which took four hours to reach Glasgow—a journey that may now be accomplished in little more than one-fourth of the time. Still more conspicuous has been the development of the facilities for travel along the western coast and among the islands. The steamboats that plyed between Glasgow and Skye formed part of the fleet which David Hutcheson was building up with great enterprise and public spirit. Few men ever better deserved a public monument than this energetic shipowner, for by him the West Highlands and Islands were finally brought well within the pale of modern life and intercourse. The vessels then in use were powerful craft, built to withstand the buffetings of the Atlantic waves round such headlands as the Mull of Cantyre and

[1] In my *Scottish Reminiscences* I have given some account of this most worthy man and his family, and of my geological doings in Skye.

the Point of Ardnamurchan. As they were the only means by which goods could be brought to the islands, or the island-produce could be carried to southern markets, they conveyed both cargo and passengers. But the cargo was the more constant and important part of the traffic, and except in the tourist season, the passengers did not count for much. Besides all the goods stowed away in the hold, the decks were often crowded with cattle and sheep, leaving but scanty room for travellers. Moreover, tedious delays would often occur, for the vessels called at many different places to discharge or take in cargo, and to transport live stock from the islands to the mainland. At these halting places there were seldom any proper piers at which ships could come alongside; hence passengers had usually to be landed in row-boats. The delays at the places of call might vary in length from a few minutes to several hours. It was therefore impossible to be certain at what time an expected steamboat would arrive. These various inconveniences have been greatly diminished, or even in some particulars entirely abolished, by the erection of proper landing piers, and still more by the introduction of swift steamers carrying only passengers and mails, and keeping to a definite and published time-table.

When I had got fairly to work in Skye, the sight of a lad carrying a bag over his shoulder and a stout hammer in his hand, wherewith he chipped the rocks as he went along, puzzled the natives of Strath. My host told me that he did not believe that a live geologist had been seen in his parish since the days of Macculloch, Sedgwick and Murchison,[1] in the early decades of the century. The crofters came to know me by the name of *Gille na Clach*, or the "Lad of the Stones." As they watched me zigzagging, aimlessly, as it seemed, over moor and mountain-side, they had, at least for a time, grave doubts of my sanity, though I seemed to be harmless. One day passing some cabins on the shore of Loch Slapin, I

[1] Macculloch died in 1835; Murchison lived till 1871, and Sedgwick till 1873.

halted to break off a fragment from a projecting rock by the roadside. As usual I looked at the chip with my lens, and having satisfied myself as to the nature of the rock, was resuming my walk, when I overheard two old crones at their doors evidently speaking about me. I had acquired only a few words of Gaelic, but the last emphatic remark, in which they both seemed to agree, caught my ear. On returning to the family circle at the manse, I was not surprised to learn that the expression I had heard meant, "He is wrong in the head."

Having greatly enjoyed my wanderings in Skye among its kindly people, and having learnt some main facts in the geology of the island, I returned to Edinburgh in October, and reported to Hugh Miller my doings, and more specially the general character of the collection of fossils which I had obtained from the Lias of Pabba. He generously referred to my work in a paper which he read at the Royal Physical Society in Edinburgh, and at his invitation I exhibited the Pabba fossils and made a short statement on the subject which was published in the Proceedings of the Society for 1854. This was my first appearance before a learned society. The kindly encouragement given to me on this occasion by Hugh Miller was of signal service. He proposed a vote of thanks to the author of the communication "of whom," he added, "if he lived, more would assuredly be heard." It may be imagined how a lad, still in his teens, would be stimulated by such words coming from such a man.

Some specially sad memories cling to my remembrance of the year 1854. In the spring, war had been declared both by Britain and France against Russia, and in the summer, amidst much popular enthusiasm, we saw the soldiers march off from Edinburgh Castle to embark at Granton and Leith for the eastern Mediterranean. The next two years brought the tragedies of the Crimean War.

In the spring of the same year, Robert Jameson died. For half a century he had held the professorship of Natural History in the University of Edinburgh. After a brief interval the vacancy was filled by the appointment of Edward Forbes, to the great satisfaction of the naturalists of the country, and to the glory of the University. But the jubilation was soon and sadly ended. Forbes had been much harassed before the appointment was made, and had to hurry north to give the summer course of lectures. But, as he said himself, the authorities, in pressing him to enter on his duties at such short notice, were killing the goose that laid the golden eggs. After a brief illness he died on the 11th of the following November. I heard him lecture, and looked forward to the great advantage of studying under him. I was privileged to be introduced to him by his college-friend George Wilson, and realised his wonderful personal charm. Little did one dream at that time that we were so soon to lose him, that Dr. Wilson would be selected as his biographer, but that in a few years he would die before the memoir was half-written, and that I should be entrusted with the task of continuing and completing it.

Carrying out my determination to pursue literary studies in addition to science, on 1st November, 1854, I matriculated at the University of Edinburgh as a student of Humanity, that is, Latin, under James Pillans, and of Greek under John Stuart Blackie. Pillans had been teaching so long in Edinburgh, first as Rector of the High School and then as Professor in the University, that Robert Chambers used facetiously to divide mankind into two sections—those who had been "under Pillans" and those who had not. I am glad to be included in the first division. The professor had now reached his seventy-seventh year, and was therefore no longer as fit as he had once been to control a large gathering of raw lads, fresh from school in town and country. But no one who was willing to learn could fail to find much that was suggestive in the prelections of the veteran professor. As he sat in his chair behind his desk, his small stature was not observable. One only saw the round bald head, the rubicund cheeks, the mild blue eyes, the hand wielding a large reading glass (for he would never consent to wear spectacles), and the shoulders wrapped round in his velvet-collared black gown. He was a true scholar, though more intent on the spirit and style of his Latin favourites than on grammatical niceties and various readings. How he loved his Horace ! and how readily he took to his heart any student in whom he could detect the rudiments of the same affection ! It was to this bond of sympathy that I owed his friendship up to the close of his life. Among the interruptions of the work of the class by a handful of unruly lads, just escaped from the subordination of school, it was impossible not to admire the patience and dignity with which he sought to restrain these ebullitions. I felt keenly for him, and was led, out of sympathy, to bestow even more care upon the essays and papers which he prescribed, than I might otherwise have given. He kept up the pleasant old custom of asking his students to breakfast, and he continued these hospitable invitations to me after I had passed through his class.

John Stuart Blackie was a man of totally different mould. He had held the professorship of Greek for only two years, having previously been occupant of the Humanity chair in the University of Aberdeen. Unlike the sedate and contemplative Pillans, Blackie was nervous, excitable, voluble ; rapidly flashing from one subject to another, gifted with a sense of humour that relieved and half-excused some of his eccentricities. His well-modelled face was lit up with sparkling eyes that seemed always alert. At this time he wore a brown wig, which a few years later he discarded, so as to allow his long scanty, grey locks to fly about, whereby he became much more picturesque. His slim figure seemed to be all on wires, and was thus a faithful index of the mental activity and versatility of the man. His restlessness of mind was characteristically displayed in the work of the class-room. The subject for the day might be a continuation of the study of Herodotus, but should any important event have been chronicled in the morning papers, such as an incident in the Crimean War, or the resignation of the Aberdeen ministry, the excitable Professor was tolerably certain to begin proceedings by at least an allusion to it, if not a more prolonged disquisition. Thus, to the great majority of his audience these extraneous dissertations, which showed the man in his raciest mood, were far more diverting than to plod through the long account of Egypt and the Egyptians given by Herodotus in the admirable second Book of his History. But to those who desired to advance in the study of Greek these improvisations on modern affairs, though often infinitely amusing, were no proper substitute for what students had come to acquire, and a certain proportion of one's time had to be regarded as lost, so far as Greek was concerned. Milton's line used to suggest itself—

"The hungry sheep look up and are not fed."

I confess to have been one of the dissatisfied sheep. And yet there was something so winning in the Professor's ways, and often what he said was so true and so impressive that I forgave him then, and when in later years I came to know him intimately in the Edinburgh Senatus, and saw much further into his nature, my transient student displeasure gave place to affectionate regard.

It was the custom of the students at the University to "drop in" occasionally and listen to the prelections of other professors than those whose classes they were attending. Following this custom, I beheld Sir William Hamilton, whom I had been taught to regard as the most distinguished philosopher of the day. Too feeble to lecture, he was wheeled into his class-room, and his discourses were read for him by an assistant who sat at his side. The impression made by his grave pale face, massive brow and look of weariness remains ineffaceably in my memory. It was his last session but one, for he died on 6th May, 1856. Not less distinct is my recollection of the assistant who read the lectures—a big-boned Celt, with a look of strength and kindliness in his large and strongly-marked features. He had gained high honours in philosophy, and eventually went to the bar. In later years I came to be intimately acquainted with him as Alexander Nicolson, one of the most genial, gifted and unsuccessful of the advocates who paced the floor of the Parliament House. He will be again referred to in subsequent chapters.

In February 1855, while in the full tide of classical studies, I was seized with a violent attack of ulceration of the throat which kept me away from college for six weeks. I had taken a high place both in the Humanity and the Greek class. Pillans had expressed his gratification with my papers, some of which, as I afterwards learnt, he said were among the best he had ever received. My class-fellows, too, used to compliment me upon my work, and before my illness the opinion was currently expressed among them that I had every chance of gaining the gold medal. In Greek, too, my position was a good one. The professor had especially praised my translations into English verse, on which he inscribed in Greek that they were "not without the Muses."

When I returned to the College, the students were hard at work preparing for examinations which were to take place in a week or two. Continued weakness prevented me from being able to recover lost ground. It was a bitter trial, but had to be borne, in the belief that it had been sent for some wise though inscrutable purpose. I came out in the prize-list of both the Latin and Greek class, but of course, not in the high place which might otherwise have been gained. Yet there were compensations. In announcing his prize-list Professor Blackie took occasion to allude to my illness, and to assure the class that had I been well I should doubtless have taken a very high place. The venerable professor of Humanity was not less consolatory. He had given out a voluntary task, as what he called an *ultimus labor*, to conclude our work under him. Each student might choose his own subject. As I loved and respected the old man, I devoted myself, heart and soul, to leave with him as finished a composition as I could prepare. I selected Horace's exquisite Ode, *Aequam memento rebus in arduis Servare mentem*, and translated it into an appropriate English metre, adding a large number of parallel passages from Latin, Greek and English poets. When the Professor announced that he had received several *ultimi labores*, he dwelt especially on mine, remarking that he was so much pleased with it, and with the spirit which dictated it, that he intended to present me with an extra prize. He thereafter asked me to come and breakfast with him, and he then gave me a special mark of his regard. He told me of John Brown Patterson, who had been his pupil in the Rector's class at the High School, where, as Dux of the school, he in 1820 carried off the medals both in Latin and Greek. This distinguished scholar passed from the High School together with the Rector to the University, the Rector having been elected Professor of Latin. Patterson was the most brilliant classic the Professor had ever had, and I was shown an interleaved copy of Horace into which he had inserted a large number of parallel passages from Latin, Greek and English poets. This precious volume Pillans placed in my hands on loan, with permission to transfer any or all of the passages quoted. I had a fitting edition of Horace interleaved, and transcribed into it every quotation which seemed to me apt and memorable—a book that for many years was my constant companion, and in the poet's own words, "laborum dulce lenimen medicumque." This was the beginning of a friendship with the Professor which increased in intimacy as years passed, up to his death in 1864.

When the College session ended I had the reputation of being one of the best classical scholars of my year and best writer of English prose and verse among my class-fellows. I naturally looked forward to a further successful career at the University. But a wholly unlooked-for catastrophe put an abrupt end to my day-dreams. Through the delinquency of a relative my father was suddenly called on to pay a large sum of money. Keenly solicitous for my success in life, he never for a moment hinted that he could no longer support the expenses of my education; but I soon saw that it would be with difficulty that he could do so without injury to the rest of the family. It was not in my power to make enough by private tutoring to enable me to pay my way. With a sad heart I resolved to relinquish my further studies at the University, and find employment of some kind until such time as an opening might present itself into the Geological Survey. *Cras ingens iterabimus aequor.*

In my perplexity I went to consult my ever kind and wise adviser, Dr. Wilson, as to what he thought should now be done. Before there was time to place my difficulties before him, he hastened to tell me that he had heard that Professor Ramsay was soon to be in Edinburgh, from whom he hoped to learn more as to the prospect of entering the Geological Survey. About the same time Hugh Miller called on my father to show him two letters. One of these was from Sir Roderick Murchison, who in May of this year (1855) had been appointed Director-General of the Geological Survey in succession to the late Sir Henry De La Beche. He lost no time in looking into the matter of the extension of the Survey to Scotland, and the chief purport of his letter was to enquire whether Miller knew of any young man qualified to be appointed to a post in the Survey. The other letter was Miller's reply. His original draft of this letter came into my possession after his death. It contained a warm recommendation of me, and a reference to his published notice of my work in Skye.

Not long afterwards Professor Ramsay came to Scotland in connection with the continuation of the geological survey of East Lothian, which he had himself begun in the previous year. At an interview with him in Edinburgh I expressed my wish that if possible the question of my appointment to the Survey might be settled soon, as circumstances had recently arisen which made this desirable for myself. He was most friendly, explained to me more fully the duties required in the Survey; set out its hardships and toil, and the slender pay which it offered. I had a strong inward conviction that notwithstanding the drawbacks, so clearly laid before me, this service provided the course which was marked out for me. It offered a practical training in geological observation such as could hardly be obtained in any other way; and after all, it might be a stepping-stone to something higher, for which this training would be a valuable preparation. So I told Professor Ramsay at the end of our interview that I believed myself able to endure the fatigues, if he thought me qualified for the work, and that I was willing to enter now. Seeing this to be obviously my ardent desire, he informed me, to my great joy, that I might consider my appointment certain; that it would only be a question of time, and probably only a short time. Meanwhile he was about to return to London, where he hoped to discuss the matter with Sir Roderick Murchison.

I believed then, and looking back across a vista of more than three score years, I believe still that

> There's a Divinity that shapes our ends,
> Rough-hew them how we will.

I recognise now, more clearly than I could see at the time, how all the various influences around me had been guided, as each unexpectedly but opportunely, and without effort on my part, came successively into play. The *Religio Medici*, of which George Wilson had given me a copy, was in those days one of my favourite books, and I found there that Sir Thomas Browne had well expressed my own conviction in his quaint words: "Nor can I relate the history of my life, the occurrences of my days, the escapes of dangers, and hits of chance, with a *Bezo las Manos* to fortune, or a bare gramercy to my good stars. . . . Surely there are in every man's life certain rubs, doublings and wrenches, which pass a while under the effects of chance, but at the last, well examined, prove the mere hand of God." And he sums up his faith in the profound remark that "Nature is the art of God."[1]

After some anxious months, it was a pleasant change and refreshment to revisit Skye, where I spent six weeks among my kind friends at Kilbride and other parts of the island. I was able to add considerably to my knowledge of the geology of the ground, and to enhance my keen enjoyment of the scenery. Having now picked up a little more Gaelic, I could often make out the gist of a conversation in which, however, my limited vocabulary prevented me from taking part. From the hospitable home at Kilbride I was franked on to one friendly household after another. It was at this time that I made my first excursion to Scalpa and into the northern regions of Skye, and had the experience narrated in *Scottish Reminiscences* (pp. 399-406).

From the West Highlands I returned in time to join the British Association, which met at Glasgow. The Geological Section, over which Murchison presided, was attended by a large muster of geologists, most of whom I saw for the first time. There were present Sedgwick, Lyell, Hugh Miller, A. C. Ramsay, John Phillips, H. C. Sorby, Nicol, Harkness, and others of less note. Murchison's duties in the chair kept him so busy that Ramsay could never fix for me the appointment with him which he had promised, but he assured me that the Chief would not leave Scotland without seeing me, and would probably be able to do so in Edinburgh. In the *Life of Murchison* (vol. ii. p. 206) I have given an account of the Geological Section at this meeting of the British Association, and of the discussion on Sutherland in which Murchison, Sedgwick, Hugh Miller and others took part. After a week or two I had the promised interview with Sir Roderick at his hotel in Edinburgh, where Ramsay introduced me to him. He made a number of enquiries about my health and walking powers, affirming that to a geologist his legs are of as much consequence as his head. In the end he informed me that I might

[1] *Religio Medici*, xvi., xvii.

consider myself appointed, and hold myself in readiness to start for the field. Although earlier in this same year the Order in Council had been issued requiring tests for fitness in candidates for appointment to the Civil Service, the arrangements for the creation of a Board and examiners were still only in process of organisation. I appear to have been the last member of the staff appointed in the old way. I never passed through the hands of the Civil Service Commissioners.

A few days after the interview with the Director-General I was directed to leave in ten days for Pembroke, there to join W. T. Aveline, who was revising the maps of that district. This order, however, was subsequently altered by another which required me to join H. H. Howell, who had begun work at Haddington. Accordingly I reported myself there on 19th October, 1855.

# CHAPTER III

## FIRST YEARS IN THE GEOLOGICAL SURVEY (1855–1860)

At the outset of this narrative of the life led by a member of the Geological Survey of the United Kingdom, it will be desirable to describe briefly the nature of the duties which he had to perform. These consisted of two main parts, I. field-work, and II. indoor-work.

I. The field-work included the following out-of-door occupations: (1) To trace on the ground the outcrops of the various rocks, and to insert them in pencil on the sheets of the Ordnance Survey. These sheets thus became geological maps, and the process of so converting them was termed "mapping." The maps being on so large a scale as six inches to the mile, there was usually room upon them for the insertion, not only of much detail in the lines traced, but also of tolerably full descriptive notes of the characters and more important exposures of the rocks. (2) To record in a note-book more detailed observations for which sufficient space could not be found on the maps, together with drawings of actually visible sections, whether natural or artificial, especially if containing material that would be required for any future printed account of the geology of the ground. (3) To arrange the work of the Collector, by noting the sections to be searched for fossils and those from which rock specimens should be obtained; in many cases also to accompany the Collector to the rocks *in situ* and start him in his work.

II. The indoor-work chiefly consisted in inking-in and

colouring the field-maps, and writing out more fully the descriptive notes made on the ground. This duty required to be attended to from day to day, so that the pencilling on the maps might not become rubbed and indistinct, but in a spell of fine weather it was apt to accumulate. It then formed occupation for the first wet day, when field-work was impracticable. In course of time, as explanatory memoirs and other office work became necessary, they were generally relegated to the winter months.

The whole work of the Survey was thus primarily a scientific investigation. Its immediate economic or industrial value was not in those days sufficiently insisted upon by the higher authorities on the staff. The most important economic service rendered by the Survey at that time lay in the careful mapping of the coal-fields and other mining tracts of ground. It was left to the individual officer to detect and make note of any material which might be of economic use. An instance of this kind occurred in my own experience. In mapping the western part of Midlothian and the eastern part of Linlithgowshire or West Lothian I traced certain bands of black shale which appeared to occur on definite horizons in the lower division of the Carboniferous series. They had never apparently been worked for any purpose, though some of them were so bituminous as to be easily kindled into flame. Mr. James Young, afterwards known as "Paraffin Young," consulted me as to the extent of these shales, and accompanied me on the ground. I was able to show him many localities where their outcrop could be seen, and to indicate to him roughly the area under which they extended. He did not say anything about the purport of his enquiry. But in a short while, having secured the right to work these and other shales over a considerable tract of ground, he began active operations for the extraction of mineral oil from them. He thus founded the oil-shale industry of Scotland from which so much wealth has since been obtained.

In the middle of last century the large-scale Ordnance

sheets had only been begun to be prepared for publication, and in Scotland were available for but a small part of the country. Consequently, as it had been determined to map the ground geologically on these sheets, areas in which the Geological Survey could begin its work were limited, and dependent on the progress of the Ordnance department. Such pressure had been brought to bear on the Government to commence the geological mapping of Scotland that it was decided not to wait for the completely engraved and finished plates, but to begin at once with those on which the topographical features had been so far inserted as to make them available for the purpose of the geologist. Though the country was to be surveyed on the six-inch scale, the geological work was intended to be published only on the scale of one inch to the mile, save where for industrial needs, such as coal-fields and other mining tracts, the larger scale might be employed. The Ordnance maps on the smaller scale of one inch to a mile were in a still more backward state. It was determined that the first of these maps to be completed and published should be that which contained the city of Edinburgh.

An officer engaged on the Geological Survey was constantly assailed with the question, "How do you know what lies below the surface of the ground? Do you dig or bore, or how?" And the mystery of his methods became still more perplexing when he was seen at work, yet without any visible apparatus or implement. He carried a thin leathern portfolio which held the requisite sheets of the Ordnance Survey cut down to a convenient size. This portfolio or "map-case," suspended by a strap over the shoulder, hung behind him inside his coat, and could at any moment be swung round and opened. The hammer was slung on the left side in a sheath attached to a leather waist-belt from which, on the right side, there was also suspended a small leather case enclosing a prismatic compass, so that both hammer and compass were concealed by the coat. The clinometer, note-book, lens, and other small objects, such as pencils, rubber, etc., went into the numerous pockets which formed a feature of the attire. A man thus accoutred, as he traversed the country on foot, showed no visible indication of his calling; and yet all his instruments, though out of sight, were at once available when required.

During the first ten years of its existence the Geological Survey was a branch of the Board of Ordnance. Its officers wore a dark blue official uniform. But a tight-fitting, well-buttoned frock coat could only be an inconvenient garment for the rough scrambling and climbing life of a field-geologist. It was accordingly at once discarded when the Survey in 1845 was placed under the Office of Woods and Forests. Each officer thenceforth chose the civilian garb that pleased him. When the military uniform was cast aside, and each member of the staff dressed in any sort of suit he chose, the nature of his occupation formed a constant source of wonderment to the rural populace. Every member of the service could draw up a curious list of the guesses which he found had been hazarded by the natives as to what he was doing. Thus, where knobs of rock projected through pasture and the surveyor examined each of them in turn, he necessarily displaced the sheep, and if a farmer happened to be watching his proceedings the geologist could not be surprised to be taken for a sheep-dealer, and to be asked what price he would be willing to give for the ewes he had been so carefully examining. If his mapping led him to the outskirts of a mining village and compelled him to thread his way to and fro among the houses, he was tolerably certain to be set down as a new inspector of nuisances, and might even overhear rude remarks about his visit. The gamekeepers were much at a loss what to make of him. Instinctively they were at first disposed to suspect him as a poacher of some sort, though he seemed to carry no gun. He learnt in the end that it was always best, in a game district, to be on good terms with the proprietors and farmers, and then the keepers, instead of sinister suspicion, often gave him valuable help in his traverses of the ground.

In recent years the conditions of service in the Geological Survey have been greatly improved in comparison with what existed when I joined. In those days the Survey offered no attraction save to men passionately fond of an open-air life, devoted to geology and to whom rate of pay was a secondary consideration. On entering the service the assistant geologist obtained the modest salary of seven shillings a day, amounting to £127 15s. per annum. A small additional grant of what was called "personal allowance" was made for each night spent at an inn. The strictest economy was enjoined. We were told that the Travelling Vote annually granted by Parliament was so small that only by the greatest carefulness could it be made to cover our current expenses. We were expected not to hire vehicles in order to drive to and from our ground, where this could possibly be avoided. Consequently, when the distances to which the mapping had been pushed from a station became so great as to involve serious loss of time in walking to and from the ground, we had to shift to some more central place. It was a further rule that where private lodgings could be had, we should not incur the expense of living at inns or hotels, save for the day or two that might be taken up in changing from one station to another. It will be understood that in many parts of the country lodgings were not to be procured, and even when they could be got, they were often far from comfortable.

The hours during which an officer should be engaged in field-work were understood to be normally from 9 o'clock in the morning, when he started out for the day, until 6 o'clock in the evening, when he got back to his station. But in practice and during the long days of summer, these nine hours were often exceeded; while in winter and in wet weather, they were necessarily curtailed. Mapping was greatly dependent on the weather. In rain and in a high wind, it was usually difficult, sometimes impossible, to open out the maps. It not infrequently happened that after a walk of some miles to the ground the weather became such as to prevent any satisfactory work from being done with the maps. By studying the sky with care, and noticing its changes from hour to hour, one became a practical meteorologist. To the geological surveyor gales of wind were almost greater enemies than heavy rains. I remember an occasion when on a breezy hill-top a map-case, incautiously opened, had its precious contents dispersed far and wide. But fortunately, it being a day of official inspection, with four members of the staff present, we scattered in all directions and succeeded in recovering every one of the pieces of the map, none of them seriously damaged. An umbrella permitted maps to be looked at, even in heavy rain, if there was no wind. But in our northern blasts that artificial shelter was sometimes blown inside out.

It may here be remarked that in the early years of the Survey's progress in Scotland it was not found necessary to suspend the work during the winter, unless the ground was covered with snow, and even then the interruption seldom lasted more than a few days. The time had not yet come when the amount of indoor work accumulated to such a degree as to make retirement into winter quarters necessary. In my own case, I was able to carry on field work during the first five winters, with comparatively few and brief interruptions on account of the weather. It was not until the sixth winter that I was withdrawn to the head office in London for three months of indoor duty.

Returning to my narrative, where it paused at the end of last chapter, I may remark that the task before my colleague and me was to continue the mapping which our Director, A. C. Ramsay, had begun in the previous year (1854). From Dunbar as a centre he had examined the ground for some distance southward into the range of the Lammermuir Hills. We now had to continue his work, by tracing the boundaries of the rocks westwards between the coast on the one side and the Southern Uplands on the other. In the course of four months we mapped a large tract of East Lothian, from the stations of Haddington, Gifford, East Linton, and Dunbar.

We were too short a time at each of these places to be able to make many acquaintances among the inhabitants of East Lothian. Through the introduction of Dr. George Wilson, however, I was received into the well-known family of the "Browns of Haddington," one of whom was the brilliant Dr. Samuel, chemist and essayist. This accomplished man of letters was then in feeble health, and confined to his room, so that I did not see him. But I was greatly struck with his venerable mother—a typical Scottish lady of the old school, with a quiet dignity of manner and a power of concise speech which revealed a sound judgment. The younger ladies of the household were much interested in spiritualistic sittings, which had been taking place in Edinburgh under the auspices of Dr. Gregory, Professor of Chemistry at the University, Robert Chambers and others. The party discussed these exhibitions, manifestations and revelations, some of the speakers expressing a conviction of their genuineness and importance, when the old lady who had been quietly listening to the talk, but repressing her opposition, rose from her seat at the luncheon-table and put an abrupt end to the conversation by demanding in the language of St. Paul, "What man knoweth the things of a man, save the spirit of man which is in him."

Some of the little towns and villages in East Lothian were in those days charmingly quiet, quaint and picturesque. Haddington had its Abbey Kirk which, though ruined, retained enough of its former glory to make good its claim to the title of the "Lamp of Lothian," by which it was known in monastic times. And, on the other side, came the prestige of the little town as the birthplace of John Knox, who directly and indirectly was responsible for the demolition of so much of the ecclesiastical architecture of Scotland. The little village of Gifford, at the foot of the long chain of the Lammermuir Hills, had kept its old church-tower, town-house and market-cross, and was indelibly fixed in my memory as the hamlet with the hostel wherein Marmion encounters the mysterious Palmer in the third canto of Scott's poem. East Linton, too, with its antique bridge across the Tyne remains with me as a pleasant recollection, and Dunbar, besides all the quaintness familiar to me in boyhood, had now gained a new and absorbing interest from the volcanic history recorded in the rocks of its shores.

A feature of these days in the field may here be mentioned. As there could be no regular meal between breakfast and dinner, it was customary to carry a small packet of sandwiches, a buttered roll or other light refection, which could easily be carried in one of the coat-pockets. Being of Juvenal's opinion "ventre nihil novi frugalius," I sometimes became so interested in the mapping on which I was engaged that the little parcel of food came back in the pocket unopened. Hence in my case arose the practice of not taking luncheon—a habit to which I have since adhered.

By the end of January 1856 I was considered competent to carry on field-work by myself, and received instructions to make Edinburgh my headquarters. The general geological features around that centre were already tolerably familiar to me, but I was now to work out their details with the minuteness required by the Survey system of mapping. I had mastered the leading principles of that system, so far at least as they were illustrated by the rocks of East Lothian, and I felt that with further practice I ought to be able to accomplish whatever might be presented by a district in which I was at home.

When I reluctantly gave up further attendance at the University it was with the firm resolve to make up for that serious loss, as far as was possible, by steady reading; especially in the subjects which would otherwise have been my study there. For several years after joining the Survey I was in the habit of rising early enough to secure an hour's reading before breakfast. Then, after my nine hours in the field, I set to work again with my books after dinner, and was seldom in bed much before midnight, and not infrequently until long after that hour. I never before had realised, as I did now, the value of what is called "spare time." Remembering Bacon's shrewd observation that "in studies, whatsoever a man commandeth upon himself, let him set hours for it; but whatsoever is agreeable to his nature, let him take no care for any set times; for his thoughts will fly to it of themselves;"[1] I constructed a time-table, on which each hour of the morning and evening had its assigned piece of work. Being alone and complete master of my time, I was able to continue this practice until it was found to require readjustment. In this way I succeeded in getting through a large amount of reading. Science had necessarily the first claim, and I tried to read widely not only in my own department but in other branches more or less connected with it. But time was also found for attention to literature. I kept up Greek and Latin reading (Horace was still a constant companion), and in the course of years made myself familiar with some of the greater writers in English literature from Chaucer to our own day, devoting particular attention to the study of style. Having a retentive memory, in which were stored long passages from my favourite poets, I used to recite these *plena voce* on the lonely moors, thus many a time, while starting a lapwing or curlew, shortening, or at least enlivening, my solitary way to or from the ground that was under examination.

I look back on this period of my life as perhaps the most studious and certainly one of the most delightful which I have been privileged to enjoy. The field-work, in the light of recent experience in mapping, was found to be wonderfully fresh and new, for the ground had never before been so minutely examined as the detail of the Survey necessitated. I was continually coming upon unexpected new facts which enhanced my interest and enthusiasm in the work. It seemed sometimes hardly credible that the rambles in the field, which had formerly been practicable only now and then as a holiday recreation, had come to form actually the business of my every-day life. Constant exercise in the open air kept me in excellent health, and even after many miles of tramping

[1] *Essay of Nature in Men.*

over hill and dale in the course of a full day, the elasticity of youth (*juventas et patrius vigor*) enabled me to sit down with undiminished zest to my books in the evening, amidst the quiet of the country villages where I lodged. It was there that I first fully appreciated the truth of Milton's line " Solitude sometimes is best society."

And yet I was not wholly without loving and devoted companionship. On one of my visits to Skye, my friends there had presented me with a fine otter-terrier—the most sagacious and affectionate four-footed creature I have ever known. I have been all my life passionately fond of animals, and this was the first dog I had ever possessed as my own. He seemed from the first to understand our relation to each other. While he had a kindly feeling towards every member of my family, and was genial also to my friends, it was for myself that the full wealth of his canine " fidelitas " was reserved from the beginning to the end. When the ground on which I was at work offered no difficulty about game, he shared my quarters in the country villages and was supremely happy. He became quite expert in recognising rock-exposures at which he knew I would have to halt, and he was never more in his element than in threading a ravine with me, when he seemed to be keeping one eye on my movements and the other on the possible appearance of a rat or other living thing. If in the evening, as he lay stretched on the hearth-rug, some sign of impatience escaped from me, as I read a book or wrote a letter, he would rise at once, put a paw on my knee and look up in my face with a wistful look of sympathy in his liquid eyes.

The first year of service in the Geological Survey is one of probation. By the close of that year I had mapped the ground on which the city of Edinburgh stands, together with part of the surrounding country. The work received the approval of the Director, who went over some of it with me, and I was encouraged by being made a permanent member of the staff.

Having now learnt the art of geological mapping, and having recognised its great practical value in geological investigation, I resolved to make use of it on the ground which I had in previous years been examining in Skye. Several visits to the district of Strath had made me acquainted with the general structure of that ground, but obviously the knowledge would be enlarged and made more exact by tracing the actual boundaries of the various rocks upon a map. No Ordnance maps of Skye were then available, but the Admiralty chart included most of the ground, and the map in Macculloch's work on the Western Isles was also useful. In the course of a few weeks in the summer of 1856 I completed the survey of the parish, and the writing of a paper descriptive of the geology formed part of the occupation of leisure hours in the following winter. Macculloch's pioneer work in the topographical geology of Scotland deserves the fullest recognition. It would probably have been more appreciated in his lifetime, but for his great self-esteem and his antagonism to the geologists who were rising into fame; some of them even venturing to examine the rocks of Scotland and publish their observations thereon. These men, who included Murchison and Sedgwick, he stigmatised as "ignorant, superficial, vain and arrogant intruders into the science." He believed that he had laid down the general distribution of the rocks of the country on his map with sufficient detail and accuracy to be the basis of any future re-survey, which, as he wrote, "will not require a refined geologist to do. The rocks are few, and it is easy to learn to recognise them; there is nothing which any man may not attain on this narrow subject, with a few weeks of experience. It will confer no particular fame on any future self-constituted geologist to have done what could have been effected by a surveyor's drudge or a Scottish quarryman."[1] Of the

[1] *Memoirs to His Majesty's Treasury respecting a Geological Survey of Scotland*, by J. Macculloch, Esq. London, Samuel Arrowsmith, 1836, pp. 17, 85. These *Memoirs* in a pamphlet of 141 pages were published after the Doctor's tragic death. Had he lived he would probably have toned down the harshness of his vituperation as the MS. passed through the press. See Ami Boué's opinion of Macculloch, given on p. 151.

true origin of the rocks, and the working out of the geological history of the earth from their evidence, he appears to have had but a feeble conception.

While surveying the range of the Pentland Hills, to the south of Edinburgh, I was much interested to find that the custom still prevailed among the peasant population of acting Allan Ramsay's pastoral play of *The Gentle Shepherd* in the midst of the very scenery which had inspired the poet. The Scottish language of the dialogue was given by the rustic actors with full Doric breadth, and even sometimes with creditable dramatic power. That the poem, which was published in 1725, should survive in the affections of the peasantry, is strong evidence of the force and fidelity of its picture of Scottish rural life. Its survival in this form has probably kept much of the old Scots tongue still in use throughout the district. Mr. Horatio F. Brown, proprietor of New Hall, the scene of the play, informed me in 1913 that an occasional performance in the old vernacular still takes place.

A new interest in the valleys and ravines of the Pentland Hills arose when I had the good fortune to discover among them, underneath conglomerates and massive sheets of lava of Old Red Sandstone age, a group of shales crowded with fossils. Some twenty years earlier Charles Maclaren had found a shell (*Orthoceras*) there, but nothing further had been done in searching the strata for more organic remains which would determine their geological age. I disinterred a good many species of brachiopods, lamellibranchs and other fossils, well preserved and identical with familiar organisms in the Upper Silurian part of the Geological Record. When subsequently the Survey Collector was put on the ground he obtained a number of additional species.

Such a "find" as this gave special emphasis to a thought which from the beginning had often been in my mind. The work on which I was now engaged, and to which I had dedicated my life, was not merely an industrial employment; the means of gaining a livelihood; a pleasant occupation for mind and body. It often wore to me an aspect infinitely higher and nobler. It was in reality a methodical study of the works of the Creator of the Universe, a deciphering of His legibly-written record of some of the stages through which this part of our planet passed in His hands before it was shaped into its present form. A few of the broader outlines of this terrestrial history had been noted by previous observers in the Lothians, but many other features still remained to be recognised and interpreted, while the mass of complicated detail, so needful for an adequate comprehension of the chronicle as a whole, was practically still unknown. I felt like an explorer entering an untrodden land. Every day the rocks were yielding to me the story of their birth, and thus making fresh, and often deeply interesting, additions to my stock of knowledge.[1] Lucretius, in a well-known passage of his great poem on Nature, has given expression to the enthusiasm wherewith a mind may be filled by the contemplation of a region of philosophical thought which no human foot has yet entered, and he warmly pictures the joy of the first comer who approaches and drinks from springs which no human lips have yet tasted. It is true that the poet confessed that he felt goaded onward by the sharp spur of fame, and by the hope that from the new flowers of the unexplored realm he would gather a splendid coronal, such as had never yet graced the head of any mortal. I had no such ambitious dreams, though I looked forward to rising in my profession to something higher than the "surveyor's drudge or Scottish quarryman" of Macculloch's sarcasm. Meanwhile the discovery of new facts in the ancient history of the land gave me in itself an ample store of pleasure, augmented by the delight of fitting them into each other and extracting from them the

[1] Virgil's prayer is one which the geologist may literally appropriate:

Sit mihi fas
Pandere res alta terra et caligine mersas.
*Æneid*, vi. 266.

connected narrative which they had to reveal. I remember the indignation with which, about this time, in reading Cowper's poems, as I loved to do, I came upon his ignorant and contemptuous denunciation of astronomers, geologists and men of science in general whose work in life was described as

> the toil
> Of dropping buckets into empty wells
> And growing old in drawing nothing up.[1]

More tolerable was Wordsworth's condescending description of the geologist, going about with a pocket-hammer, smiting the "luckless rock or prominent stone," classing the splinter by "some barbarous name" and hurrying on, thinking himself enriched, "wealthier, and doubtless wiser than before."[2]

Samuel Johnson is recorded by his biographer to have expressed his belief that "there is no profession to which a man gives a very great proportion of his time. It is wonderful, when a calculation is made, how little the mind is actually employed in the discharge of any profession."[3] Whether the active clergyman, lawyer or medical man would subscribe to this statement may be questioned. But I found that the profession which I had chosen demanded all the time and thought which I could give to it, both with body and mind.

The discovery of a large group of organic remains, previously unknown to exist where I found them, might have been made by anybody, and conferred no particular credit on the finder. But it made a profound impression on my mind, as its full meaning in the unravelling of the geological history of the district gradually revealed itself. I lost no time in including it, with an account of my recent doings in Skye, in a letter to my kind friend Hugh Miller. His note in reply, from its interest in reference to his own field-observations, which he never lived to publish, may be inserted here.

[1] *The Task*, iii. 150, 197. [2] *Excursion*, book iii.

[3] Boswell's *Life of Johnson*, April, 1775.

Wednesday evening 9th October, 1856

My dear Sir

Could you drop in upon me on the evening of Saturday first, and share in a quiet cup of tea. I am delighted to hear that you have succeeded in reading off the Liassic deposits of Skye, and shall have much pleasure in being made acquainted with the result of your labours. Like you, I want greatly a good work on the English Lias, but know not that there is any such. My explorations this season have been chiefly in the Pleistocene and the Old Red. I have now got Boreal shells in the very middle of Scotland, about equally removed from the eastern and western seas. But the details of our respective explorations we shall discuss at our meeting.

Yours ever

Hugh Miller.

Having been fortunately able to accept this invitation, I spent a couple of hours with him in his home at Portobello. He was interested in the discovery of so large a series of well-preserved fossils in the basement rocks of the Pentland Hills, and not less pleased with the details of my work in Skye. He had spread out on the table his trophy of northern shells, which enabled him to affirm that, at a late date in geological time, Scotland was cut in two by a sea-strait that connected the firths of Forth and Clyde. He had found marine shells at Buchlyvie in Stirlingshire, on the low ground between the two estuaries. Finding me not quite clear as to the precise location of Buchlyvie, he burst out triumphantly with the lines which Scott puts at the head of the immortal chapter of *Rob Roy* wherein are told the adventures at the Clachan of Aberfoyle :

Baron o' Buchlyvie,
May the foul fiend drive ye,
And a' to pieces rive ye,
For building sic a toun,
Where there's neither horse-meat, nor man's meat, nor a chair to sit doun.

Two months later I had another memorable evening with him, and had then the opportunity of showing him the maps of the Geological Survey of Midlothian, as far as the work had gone. He was especially struck by the details of the large-scale map of Arthur Seat, and the volcanic history to which they pointed. Having realised the successive phases of this record, he with characteristic rapidity passed from the details which I had given him to the sequence of events in the life of the volcano, and turning to his daughter, who with his wife was sitting near, he exclaimed to her, "There, Harriet, is material for such an essay as has been prescribed to you at school." Then in a few graphic sentences he drew a picture of what he believed to have been the history of the eruptions.

It was my last interview with this great man. Next week the country was startled to hear that he had been found dead in his bath-room, with a pistol lying by his side. To me the shock was overwhelming. Before I knew him personally, his writings had filled me with admiration of his genius, and I looked up to him with the enthusiastic veneration of ardent boyhood. But after my first introduction to him, he had by so many kindly acts shown his warm interest in my welfare that he had completely gained my affection. I had been with him only six days before his death, and had found him as lively and friendly as ever. This sudden bereavement was the deepest sorrow I had yet known. On the day of the funeral at the Grange Cemetery in Edinburgh, I saw the remains of my friend laid to rest next to the grave of Thomas Chalmers—two of the most illustrious Scotsmen of my time.

Mrs. Miller, whose love of literature was the first link that connected her with her future husband, the stone-mason of Cromarty, lost no time in collecting his essays and lectures, and publishing them in a succession of volumes. In this pious duty she asked me to assist her. The little help I was competent to give was willingly rendered in grateful remembrance of her husband's kindness.[1] Years afterwards, when the centenary of Hugh Miller was celebrated at Cromarty in 1902, an ampler opportunity came to me to acknowledge my admiration for his genius, my gratitude for his invaluable aid as a friend, and my life-long indebtedness to his example.

With Edinburgh as headquarters, it was pleasant to be able to see Professor Pillans from time to time. He continued to interest himself in my progress and to invite me to lunch or dine with him. As a confirmed *laudator temporis acti*, he hoped that my scientific pursuits would not extinguish my love of the Latin classics. I remember one conversation with him in which he told me that in his opinion the taste for what he called "elegant literature" had gone down. "Men nowadays," said he, "care only for three things—money-making, mechanical inventions and spasmodic poetry. I have been all my life fond of poetry, and I find solace in it still. But I must go back several generations for what really interests and pleases me. There is Tennyson, and another writer, Browning, that I hear people raving about. I have tried to read them, but I confess I cannot understand much of them, and they give me no real pleasure. When I want to enjoy English verse, I go back to the masterpieces of Dryden and Pope."[2]

On another occasion I had an interesting talk with him on his experiences as an educational reformer, especially in the management of infant schools. He had energetically combated the system of teaching by rote, and of compelling young children to burden their memories with genealogies and dates. "I was at an infant-school," he remarked, "and you won't guess what question I heard put to a class of little tots, not more than four or five years old—'How long did Jeroboam reign over Israel?'"

[1] Her acknowledgment of this assistance recalled his generous forecast, and she stated her belief that he looked upon me "as the individual who would most probably be his successor as an exponent of Scottish geology." Preface to *The Sketch-book of Popular Geology*.

[2] Quoted in *Scottish Reminiscences*, pp. 174, 357.

Pillans was not without a vein of dry humour. There was, and still is in Edinburgh, an exceedingly pleasant dining confraternity called the Royal Society Club, of which he was a member, to which further reference will be made in a later chapter. Like its prototype in London, it was a convivial gathering, meant not only to promote good-fellowship among its members, but to form a nucleus of attendance at the meetings of the Royal Society of Edinburgh which were held at a later hour of the same evening. Pillans would come to the dinner, but avoided the meeting that followed it. He would say to those who rallied him on this procedure: "I much enjoy the *play* (meaning the dinner); but I can't stand the *farce* (F.R.S.) which comes after it."

It was at one of the Professor's luncheon parties that I first met his life-long Edinburgh associate, Leonard Horner, who thenceforward became one of my kindest and most helpful friends. As will be told in later pages of this narrative, when I first went to London for the winter indoor-work of the Survey, he launched me into the scientific world there. Through him I became intimately acquainted with his son-in-law, the illustrious Sir Charles Lyell, and with other geologists resident in the metropolis, and at various evening receptions he introduced me to people of mark outside the ranks of science.

During the years 1856 to 1859 I was engaged in surveying the western half of the county of Midlothian and the northern half of West Lothian up to the borders of Stirlingshire. To me the most novel and outstanding feature of this region was the remarkable abundance of the traces of former volcanic action which were revealed as the mapping advanced. Charles Maclaren, as already mentioned, had published an excellent description of the structure and history of the volcanic mass that forms Arthur Seat, and of the nature and arrangement of the ancient lavas of the Pentland Hills. But I was wholly unprepared for the number of distinct volcanic vents that could be recognised, and for the evidence that a great deal of the igneous material which had been looked upon as intruded into the terrestrial crust, like the classical example in Salisbury Crags at Edinburgh, really flowed out at the surface as streams of lava, or was ejected into the air in the form of dust and loose fragments. Moreover, I found that it was possible to recognise a long succession of eruptions, indicating the persistence of volcanic activity in this part of Scotland for a protracted series of ages. The oldest lavas and tuffs in the region which I had surveyed appeared to be those intercalated in the Old Red Sandstone of the Pentland Hills. Another younger and widely traceable series could be detected on successive stratigraphical horizons, from near the base of the Carboniferous formations up at least as far as the base of the Millstone Grit. And later than all these were the massive basaltic dykes which could be followed for miles, with an east and west trend, cutting through the earlier volcanic ejections.

This mass of evidence was almost entirely new to science, and the gathering of it formed one of the chief sources of the enthusiasm with which the field-work filled me. I grew more and more engrossed by the innumerable well-preserved details of ancient volcanic action, and by the part played by the erupted material in the building up of the framework of the Lothian landscapes. From these first beginnings of the study, the wide subject of volcanism became ot increasing interest. It grew to hold the dominant place in my geological enquiries, leading me ultimately to pursue it far and wide over the whole of Scotland, and into every tract of England, Wales and Ireland where relics of volcanic history have been preserved. The study further impelled me in later years to visit the volcanic regions of Auvergne, the Eifel and Italy, and eventually to explore the great lava-fields of Western America. The spur which drove me through all these wanderings was unquestionably the enthusiasm for the subject first kindled within me by the unravelling of the volcanic history of my native county.

It was evident that in the geological history of the area of Central Scotland there had been at least two great periods of volcanic activity, each marked by its own distinctive features. Further research might discover evidence of still older as well as of younger eruptions in the same area. The possibility of the existence of traces of a vast chronological succession of volcanic periods in Scotland as a whole, engaged my thoughts and directed my holiday excursions for ten years, until at the meeting of the British Association at Dundee in 1867, I was able to present the first attempt to offer a sketch of volcanic history in Britain.

One of the difficulties attendant on the study of this history arose from the want of any convenient method of determining the petrographical composition of the different varieties of volcanic rocks. They had long been known by the vague epithet of the *Trap* rocks, a name borrowed by us from abroad, where it was assigned to igneous masses that occur in sheets with stair-like escarpments. But they obviously comprised many kinds of material, differing greatly from each other in composition and internal structure. Chemical analysis was a lengthy and expensive method of discrimination, and quite impracticable in the roving life of a field-geologist. The preparation of thin slices of the rocks for examination under the microscope had not yet come into use. In my perplexity I appealed to my old acquaintance Professor Fleming, who, as a Wernerian and a mineralogist, might possibly know of some method that would assist me. He replied that he had been in the habit of breaking fresh pieces of "trap" into smaller fragments which he placed in the bowl of a clay tobacco-pipe, and then thrusting the pipe into a brisk fire. The fragments were not melted, but their component minerals could then be more or less distinguished from each other. He concluded in his pleasantly sarcastic manner by congratulating me on being better "up to trap" than he could pretend to be. But I need not say that his process gave me no real assistance. A few years later, after the publication of Sorby's great paper, I was put in possession of the true key to the solution of these petrographical problems.

Some of the little villages or towns of West Lothian which were successively my headquarters during these early years, like those of East Lothian already mentioned, dwell pleasantly in my memory. South Queensferry was then a quiet sleepy place, with its Hawes Inn, much in the same state as when Jonathan Oldbuck frequented it, the quaint main street, the primitive little harbour and the tiny ferry steamboat that crossed so many times in the day to the north side of the Firth. The Provost at the time, Robert S. Wyld, whose acquaintance I soon made, proved to be one of the most genial and cultivated men I ever met, and our intercourse ripened into a warm friendship with him and his family. Dalmeny Parish, to which the house in which I lived belonged, has in its church one of the most beautiful and perfect relics of Norman architecture in Scotland, and I used to be glad to attend the Presbyterian Sunday service in so impressive a sanctuary. Queensferry unhappily is now no longer what it was then. The Forth Bridge, the railway, and many other modern innovations have destroyed its antique quaintness and somnolent quietude.

The town of Linlithgow was another charming resting-place for some months. The windows of my quarters there looked over the Cross Square to the old Palace and St. Michael's Church, and the fall of the water from the grim sculptures of the central fountain made a perpetual soothing lullaby. Here, as in an appropriate setting, I read the poems of Sir David Lyndsay of the Mount, Lord Lyon King at Arms, and could sometimes almost fancy the antique days back again. Among the old-fashioned streets, imagination carried one three centuries into the past; the noise of my hobnailed boots on the stone pavement sounded like the clank of spurs; I seemed to be elbowing burghers in doublet and hose, friars, heralds, pursuivants, rough soldiers, and fair ladies in damask and satin, tripping to the Palace gate, with page behind. Lyndsay's panegyric seemed none too high-sounding:

> "Adew Lithquo whose palice of plesance
> Micht be ane patrone in Portugal or France."

During the months spent in West Lothian one of my evening occupations was the preparation of an account of the geology of Strath in Skye. Dr. Robert Christison, when he heard of this paper being in progress, advised me, if I intended to keep Edinburgh as my field, to read my papers before the Royal Society of Edinburgh. He offered to read them for me himself, "unless they were very profound, in which case he would get a very profound geologist to read them." His advice was sound, and I acted on it as long as Edinburgh remained my centre. But in the present case there were cogent reasons why the paper should be read before the Geological Society of London; and it was accordingly read there on April 22, 1857. Professor Ramsay, who communicated the paper, sent me a gratifying account of its reception at the meeting, where Murchison, Lyell, Egerton, and Leonard Horner were among the auditors.

For several years after the publication of this paper, visits to Skye and the neighbouring islands continued to be the occupation of my holiday weeks. Particular attention was devoted to the island of Raasay, of which I carefully mapped the geology on the large Admiralty chart. A considerable collection of fossils was also made from the various subdivisions of the Jurassic series and a detailed paper was intended to be written recording the observations. But the palaeontologist, to whom the fossils were sent for determination, delayed to examine them, and eventually died; and the specimens were never recovered. Thus to my great regret the proposed paper never was written. But the weeks spent in Raasay remain in my memory as a bright and joyous time. The geology was full of novel interest, the weather exceptionally fine, and the landscapes in every direction singularly impressive. The effects of light and colour upon the lofty

Skye mountains on the one side and upon the heights of Applecross on the other, were such as I had never seen before. The sunsets were especially glorious, and owing to the prolonged daylight of mid-summer their tints seemed hardly to fade from the sky until they melted into the hues of sunrise. Truly one realised the full meaning of Juvenal's allusion to the short nights of Britain (minima contentos nocte Britannos). There was every inducement to linger on the heights until long after supper-time, and to return to the hospitable shelter of the gamekeeper's cottage only when the shadows had begun to deepen in the valleys.[1]

The exuberance of my joy in the active and continuous prosecution of geological investigation, alike on official duty and holiday excursions, could not find adequate vent in solitary country quarters. It filled me with an eager desire to interest others in a subject which, though lying open to all, attracted comparatively so few, but had completely fascinated myself. This effervescence of youthful enthusiasm led to the writing of a little volume, with the title of *The Story of a Boulder*, which was published in 1858. It met with a favourable reception from the press, and eventually took the fancy of Alexander Macmillan, who purchased the remaining stock from Thomas Constable of Edinburgh, the original publisher, and substituted a new title-page with the name of his firm as its publishers. It was never reprinted.

The desire to share with others the deep joy which the study of Nature had brought to myself, found ampler vent in later years. It formed one great source of the pleasure of my professorial life in Edinburgh, when each session brought a fresh band of eager youths to be for a time my companions and fellow-workers in the study. And it led to the production of a series of educational works which, as will be told in a later Chapter, have carried my message to thousands of readers in all parts of the civilised world.

[1] Some incidents of my visits to Raasay and the north of Skye are given in my *Scottish Reminiscences*, pp. 93, 227, 399.

My first visit to London after entering the public service occupied two weeks in February 1859, and gave me an opportunity of making the acquaintance of those of my colleagues in the Survey who were then at the Jermyn Street Museum, while through Leonard Horner and Sir Charles Lyell I saw a good deal of society in the short time. From this brief peep into the big world I returned with renewed alacrity to the completion of the first or Edinburgh Sheet of the Geological Survey map of Scotland. The portion of that sheet assigned to me was nearly finished south of the Firth of Forth, but it included also that part of the county of Fife which projects southward at Inverkeithing and North Queensferry. This strip of country had still to be surveyed, and in order to fit accurately into the sheet to the north, the geological lines had to be prolonged for a mile or two beyond the boundary of the Edinburgh sheet. It was my first acquaintance with the geology of Fife, and furnished me with many fresh and striking manifestations of volcanic phenomena—a foretaste of the rich harvest which the county was afterwards to yield in the same field. It was also an interesting district from the character of its old buildings and its historical associations, for it included at the one end Dumfermline, with its noble Norman abbey and the tomb of Robert the Bruce, and at the other the quaint little burgh ot Inverkeithing, with its curious natural harbour. Moreover, the islands in the Firth had to be visited and mapped, from Inch Garvie, between the two Queensferries, down to Inchkeith. Thus the completion of the sheet took up most of the earlier half of the year.

In another direction the progress of the mapping brought me back into East Lothian, where some of the boundary lines which we had run into the flanks of the Lammermuir Hills required to be carried well to the south, before the publication of the Haddington sheet could be accomplished. These tracts form part of the range of the Southern Uplands, stretching across Scotland from the North Sea at St. Abb's Head to the south-western coast of Ayrshire and Galloway—a lonely region of smooth rounded green hills, traversed by many narrow valleys, each with its clear river and scattered farms.

> The grace of forest charms decayed,
> And pastoral melancholy,

which Wordsworth so truthfully sang, greatly impressed me at this first visit, and their fascination has lasted ever since. Lodging in the little hamlet of Longformacus, I made the acquaintance of the Darlings of Priestlaw, of whom I have elsewhere given some account.[1] With no little interest I learnt from the elderly Miss Darling that when she was a girl she had accompanied Sir James Hall and Professor Playfair in their excursions up the Fassney Water in the last decades of the eighteenth century. She said that she had seen no geologist since then, and would fain hear something of what was thought now about the history of the earth. Besides possessing the characteristic family charm, she was noted for having collected and stored in her memory all the legends and traditions of her native vale which she had been able to recover. We agreed to exchange wallets, I giving her such information as I had been able to gather regarding the geology of Lammermuir, while she retailed to me with graphic force a most interesting collection of tales connected with the landscape before us, as we sat in her garden at the farm of Priestlaw, and looked over to the valley of the Whitadder and the heathy slopes beyond. It was arranged that after I had taken a few weeks of holiday, she should have a collection of stones brought up from the river, in order that I might discourse to her from them, while she on her part promised to continue her stories and legends. When the holiday was over I returned to Priestlaw, but found, to my infinite regret, that in the interval Miss Darling and one of her brothers had been laid in the grave—martyrs to their generous care of a poor tramp from whom they caught the fever out of which they had nursed him.

[1] *Scottish Reminiscences*, p. 206.

Part of the holiday just referred to was devoted to attendance at the meeting of the British Association, held this year (1859) in Aberdeen. Over the Geological Section Sir Charles Lyell presided, and there was a good company of geologists present, including Murchison, Daubeny, Nicol, Pengelly, and others. Pengelly read one of his earliest papers on the bone-bearing caves of the south coast. Sir David Brewster created some astonishment by communicating a paper on "A Horse-shoe nail found in the Old Red Sandstone of Kingoodie." He was too grave a man of science to be capable of a joke against Section C. The geologists kept their gravity with praiseworthy patience, but only the title of the paper was allowed to appear in the official Report of the meeting. It was on this occasion that I read my first paper on the subject that was engrossing so much of my time and thought—the chronology of the volcanic outbreaks in the geological history of Scotland. Among the auditors was my old friend Professor Christison, who reported to his son:—"I had only time to hear the close of a paper by young Geikie, and very warm commendations bestowed on it from all quarters."[1] I had noticed also among the audience Professor Blackie, standing against a wall not far from the platform, with his brown plaid wrapped round his shoulders, and his soft felt hat crushed between the hands that rested on his walking stick. After the close of the discussion on my paper he came up to me and abruptly put the question, "Weren't you in my class?" On my answering in the affirmative, his next words were "Well, come and dine with me." Having for so many years held the Chair of Humanity in the Aberdeen University, he had many old friends and acquaintances in the town, and it was with one of these families that he was now a guest. When I was ushered into the drawing-room next evening he came quickly forward, took me round the neck and introduced me with a little complimentary speech to every one of the ladies in the room, none of whom had I ever met before.

[1] *Life of Sir Robert Christison*, vol. ii. p. 276.

I knew it was "pretty Fanny's way," but this first personal experience of it, although most kindly meant, was a somewhat trying ordeal.

The Prince Consort being its President this year, the British Association was invited to Balmoral where Queen Victoria was then in residence. I remember with special pleasure that on the top of the omnibus or stage-coach which conveyed some of us up Deeside there was seated next me the artist E. W. Cooke whose tongue and pencil were wonderfully active during the whole of the excursion. I had learnt greatly to admire his paintings, which seemed to me to unite a faithfulness of detail with a breadth of treatment such as few other of our living landscape painters equalled. Before the end of the day we laid the foundation of an agreeable friendship which lasted as long as he lived. When in after years I was much in London, I spent many happy hours at his picturesque home on the south side of Hyde Park, and watched with no little amusement the gradual growth of his collection of grotesque and ludicrously impossible animals constructed by himself out of portions of shells, crustacea, and other organisms. On our way to and from Balmoral he astonished me by his mastery of the pencil in sketching. The items of amusement arranged for the members of the Association at the royal residence included Highland games, and a muster of Highland clansmen in full military attire, with bagpipes in ample blast. As the kilted ranks deployed, Cooke was able to transfer them to his sketch-book with amazing rapidity, giving the impression of number, as well as of the characteristic martial strut of the men. Another example of his rapid sketching power was shown in his delineation of the great granite quarry near Aberdeen, where, in honour of the Association, a huge mass of solid rock was brought down by an abnormally large explosive charge. He drew the scene just after the explosion, and with a touch of his characteristic humour, he represented the geological excursionists clambering over the fallen debris in search of specimens of the rock, some of them creeping towards inaccessible crevices like so many black beetles. This sketch he presented to Sir Roderick Murchison, from whose possession it came into my hands. As it stands on one of the tables in my library it continually reminds me of the artist and of my Chief.

A grievous loss befell me in the autumn of this year. On 22nd November my ever constant friend, George Wilson, died. His health, always frail, gave way somewhat more rapidly towards the end, but seeing him now much less frequently than when I lived in Edinburgh, I did not realise that the end was so near. His character was perhaps the most lovable I have ever known. Though fighting with constant bodily suffering, he maintained a brave cheerfulness and even buoyancy of spirit that were wonderful to see. His truly Christian patience and genuine goodness, his full-hearted sympathy that would so often overflow into active helpfulness; his keen love of science, the clearness and grace with which he discoursed on scientific subjects, and the singular charm which he brought with him wherever he came, were to me a memorable lesson. His old college friend, Edward Forbes, who well knew his worth, used to speak of him as "a splendid jewel in a shattered casket." For years he had been my most intimate friend. To him more than to any other was I indebted for guidance and encouragement at the beginning of my career, when the future seemed so dark and uncertain. As he watched my early progress he would sometimes say that a Professorship of Geology was sure to be established in the University of Edinburgh, and that I must keep that chair in view as the goal for which I should prepare myself by assiduous devotion to the pursuit of science.

The same month of November witnessed the most important event in the history of Natural Science during my time—the publication of Charles Darwin's *Origin of Species*. Geologists in this country were perhaps somewhat slow in appreciating the bearings of this remarkable treatise on their own branch of science. It taught them a new method of interpreting the crust of the earth and showed them that what they had presumed to be a fairly continuous and complete record, was full of gaps, even when these could not be detected by any visible stratigraphical breaks. I well remember the deep impression made on my mind by the reading of his two chapters on the "Imperfection of the Geological Record" and the "Geological Succession of Organic Beings." It was a new revelation of the manner in which geological history must be studied.

The amount of Survey indoor-work had now accumulated to such an extent that time had to be found for it in the winter by abandoning field-work for a while. No proper office had yet been provided in Edinburgh for the Survey, though temporary accommodation had been obtained in a building which was to be pulled down in order to make way for the present Science and Art Museum. The preparation of the maps for the engraver, the duplication of the 6-inch field maps, and the writing of the Memoir descriptive of the geology of the Edinburgh district could be more conveniently done at the head office in London. Henceforth for some years I spent two or more months of each winter in London.

This change of winter quarters led to the growth of a close friendship with Alexander Macmillan, whom, as already stated, I first met in Edinburgh at the house of George Wilson. He had not yet settled in London, but still lived at Cambridge. It became an agreed arrangement between us that when winter brought me to London, I should pass the Christmas recess at Cambridge, as a member of his family circle. Through him I made many friends both at Cambridge and in London. The Cambridge visits were always full of interest, for my host's circle of acquaintance among the University men was extensive, and he launched me into it with all his ardour. Among interviews which more particularly dwell in my memory one of the earliest and most memorable was an unforgettable evening spent with Professor Adam Sedgwick in his rooms at Trinity College. He happened to be in excellent spirits, and full of reminiscences of his early years. Among his narratives there was one of a Cumberland picnic, which he gave with much humour. The fore part of the day of the festival proved to be so wet that the luncheon, instead of being taken outside, was served in the house. Somewhat later, as the rain ceased and the sky cleared, the company resolved to do something in the open air. There had been some talk of an owl's nest in a tree near at hand, and Sedgwick, then a young man, proposed that they should pay a visit to the nest. As he told the story, sitting in front of the fire with his feet on the fender, he struck his thigh with the palm of his right hand, exclaiming at the same time, "I was a good deal suppler in the legs than I am now; so proposing to climb the tree and look after the owl, I asked if any of the young ladies would accompany me, but none of them volunteered for the ascent." He climbed the tree alone and found that sure enough there was a nest. The next proposal was to call at the abode of an old woman who was reputed to be a witch. "When we reached her," said the Professor, "she proved to be the veriest old hag I had ever set eyes on. Her dark skin, wrinkled and shrivelled, looked like an old pack-saddle, and I actually took out my pocket lens to have a better sight of it. One of the young ladies wished me to ask the woman how she had lived in her youth to be able to attain such longevity. Well, I put the question pretty much in the words of the questioner, but the creature could not understand what was asked; whereupon I put it in good broad Cumberland speech, to which she promptly replied in the same dialect, "O is that what the leddy wants to ken; tell her I married a tinker and lived for sixty years under the hedges!" And so the chat went on till late. I have always regretted that I did not write down at the time all that I could remember of it. But it was a rare experience to see old Sedgwick's face kindle and those eagle eyes of his sparkle in the lamp-light as he recalled his recollections of the past. I do not think that much geology entered into our talk. No allusion was made to Murchison or to the disagreement that had arisen between him and his old and intimate friend the Woodwardian professor.

In the course of holiday rambles in those years I had made out a new and rather important feature in the history of the Old Red Sandstone of at least the southern half of Scotland. I found the upper division of that formation to be separated from the lower portions by a strong break or "unconformability," pointing to the lapse of a long interval of time unrepresented in the region by any intervening strata. This observation formed the subject of a short paper which I read to the Geological Society on January 18th, 1860. But I had not forgotten Dr. Christison's advice to make communications to the Royal Society of Edinburgh. Expanding the paper which had been laid before the British Association at Aberdeen, I presented it the same winter to that Society, where, as a memoir "On the Chronology of the Trap-rocks of Scotland," it was read for me by Robert Chambers. Though not free from errors, it was the first attempt to arrange in sequence of time the different volcanic periods in the geological history of Scotland. The memoir appeared in vol. xxii. of the Society's *Transactions*.

The death of Professor George Wilson left the question of his unfinished *Life of Edward Forbes* to be considered. His sister, who was his executrix, wished that it should be entrusted to me, but Mrs. Forbes and some of her husband's friends naturally objected to place the task in the hands of a young and untried man. During the correspondence which ensued I received an invitation from Mr. Robert A. C. Godwin Austen to visit him at Chilworth, near Guildford, in Surrey. One of the ablest geologists of his time and one of the friends of Forbes, he had been entrusted with the task of completing and publishing some of that eminent naturalist's unfinished manuscripts. He had the reputation of being a caustic critic and, as Ramsay told me, he had made some strictures on my Skye paper at the Geological Society. It was therefore with not a little trepidation that I went to him, at his picturesque old manor-house in Surrey. We discussed the subject of the biography, and he proved most friendly and sympathetic. I parted from him with his blessing, and every good wish for the success of my undertaking. My colleague in the Survey, J. B. Jukes, also an old associate of Forbes, withdrew his opposition, and it was amicably arranged, about the middle of February 1860, that the biography of the eminent naturalist should remain in my hands.

When the documents for the preparation of the work were transmitted to me I found that Dr. Wilson had brought the narrative only to the point at which Edward Forbes abandoned the intention of making medicine his profession, and resolved to devote himself to natural history as his permanent vocation. His student life had been fully narrated, but the whole of his subsequent public career remained untold. It was therefore with no little diffidence that I entered upon this interesting, but obviously delicate and arduous labour. So much encouragement, however, came to me from some of Forbes's friends, and my admiration of the man himself and of his scientific achievements was so great that I ventured to embark on the undertaking. As I could not allow it to interfere with the active prosecution of the work of the Survey, either in the field or in the office, the biography, which was begun early in March 1860, advanced at first slowly and at irregular intervals, amid changes of abode, and often at a distance from libraries. But in little more than a year it was completed, and was published in the middle of June 1861 as an octavo volume of nearly 600 pages. From the artistic point of view it was unfortunate that the early years of the naturalist had been treated at such length as to fill about a third part of the book; but this part was so fully and admirably done that it must needs be published as the author had left it. The rest of the biography had necessarily to be planned on a less voluminous scale. The writing of it was to me a labour of love; I thoroughly enjoyed it while it was in progress, and was sorry when it came to an end.

My field-work in 1860 lay chiefly in the west of Fife with Dunfermline as headquarters—a neighbourhood as full of interesting geology as it was of historical associations. The old abbey and the ruins of the royal palace were favourite haunts of mine, and I often spent the week-end at Pitliver, an old-fashioned mansion in the neighbourhood, where at that time lived Allan Maconochie Welwood, formerly Professor of Scots Law in the University of Glasgow, and his wife, Lady Margaret Dalrymple, daughter of the Earl of Stair. The retired professor possessed a wonderful vivacity, and had a fund of anecdote of his legal and other experiences at home and abroad.[1]

The ground which could be mapped from this centre presented to me some novel features. It included a portion of the Fife coal-field, and thus brought me for the first time in contact with underground operations and mining-plans. Its display of volcanic rocks was absorbingly fascinating, and from this time onward I learnt more from the county of Fife than from any other part of Scotland, regarding the details of volcanic action as preserved in the terrestrial crust. It was not, however, until many years later that these details, accumulated in a series of note-books, were arranged, and an account of the volcanic eruptions in this part of Scotland during the Carboniferous period was published in the official *Memoirs of the Geological Survey*.[2]

In the summer of 1860, with a view to the publication of the East Lothian sheets of the Survey, it was found necessary to extend the mapping further south into Berwickshire. Being then within easy reach of the Cheviot Hills, I took the opportunity of spending a couple of long days in that region with my friend Dr. Paxton of Norham, who knew a sheep-farmer of the type of Dandie Dinmont in one of the valleys. Our rambles took us through

[1] An account of one of his adventures is given in my *Scottish Reminiscences*, p. 169.

[2] "The Geology of Central and Western Fife and Kinross," 1900. "The Geology of Eastern Fife," 1902.

Kirk Yetholm, the headquarters of the gypsies in Scotland. One had often come upon small encampments of this wandering tribe, sometimes even in remote Highland glens, where they usually looked ragged and unkempt, and were regarded as little better than thieves and vagabonds :

> Loud when they beg, dumb only when they steal.

It was a surprise to see them at their chief centre, presided over by their queen, by no means the haggard tatterdemalions one had been used to meet. The characteristic gypsy features were here well shown—olive-coloured skins, brilliant black eyes, dark hair and frequently good looks. Some of the young women were pretty, and the men included a few handsome fellows.

From the prosecution of my ordinary surveying I was most unexpectedly called off to the Highlands in the autumn of the same year. Sir Roderick Murchison was at that time engaged in following up the conclusions which he drew from Charles Peach's discovery of recognisable Silurian [Cambrian] fossils in the limestones of north-west Sutherland, and he now proposed to extend his application of these conclusions to the central and southern Highlands. He had already been accompanied by A. C. Ramsay, with whose assistance he mapped the geology of western Sutherland. This year his former companion suffered a serious breakdown in health, which necessitated an absence from duty for six months. The Director-General now requested me to assist him in a prolonged series of traverses through parts of the Highlands which he still had to visit. The request was of course a command. But I was ill-equipped for work among the crystalline schists, of which I had little experience, and though it was a mark of his confidence that he should select one of the youngest members of his staff, I complied with some doubt as to my competence. But the opportunity of seeing much new ground, both in regard to geology and scenery, was a chance to be prized and used; and if I could succeed in being of real service to the Chief, it might prove a help to future progress.[1]

The great review of troops by Queen Victoria at Edinburgh, in August 1860, led Sir Roderick at the last moment to change his plan of travel, and postpone the start for some days. Having got as far as Glasgow before news of this change reached me, I returned to Edinburgh and was able to witness the magnificent review from the slopes of Arthur Seat. On the 6th August I met my Chief at Greenock, and we sailed next day for Islay.

Murchison's large circle of acquaintance enabled him to bespeak the hospitality of friends over a considerable part of the region which he intended to visit, where there were few or no inns. Thus at the very beginning of the journey we were received in the friendliest way by Mr. Ramsay of Kildalton, Islay, from whose commodious house, embowered in masses of fuchsia, then in full bloom, I made my first acquaintance with the quartzite mountains of this region. The Chief was no longer able for much climbing, though he could still take a long walk on a level road; hence I undertook all the observations which required the ascent of rough and mountainous ground. From Islay we crossed to the neighbouring island of Jura, where Campbell of Jura entertained us. This island proprietor was a born deer-stalker, and so keen on the sport that he used to live, for a day or two at a time, in one of the numerous caves on his western shore. Five and thirty years after this time, on returning to Jura, I was there told an anecdote of him. A party landed from an English yacht, then at anchor in Loch Tarbert, and in passing the mouth of a cave, noticed the laird inside, whom they took to be a hermit, retired from the vanities of this world. Pitying his forlorn condition, and the necessarily scanty supply of food which he could scrape together in so wild a place, they kindly made up a basket of provisions and sent it ashore for his sustenance. Next morning before the anchor was weighed, a boat came

[1] More detailed allusions to this Tour will be found in my *Life of Murchison*, vol. ii. pp. 229-239.

alongside of the yacht, with a gamekeeper who brought a haunch of venison with the thanks and compliments of Campbell of Jura.

The laird was good enough to give one of his keepers to accompany me to the summit of one of the famous Paps—a rough climb, but one long to be remembered on account of the views from the top. For the first time I could see the whole range of the outer Hebrides bounding the western horizon, while the southern half of Scotland from Ben Lomond to the far uplands of Galloway lay clearly defined, and the north coast of Ireland looked wonderfully near. But the most novel and astonishing landscape was that displayed by the hills and glens which lay at my feet in the heart of Jura. I shall never forget the impression made on my mind by this first display of the white quartz-rock of the Highlands. I stood in silent wonder for a while, tracing with the eye the clearly marked lines of stratification along the sides of the glens, and the singular contrast of the dark basalt-dykes climbing the cliffs from bottom to top. What a picture of the effects of age-long waste—some of the oldest rocks of the Highlands pierced by some of the newest, and all of them carved out into the present rugged scenery!

At Glen Quoich, the Highland home of Mr. Edward Ellice, M.P., we dropped into a large house-party in which one of the most interesting members was the father of the host, well-known as "Bear Ellice," then in his seventy-ninth year. I had many pleasant talks with him, the subject of them being not infrequently the virtues of his daughter-in-law, our hostess. It was touching to watch his devotion to her, and she was evidently well worthy of his affection. One of the company was Prosper Mérimée, whose chief occupation out of doors seemed to be sketching the scenery in body-colour on coarse brown paper. He showed me some of his drawings, which filled me with surprise that so accomplished a man of letters should spend his time in the production of such crude and inartistic efforts to represent the noble scenery around him. Doubtless the employment was at least a relief for a time from the turmoil of French politics; but he might be also making mental notes, not only of the scenery, but also of the company at Glen Quoich, and the manners and customs of modern life in the Scottish Highlands during the shooting season, to be embodied in some brilliant contribution to the *Revue des Deux Mondes*.

We made an excursion down Loch Hourne as far as Barrisdail—the grandest of the West Highland sea-lochs. Barrisdail House, at which we called, brought up memories of the '15 and '45, when Macdonald of Barrisdail played a not inconspicuous part in that troubled time, suspected of treachery to both sides. He was said to be the inventor of a "racking machine," which he made use of to extort confession from thieves, and "in which one could not live above one hour." He was eventually seized and imprisoned in Edinburgh Castle where, although condemned to death by the Lords of Justiciary, he remained for some nine years, when he was liberated and "took the oaths of allegiance and abjuration." Mr. Ellice had acquired the extensive estate of Glen Quoich not long before our visit, and was making new roads through it, thereby laying bare many excellent sections of the rocks of the district. As these were the first fresh exposures of the Highland schists which I had seen, I studied them with all the zeal of one who had everything to learn about those ancient and puzzling parts of the earth's crust.

At Inverinet, on Loch Duich, we spent three or four days. The district of Kintail having been the country of Murchison's ancestors, he could not, in spite of the paramount claims of its geology, resist the temptation to revisit some of the places where Donald Murchison, acting as factor on behalf of the Earl of Seaforth, collected the rents, baffled the King's troops sent against him, and carried the money to the Earl in France. Standing on a knoll in the Bealloch of Kintail, his left thumb thrust into the arm-hole of his waistcoat (a characteristic attitude) and his right hand holding a stout walking-stick, Sir Roderick pointed out the leading features of interest, historical and geological. His face kindled with the old martial fire as he went over some of the more famous deeds of the '15. It was while this mood lasted that he asked and obtained from the proprietor of the ground, leave to choose a site for a monument to commemorate the exploits of his illustrious kinsman. We went by boat on this pious quest. Creeping in and out among the little bays, islets and promontories of Loch Duich, he fixed at last upon a knoll of rock amid the heather and bracken, whence we could look over to the ruined Eilan Donan Castle, and away up into the mountains of Kintail and Glenelg on one side, and across to the peaks and glens of Skye on the other. On that spot the monument was afterwards raised, where it is seen by every vessel that passes through these narrows.[1]

Another reminiscence of this visit to Loch Duich occurs to me as I write. It was of consequence, in our rapid progress over the ground, to see the series of rocks exposed along the opposite shore of Skye. The parish minister of Glenelg, Mr. Macrae, hearing of my wish to land on that coast-line, offered to bring his yacht to Inverinet and let me explore these shores to my heart's content. I had met him at Kilbride, and knew him to be an accomplished yachtsman, who almost lived on board his little vessel. Mr. Mathieson, our host at Inverinet, shook his head, and said that my goal would be sooner reached by driving to the coast and crossing by the ferry to Skye at Kyle Rhea. But the worthy minister was insistent. Next morning his yacht duly appeared, and we started on our short cruise in the finest weather. A slight breeze carried us gaily down the loch, but it soon fell away and, in spite of the clergyman's whistling incantations, would not come back. Eventually the tide turned, and we slowly drifted back again to Inverinet, where we arrived just in time for dinner.

From Loch Duich Sir Roderick's planned course led us north by Balmacarra, across Loch Carron to Jeantown and thence to Shieldag on Loch Torridon. On

[1] *Life of R. I. Murchison,* vol. ii. p. 232.

boating to the head of that singularly picturesque sea-loch, we found that the carriage ordered from Kinlochewe had not come, and no conveyance could be had to carry us and our baggage on to Loch Maree. We had accordingly to walk the twelve miles through Glen Torridon to Kinlochewe Inn—a notable feat on the part of the Chief now in his sixty-eighth year. This walk was in some respects the most instructive I had ever enjoyed, for the glen abounded in admirably fresh glacier moraines, perched blocks, and a truly magnificent display of mountain-structure on both sides of the long hollow. One mountain seemed to me more especially worthy of close examination. Starting accordingly next morning alone, I had a long day of the keenest enjoyment.[1] From the highest peak I looked across a tumbled sea of mountains far away to the blue Atlantic and the faintly seen range of the Outer Hebrides. But the geological structure which had caught my eye from the bottom of the valley proved to be far more remarkable than was expected, for it showed the existence of great dislocations by which the dull red sandstone and its overlying white quartzite were thrown into inverted positions. Little did I know at the time that these fractures were only part of a gigantic system of displacements, by which the whole of the north-west Highlands have been so greatly affected that the very bottom rocks have been thrust up from below, and pushed for miles over some of the younger formations, yet with a junction-line often so apparently normal and undisturbed as to deceive even the most experienced geologists. Murchison, when I showed him the drawings which I had made of the sections actually visible, was rather nonplussed, but thought they were merely local phenomena.

Time passed so quickly on this day of exciting discovery that the shades of evening were beginning to gather in the valleys before I reluctantly turned back. A good many miles of a rough tract of mountains, glens, rivers and bogs, across which there was no track, lay

[1] *Scottish Reminiscences*, p. 128.

between me and Kinlochewe. Fortunately the moon rose at last, and I arrived at our quarters towards midnight, to find Sir Roderick in a condition of great uneasiness. As hour passed after hour, he insisted with increasing vehemence that some of the people of the inn should turn out with lanterns to seek me. But his remonstrances met with only sullen indifference. "It was the Sabbath day," they said, "the gentleman shouldn't have gone out to walk on the Lord's Day."

Two days later I procured a guide and walked across the mountainous country from Loch Maree to Ullapool on Loch Broom—a tramp of some thirty-two miles through ground of the most varied and striking scenery and full of geological interest. But the key to the interpretation of the great geological problem of this region had not yet been found, and like previous observers I was deceived by the singular manner in which Nature had concealed it. In later years, after the secret had been discovered, I went over all this region with my colleagues in the Survey, whose admirable mapping allowed me to see the ground with fresh eyes and redoubled enjoyment.

My Chief parted from me for a few days, while I crossed to Skye for the purpose partly of seeing the rocks in Sleat which, in spite of the kindness of the minister of Glenelg, I had not been able to visit, and partly of making a brief stay at Kilbride, where Dr. Donald Mackinnon, the eldest son of my former host, was now resident as minister of the parish. I rejoined Sir Roderick at Balmacarra, where we took the steamer to Arisaig. Thence we made the long traverse to Banavie, halting at frequent intervals in order to look at the rocks on the way. Driving up Glen Coe we reached the Black Mount. The Marquess of Breadalbane, who received us there, took an interest in mineralogy. With great public spirit he kept the mines at Tyndrum going, though they probably hardly paid the expense of working them. Murchison had told him that the old gneiss of the north-west of Scotland was often strongly hornblendic, which the younger gneisses of the Central Highlands were not; but the Marquess assured the geologist that he had hornblende on his property at no great distance. We walked to the place with our host, who showed us what was beyond question a hornblendic rock.

Posting from Killin to Dalnacardoch, where there was then no railway, we drove along the great Highland road to Blair, where Sir Roderick wished to look at the classic section of the rocks in Glen Tilt. This noted glen was only open to the public on certain days of the week, and the day of our arrival happened not to be one of them. But as time was now pressing, Murchison thought it best to drive to Blair Castle and ask permission to enter. When we reached the door, the Duke of Atholl and a company of friends and gillies were standing there. On receiving Sir Roderick's card, the Duke came to the carriage and insisted on our remaining over the night with him. He sent down to the village for our luggage, which had been deposited at one of the two hotels, and allowed his head forester to accompany me wherever I liked to go for the rest of the day. I had thus an opportunity of seeing the famous section in the channel of the Tilt, and also some of the vast quartzite masses which tower on either hand above the glen. In the evening the Duke appeared at Sir Roderick's room, somewhat dusty from a long but happily successful search for a copy of the classic memoir of Playfair and Lord Webb-Seymour on Glen Tilt, which he knew was somewhere under his roof.

This prolonged Highland tour ended at Edinburgh on the evening of 1st October, and I returned to field-work in Fife with my centre at Dunfermline. Looking back now from the experience gained in subsequent years, I have to admit that this expedition into the Highlands, though it obtained a number of interesting details of geological structure, was a premature attempt to solve problems which have not yet been all solved, in spite of the amount of investigation which has been given to the subject. It led, indeed, to some highly probable suggestions, such, for instance, as the identification of the quartzites of the Central and

South-Western Highlands with those of Sutherland, the geological age of which had been established on solid palæontological evidence. But the true structure of the Scottish Highlands was far too complicated to be unravelled by desultory and hasty traverses of the ground. It required to be patiently worked out by detailed mapping, such as has revealed the complicated grouping of the rocks in Sutherland—a complication which had so completely misled Murchison and his contemporaries.

For myself, however, the expedition was an experience of the utmost importance. Besides the invaluable practice in field-observation among rocks that were new to me, it gave me many impressive and memorable lessons on Denudation, and laid the foundation of a clear recognition of the stupendous part which surface-erosion has played in the production of the present landscapes of the country. Personally also it cemented the friendship which Murchison had extended to me, and which was thenceforth to lead to so close and fruitful an intimacy.

# CHAPTER IV

## LECTURING AT THE SCHOOL OF MINES. FIRST JOURNEYS ABROAD. DIRECTORATE OF GEOLOGICAL SURVEY OF SCOTLAND

(1860–1867)

For the first few years the staff of the Geological Survey in Scotland numbered only two—my colleague, H. H. Howell, and myself. To him was assigned the mapping of the Midlothian and Fife coal-fields, and when this work, in the course of a few years, was completed he was recalled to the staff in England, whence he had been transferred. Thus for some time I remained the sole representative of the Geological Survey in Scotland. My immediate duty was to extend the survey of the Lothians southward into the counties of Berwick, Peebles and Lanark, and westward into the borders of Stirlingshire. Thereafter I was to cross the Firth of Forth into the west of Fife. The variety of geological interest in these areas, and the patient labour required for the detailed mapping of them, made them an admirable training ground for a field-geologist. It was for me a piece of great good fortune that my " prentice years " were passed in such a diversified region.

The indoor-work referred to in the foregoing chapter continued to increase, as the area of survey widened, and the winter recess grew gradually longer. There being no adequate office for this work in Scotland, it continued for several years to be done at the headquarters of the Survey in London. At these times Christmas always found me with Alexander Macmillan and his family at Cambridge. The Christmas of 1860 was marked there by the keenest frost I had ever experienced. The thermometer was said to have gone down several degrees below zero. An odd result of this low temperature was seen on the face of a College don whom I met in the street. Every individual hair of his black beard was coated round his mouth with a thin veneer of hoar-frost, so that,

> Fringed with a beard made white with other snows
> Than those of age,

he was partially converted into a venerable patriarch.

My sojourn in London this winter was prolonged for three months and a half by a wholly unexpected event. Professor Ramsay's health gave way so much that he felt unable for his course of lectures to the students at the School of Mines, and he asked me to relieve him of them. The lectures were to begin on 11th February. As the matter was only settled on 31st December, there remained barely six weeks in which to prepare a full course of systematic lectures. The opportunity, however, was one which obviously should not be declined, if the duty could be satisfactorily performed. It was an occasion to " grasp the skirts of happy chance " ; and although with little experience in public speaking, having only given an occasional lecture in country towns, I determined to avail myself of it. Professor Ramsay considerately placed his lecture-notes at my disposal ; but I soon found it more practicable, while keeping to his general scheme of arrangement, to work out each section in my own way. It was unfortunate that there were already on my hands a number of undertakings, most of which could not be set aside.[1] The six weeks for preparation came to an end

[1] The subjoined list shows the various employments which filled up the three and a half months spent by me in London during the winter of 1860-1 :

*Geology of the Neighbourhood of Edinburgh,* the first published volume of Memoirs of the Geological Survey of Scotland.

Course of Systematic Lectures on Geology to the students at the School of Mines.

Short course of Popular Lectures on Geology to Working-Men, at School of Mines.

Conjoint paper with Sir R. Murchison on the Geology of the

before much advance had been made. The outlines of the first few lectures were carefully worked out, but after these were exhausted, I was living "from hand to mouth." Each day's lecture had to be thought out, and notes for it to be jotted down, in such odd intervals as could be snatched for the purpose during the day, but very largely in the evening. With the help of a pot of strong tea, I was often able to continue my labour till two or three in the morning. It was a novel sensation to address an audience, most of whom were busy all the time taking notes of what was said. At first some effort was needed to overcome one's nervousness, but I succeeded in the end, and was able to complete the course, to my great relief.

While preparing these rather detailed discourses to the students, I was suddenly confronted with a request to give the course of popular lectures to working-men which were also delivered in the theatre of the School of Mines. Professor Ramsay came to me on the morning of 4th February with a request to take his lecture that same evening. I agreed to do so, and after office-hours took a long walk to Chelsea to think over the subject and arrange it in my mind. On entering the theatre and finding the place packed from floor to ceiling, many standing in the passages, all men and mostly young, I confess to have been somewhat nervous, and not to have succeeded as I wished, although the kindly audience frequently interrupted me with applause. I was able to do better in the succeeding evenings. The working-men were most attentive and intelligent, and it was always pleasant to remain at the end of the lecture, to hear and answer their questions. In the following winter Professor Ramsay

---

Highlands and Islands, giving an account of our work: *Quarterly Journal of Geological Society*, vol. xvii, 171-240.

First Sketch of a new Geological Map of Scotland (in conjunction with Murchison) published by W. & A. K. Johnston.

*Life of Edward Forbes*, completed in spring and published in June 1861.

Articles in *Good Words* for January and February.

Article in *North British Review* on Recent Discoveries in Scottish Geology.

Occasional articles in *Saturday Review*.

was able to give his course of systematic lectures to the students; but I relieved him of part of his duties by delivering the popular lectures in the Jermyn Street theatre on The First Principles of Geology.

After so long a spell of indoor work in London, it was a welcome relief to resume mapping in the field, across the western half of Fife, from the Saline Hills on the west to the heights of Burntisland on the east. The volcanic geology of that region had now become so extraordinarily engrossing that I resolved to visit the area of Central France, where, among a group of recent but extinct volcanoes, which had been admirably described by G. P. Scrope, I believed that light might be thrown on the more ancient and partially buried Carboniferous volcanic rocks of Scotland. Of any well educated man it used to be said that

> It would be great impeachment to his age,
> In having known no travel in his youth.

But to a geologist travel is as the breath of his nostrils, and can hardly be begun too soon. The trip to France would be my first visit to the Continent; but if life lasted it was my intention to take every opportunity that offered to see other countries than my own, and thereby to widen the experience that can be gained in the British Isles.

At this time I had not met Mr. Scrope, but when he heard from Mr. Leonard Horner of my intended excursion, he sent me a long letter, containing much information that would be useful in Auvergne. He also presented me with a copy of his classic work—*Geology and Extinct Volcanoes of Central France*, which was my constant handbook on the ground. Mr. Scrope was not only a skilled observer in volcanic matters, but also one of the leaders of the small company of geologists who saw and proclaimed the potent influence of the agents of denudation in shaping the landscapes of a country. The letter which he sent me on 16th June, 1861, was the beginning of a correspondence that lasted as long as he lived. Further reference to him will be made under the year 1870. As a fellow-traveller, also for the first time abroad, John Young, my old school-fellow, both at Mr. Black's seminary and at the High School, was induced to accompany me. Although our paths in life had diverged, we both cherished the recollection of our early years, and never had lost sight of each other. Having chosen medicine for his career, he had graduated M.D. at the Edinburgh University, and was now in the practice of his profession at the Morningside Asylum, Edinburgh. But he had not found the medical career to be to his liking. He used sometimes to accompany me in the Survey work in different parts of the Lothians, and grew so much interested in the pursuit of natural science in an open-air life that he eventually applied for a post on the Geological Survey. In this application he was successful, and had received his appointment only a few months before our start for France.[1]

A pleasant and instructive time was spent by us among the admirably well-preserved puys and lava-streams of Auvergne. In a region where the volcanic energy was active in comparatively recent time, though probably far beyond the furthest dates of history, and where, therefore, the superficial features displayed by that energy had not yet been effaced by the influences of denudation, it was an experience of deep interest to a student of the far more ancient Scottish Carboniferous volcanoes, to see in this French area what was probably the general aspect of many of the little eruptive vents and their lava outflows in the basin of the Firth of Forth. This experience served to vivify all my subsequent study of the volcanic geology of Scotland. Another valuable influence of the excursion came from the impressive evidence of the potency of subaerial erosion in sculpturing the present landscapes of a country. The lesson taught by the scenery of the Scottish Highlands was repeated here, but with even greater clearness. Nothing I had read in geological literature, not even Scrope's own classic

[1] He left the Survey in 1866 to become Professor of Natural History in the University of Glasgow.

descriptions and drawings of this very region, prepared me for the contemplation of changes so stupendous as those of the erosion of the ravines and opener valleys of Le Puy. The long level lines of the undisturbed stratification, so plainly traceable everywhere, afforded a measure by which the extent of the denudation since the close of Tertiary time might be estimated. Each new landscape that met the eye, as we journeyed through the region, enforced with fresh insistence the same lesson, and I left Auvergne more firmly persuaded than ever of the cumulative energy of the seemingly feeble subaerial agencies of waste. These agencies are the sculpture tools wherewith the scenery of a country has been slowly carved out, during the vast lapse of time in which the work has been in progress. I returned to my native country with fresh eyes and much acquired knowledge regarding both the history of the surface of the land and the part played on that surface by volcanic action.[1]

From Auvergne Young and I made our way southward to Marseilles, skirted the vast Étang de Berre, and crossed the Rhone delta, halting at Arles and Nismes. The enormous expanse of well-rounded shingle in the delta gave another striking proof of the results of long-continued subaerial waste, and of the transporting power of the rivers, once doubtless greatly more voluminous than now, which swept such coarse material across the plain to so great a distance from the parent rocks. The Roman antiquities at the two towns on the delta were duly visited and admired.

During these early years of Survey life two brief holiday excursions, one to Ireland, the other to the Inner Hebrides, had a lasting influence on the progress of my volcanic studies. In the spring of 1866 my colleague, J. B. Jukes, Director of the Geological Survey of Ireland, invited me to spend a week with him in Antrim, for the purpose of examining the remarkable sections of the

[1] An account of the journey in Central France appeared in Francis Galton's *Vacation Tourists*, 1861, which was reprinted in my *Geological Essays at Home and Abroad*, 1882.

Tertiary eruptive rocks displayed along the northern coast of that county. After two pleasant days in Dublin we made for Portrush, working our way thence along the shore, by the Giant's Causeway and Cushendall, to Belfast. It was a short but instructive excursion, which for the first time gave me detailed acquaintance with the Tertiary basaltic plateaux of our islands, and I then determined to lose no time in attacking the corresponding rocks of the Inner Hebrides which Edward Forbes thought were probably of Jurassic age (an error adopted by me from him), but which I now felt tolerably certain must belong to the same Tertiary volcanic episode as those of Antrim.

Accordingly the summer vacation of the same year was devoted to a descent upon the island of Mull. It was more especially desirable to examine with care the locality at Ardtun Head where lie the "leaf-beds," made known by the Duke of Argyll, in which well-preserved remains of Tertiary plants had been found. Having an introduction to the Duke of Argyll's Chamberlain, I received every assistance, including the labour of two men to blast the cliff, and lay bare a portion of the leaf-bearing zone, together with a pony for my own use.

I shall never forget the surprise and delight which accompanied this little bit of simple exploration. It could be seen at a glance that the "leaf-beds" lie beneath a vast pile of lava-sheets, three or four thousand feet thick, which rise, one above another, up to the summit of Ben More (3185 feet). The occurrence of Tertiary plants at its base showed that this great accumulation of volcanic material cannot be so old as Jurassic, but must certainly be younger than the "leaf-beds," which are shown by their plants to be of older Tertiary age.

It was my good fortune at this time to visit the southern coast of Mull and to find beds of chalk-flints lying beneath the basalt-cliffs—a further proof, if any were needed, of the geological age of the volcanic eruptions. But geology was not the only attraction of these interesting shores. At Carsaig I had a most friendly welcome from the Maclean of Pennycross, and formed with him and his family a cordial intimacy, which lasted until gradually broken up by death.

Now and again into the daily routine of official life there would come some unexpected message or other communication from the outside world. Thus in the summer of 1861, while deeply immersed in the volcanic geology of western Fife, I received, through Sir Roderick Murchison, the offer of the post of Geologist to the Maharajah of Cashmere, with a good salary, and outfit and travelling expenses paid. It required no long consideration to decline this tempting proposal. I had no very definite plans in my mind as to what my future would be or what I should strive to make it. Francis Bacon, whose essays were favourite reading in those days, had written—" If a man look sharply and attentively, he shall see Fortune; for though she be blind, yet she is not invisible." I cannot say that I saw the goddess at this time, but I very clearly realised that the scene on which I had to work lay in my own country.

" Better fifty years of Europe than a cycle of Cathay."

My letter to Murchison declining the offer, together with the rest of my correspondence with him, came into my hands after his death, and I found that in preserving it he had docketed it on the back, " Geikie refuses to go to Cashmir, salary £1000 per annum."

In the spring of the same year the agreeable intimation reached me that the Royal Society of Edinburgh had elected me into its fellowship. As this Society, venerable among the scientific institutions of Scotland, counted among its founders in the latter half of the eighteenth century James Hutton and John Playfair, it was the early home of geology north of the Tweed. If Edinburgh was to be the centre of my future life, this Society would play a not unimportant part in it.

Another event, which came as a great surprise, was my appointment to be Examiner in Geology to the University of London—a post which necessitated one or two brief visits to London in the course of the year. Such marks of recognition from the outside world are naturally prized by everyone, but by none more sincerely than by those whose lives are chiefly spent in quiet work in remote parts of the country.

When the indoor winter work was done in Edinburgh, or when, during the year, an occasional visit had to be made to that city, I was able to keep in touch with old, and to make there new, friends. On one of these occasions in December 1861 I found my old preceptor, Pillans, though in his eighty-third year, almost as fresh as ever. He was quite able to entertain a dinner-party at which I had the gratification of meeting again the venerable Charles Maclaren, one of the best of the Scottish geologists of his day who, though he had long ceased from the active prosecution of geology, kept up his interest in the progress of the science. In a pleasant retreat, on the southern outskirts of Edinburgh, he passed the last years of his life, where from time to time I spent enjoyable hours in his company. As an admirer of good style he used to discuss this subject with me, giving valuable hints as to what to avoid. While in Edinburgh at this time I accepted, for practice in public speaking, some of the invitations which came to me to give popular lectures on the subject of my favourite science, and delivered two addresses before the Philosophical Institution of Edinburgh. It was on one of these occasional visits to the Scottish capital that I had the pleasure of listening to Jenny Lind for the only time.

Although the geologists of the Survey had instructions to map only the solid rocks under the superficial looser accumulations known as "the Drifts," it was not possible to shut one's eyes to those later deposits, and to the extraordinary problems presented by them. From the beginning of the life on the Survey, I had been noting, on the maps and in my note-books, many of the features of these deposits, but in the year 1861 I deliberately set myself to undertake a serious study of them, with the view of trying to make out the history of the events of which they are the record. They were commonly regarded as various marine sediments, spread over the country when it was submerged under a sea on which icebergs and rafts of floating ice transported rock-debris from northern lands. I was led independently first to doubt and finally to discard this generally accepted explanation. The more one studied the smoothed and striated rock-surfaces which lie immediately below the "Drift," and noted the varying characters of the boulder-clay which rests on these surfaces, one became the more convinced that the phenomena could not all be accounted for by floating ice, but demanded the former existence of a great terrestrial ice-sheet or sheets, as Agassiz had insisted, twenty years before. My note-books were full of sections and local details from all the areas which I had mapped. In 1862 I spent my summer holiday in traversing the southern counties of Scotland for the special purpose of studying the "Drift," and the ice-moulded rocks beneath it, as far as these can be seen in that region. I remember that at this time I was guided to a fine recently-exposed section of boulder-clay on the Slitrig Water by Mr. J. H. A. Murray, who was then a schoolmaster in Hawick, and afterwards rose to fame and knighthood as Editor of the *Oxford English Dictionary*. This excursion took me into hitherto unvisited parts of the country. For the first time I wandered on foot down Liddesdale, Border ballads on my lips, and with eyes on the outlook, not only for sections of "Drift," but also for every old Border peel or tower within sight or reach. The Solway shores seemed not unfamiliar to one who remembered the pages of *Redgauntlet*. The Kirkcudbright coast, with its rocky headlands and little bays, brought to one's recollection scenes in *Guy Mannering*. Walking up Nithsdale and diverging into its side valleys in search of fresh geological sections, I found that Walter Scott had waved his magician's wand over that vale too, where the deep and narrow ravine of Crichope Linn will to all time be associated with Balfour of Burley in *Old Mortality*.

Crossing Ayrshire in continuation of this quest of the

Glacial Drift, I passed over into Bute, where I had arranged to spend the winter months, clearing off indoor work of the Survey, and taking an occasional excursion along the Kyles and to the sea-lochs, under the guidance of my much esteemed friend the Rev. Alexander Macbride, Free Church minister of North Bute. One of the most modest of glacialists, he had never published his observations—yet for years past he had been scouring not only the shores of Bute, but those of Argyllshire, and had discovered many new places where the Arctic shell-beds have been preserved. He conducted me to some of his numerous "finds," and thus enabled me to collect much of the local detail which, together with the many observations made in the course of years in the work of the Survey, and the conclusions founded on them, was embodied in my memoir "On the phenomena of the Glacial Drift of Scotland." This essay formed a volume in the *Transactions of the Geological Society of Glasgow*, published in April 1863. It contained, I believe, the first attempt to present a connected view of the sequence of events in the history of Scotland during what is known as the Glacial Period or Ice-Age. It collected evidence to show that the boulder-clay was mainly the product of the action of land-ice, and not of drifting bergs or floes. And it brought forward, for the first time, proofs that the Ice-Age included intervals when some parts of the country, up to heights of 800 or 900 feet above the present sea-level, were free of ice and clothed with vegetation.

At the time of my wintering in Bute, one of the most familiar of the vessels afloat on the Firth of Clyde and in Rothesay Bay was the little sailing yacht of the veteran James Smith of Jordanhill, known to the world at large as the author of an admirable volume on the *Voyage and Shipwreck of St. Paul*, but honoured among geologists as the man who first proved, from the occurrence of Arctic shells in the upraised sea-bottoms along the shores of the Firth of Clyde, that Scotland, at a comparatively recent geological period, had a climate like that of northern Norway. Though he had reached his eightieth year, he was still vigorous, and passed most of the summer in his yacht, taking an active interest in the aquatic life of the busy Firth. When I visited him on board his trim little vessel, he looked so hale that one could almost believe him to be fit to start off any day for another cruise in the Mediterranean waters which he once knew so well.

About the time of the publication of my memoir on the Glacial Drift an important change was introduced by the authorities at headquarters into the mapping of the geology of the country. The Drift had never been mapped, and indeed was commonly regarded as a serious impediment to the mapping of the older formations, which it covered and concealed. It was forgotten, however, that these superficial accumulations are of great agricultural importance, for upon their composition the character of the soil that covers them largely depends. This economic interest of the much maligned Drift having now at last been recognised, orders were issued from headquarters that thenceforward the areas of the superficial accumulations were to be traced upon the field-maps in addition to the older solid rocks beneath and further, that the tracts of the country which had already been surveyed were to be revised, in order to have the Drift inserted upon the maps. This rearrangement necessitated an increase of the staff. The transference of my colleague, H. H. Howell, to England having left myself sole representative of the Survey in Scotland, it was determined, in the course of the year 1861, to add two assistant geologists. These were my old school-fellow John Young, already mentioned, and my brother James. In the following year a third assistant, B. N. Peach, an able student from the School of Mines, was appointed.

It was unfortunate that the Drift mapping should be the first kind of surveying work assigned to these fresh recruits. Not always an easy task, it is apt to be thought monotonous, until its special interests are discovered by patient work. As senior member of the staff it fell to me to train the new men. Eventually the Local Director,

Professor Ramsay, finding it impossible for him to inspect the details of the mapping in Scotland as frequently as he wished, constituted me his "lieutenant-general," to use his own phrase, with instructions to visit the new men from time to time and report to him on the progress of the work.

During this period of inspecting duty I first saw Loch Skene (1680 feet above sea-level) and recognised it as a tarn still dammed up by the moraines of a glacier that filled a picturesque recess in the high grounds of White Coomb in the uplands of Dumfriesshire. Young was stationed at the shepherd's shieling near the foot of the waterfalls of the Talla, a tributary of the Tweed. When I visited him there, we climbed over the watershed above the Linn, and were astonished to find such perfect relics of a glacier, of which neither of us had ever heard, though it had been detected some years before by Robert Chambers. We followed the track of the glacier southwards and rested for the night at the well-known little hostelry of Tibbie Shiels. On getting up next morning we found the whole countryside buried under several inches of snow which had gathered into deep drifts that made our return into the Talla valley impracticable. The household was engaged in churning butter, an operation in which we were glad to assist. This snowstorm taught me a memorable lesson on the physiography of a landscape. The ground being covered with a continuous mantle of uniform white, the distracting influences of colour were removed, which so obscure the minuter irregularities of the surface. One could thus see, as in a piece of statuary, the detailed lineaments of every slope. Each little dimple had its pale violet shadows, and the surface of the declivities all around could now be seen to be infinitely more varied in their surface contours than would be suspected when they are beheld, in the full light of day, clothed with all the forms and tints of vegetation.

The increased attention now being paid to the Drift deposits of the country led the Local Director in 1863 to ask me to accompany him in an examination of the exposures of these accumulations along the coast of the north-east of England from Berwick-on-Tweed to the Humber. Parts of this shore-line reveal with great clearness the composition and sequence of some of these deposits. At many places we took percentages of the various types of stones in the boulder-clay, and gained from them a good deal of information respecting the probable direction of movement of the ice-sheet. I filled many pages of my note-book with these percentages, and with sketches and detailed descriptions of the sections, in the expectation that my companion would write an account of the tour, and that my notes would be of service to him. But in the multiplicity of his duties, and with his impaired health, he never succeeded in preparing such a paper, and my notes have remained unused.

What with my own studies and superintending the Drift work of the Survey, my mind was rather obsessed at this time with glacial questions. Summing up the knowledge which had now been gathered together, I wrote a descriptive account of the Ice-Age as it had left its records on the surface of Scotland, which appeared in the *North British Review* for November 1863.

In the progress of the Geological Survey in Scotland, which partly depended on that of the large scale Ordnance maps, it was now decided to break new ground on the west coast of the country, beginning in the south of Ayrshire, and working thence north and east into the coal-fields. Accordingly in October 1863 I proceeded to the ground assigned to me in Carrick. When my friend Mr. Richardson, the Free Church minister of Dailly, heard of our intention to begin in the south of Ayrshire he insisted on my taking quarters with him and his wife at their manse. The situation was exactly suited for the work to be done, as it lies in the vale of the Girvan Water at the foot of the Carrick Hills, a few miles from the coast, with which it is connected by the line of railway that runs northward to Glasgow. Close to it one of the smallest coal-fields of Scotland was in active working.

The genial society of the manse made my stay under its roof memorably pleasant.

I had not been long established there when a letter came from my constant friend and correspondent Leonard Horner, containing the following paragraph: "You are now in the immediate neighbourhood of Dalquharran Castle, the residence of my friends Mr. and Mrs. Kennedy, and as their acquaintance may make your stay at Dailly more agreeable, I enclose a note of introduction to him. Dailly was the birthplace of my dear friend Mr. Thomas Thomson, advocate, of his brother John, the minister of Duddingstone and admirable painter, and of the most excellent wife of Professor Pillans, whom he lost many years ago. Their father was the minister of the Parish."

This timely introduction not only brought me the friendship of the family of Dalquharran, but through them opened the way to a wide circle of acquaintance all over the county. Perhaps no other shire in Scotland could furnish an ampler list of landed proprietors, both of the old stock and the newcomers. The former, in addition to titled possessors of large estates, included many smaller landholders, some of whom could trace their genealogy to a remote ancestry. The Right Honourable Thomas Francis Kennedy of Dunure was an excellent type of these lesser landed gentry.[1] Educated in Edinburgh under Pillans and Dugald Stewart, he was from his youth associated with the brilliant literary coterie which then flourished in the Scottish metropolis. He used often to recount to me his reminiscences of the men and clubs of that time. As he was born near the town of Ayr, and had passed much of his life in the county wherein he owned considerable landed property, he retained a lively recollection of the state of the south-west of Scotland in the closing years of the eighteenth and the early part of the nineteenth century. As my intercourse with him grew more intimate, I often heard him contrast the present condition of his district with what it was in his boyhood.

[1] *Scottish Reminiscences*, pp. 190-193.

In those days the hardships of the small upland farmers and peasantry were, in some prolonged winters, almost incredibly severe. He told how these people were sometimes compelled to bleed their cattle and mix the blood with oatmeal to keep themselves in food. He used to describe the Scottish diet of his early days, which included a number of dishes now seldom or never seen, such as solan-goose from Ailsa Craig, white and black puddings, crappit-heads, singed sheep's head, and sundry other preparations in the national cuisine, long since banished from the tables of polite society. His account of the appearance of the solan-goose on the table was given with great humour, and possibly with a little heightening of the colours. First, the bird was buried in the earth for a certain number of days. It was then dug up and roasted. So pervasive and nauseous was the odour that spread from the kitchen through all the castle, that the windows were kept open to let it escape. He never entertained me to this example of the ancient dietary, but he not infrequently introduced other dishes which were quite palatable. I finally struck at the black puddings.

Kennedy of Dunure was a gentleman of an antique cast, courteous and stately in his manners, proud of his descent and of his ancestral possessions, and tenacious of his rights, which he was sometimes thought to insist upon more than he need have done. He looked carefully after his breeds of cattle, and was keenly alive to the use of new inventions for the improvement of agriculture. Part of the little Dailly coal-field was on his estate of Dalquharran, and he worked the mineral according to the best known methods.

Though now a retired country squire, he had been an active politician in his time. For sixteen years he sat in Parliament as member for the Ayr Burghs, and was closely associated with Cockburn, Jeffrey, Horner, and the other Whigs who brought in the Scottish Reform Bill. He had retired from public life in 1854, and thereafter lived entirely at his Ayrshire home, save that for some twenty years he used to come up to London for the season. When I made his acquaintance he was in his seventy-fifth year, but still vigorous both in body and mind. Though he seldom referred to his political life, and never to the circumstances in which it ended, I understood that he had not been well treated by the chiefs of his own party, and that one of his friends, Lord Murray of the Scottish Bench, in indignation at this treatment and out of personal regard for him, settled a large pension on him for life. In the course of frequent visits to Dalquharran during the next few years, I learnt from him that he possessed a voluminous correspondence, particularly with Lord Cockburn, extending over most of the first half of last century and relating mainly to the strenuous political life of that period in Scotland. Eventually he collected these letters, and published them in 1874 in a volume of nearly six hundred pages.[1] But the book had no explanatory notes or connecting narrative, and worst of all, no index, so that save to undaunted students of political history, it could hardly be called readable. The copy which he gave me was specially acceptable for the fine engraving of his portrait (painted, I think, by Raeburn) when he was thirty-four years of age and M.P. His handwriting was the most illegible in all my correspondence, resembling nothing so much as the track of a sparrow across a sheet of paper after its feet have been dipped into an ink-pot.

Mrs. Kennedy, daughter of Sir Samuel Romilly, Master of the Rolls, was a singularly gentle and gracious lady, fragile and beautiful in her old age, cultivated and keenly intelligent. She had been married twenty years before her only child, a son, was born. Mr. Kennedy used to remark on the curious coincidence that he himself was also an only child, born after twenty years of wedlock.[2]

[1] *Letters chiefly connected with the Affairs of Scotland from Henry Cockburn, Solicitor-General under Earl Grey's Government, afterwards Lord Cockburn, to Thomas Francis Kennedy, M.P., afterwards the Right Hon. T. F. Kennedy, with other letters from eminent Persons during the same period, 1818-1852.*

[2] *Scottish Reminiscences*, p. 193.

Dalquharran Castle is a large modern mansion, built in a massive but rather tasteless style, in curious contrast to its predecessor, the older castle, which now stands as a picturesque, ivy-clad ruin a short distance off, near the river. The laird remembered when this ruin still had its roof on and was partly habitable. Dalquharran became a second home to me during the years of my residence in Ayrshire, and not infrequently afterwards, as long as my kind friends lived. After my marriage my wife accompanied me. Her youth and her songs at once gave her a place in Mrs. Kennedy's affections.

One of my most vivid and delightful recollections of Dalquharran is the series of family portraits, some by Raeburn, that hung in the dining-room. One could never tire of gazing at them and marvelling at their charm and masterly execution. Scotland was fortunate to have so consummate a painter, and her gentry did well to employ him so freely as they did. It is a great source of satisfaction to see how steady in recent years has been the growth of the public appreciation of Raeburn.

The year 1864 is memorable to me, for in it my friends Professor Pillans and Leonard Horner both passed away. Of the Professor I was glad to have seen more in recent years under his own roof. While the recollection of him as he appeared in the class-room will never fade from my memory, I delight also to picture him as he was in later years when he had become a personal friend. Of Leonard Horner, to whom Pillans first introduced me, I wrote at the time a short appreciation which was published in the *Scotsman* newspaper. It was unsigned, but his daughters in London guessed its authorship, and they sent me a letter of thanks for the "deep feeling of gratification" which my picture of their father's character and work had given them. At the request of the Royal Society I wrote the obituary notice of him for the *Proceedings*. Leonard Horner had a remarkably grave manner, beneath which, however, lay a most kindly and responsive human heart. It used to be said of him that he was "born to fill chairs," and he certainly often presided at the societies and associations with which he was connected, always doing so with remarkable calmness and deliberation. Somebody remarked of him that he "looked wiser than any man could possibly be." Before he finally settled in London he had made his mark in Scotland as an educational reformer. He was the founder of the School of Arts in Edinburgh—an institution for the education of working-men, which led the way in this department of educational effort and has flourished with vigour. Out of it has grown the present excellent Heriot Watt College. His contributions to geology were few, but sufficient to entitle his name to be remembered among the writers in the infancy of that science in Britain.

At this time (1863-64) such leisure as I could secure was in great part spent in the writing of my volume on *The Scenery of Scotland viewed in connection with its Physical Geology*. The Royal Society of Edinburgh requested me to give them an address on the progress of the Geological Survey in Scotland, which enabled me to put before the public a statement of what had been done in the first ten years of its work in the country. In two years from the time when my own field-work had been transferred to the western side of the country, I had completed the mapping of the high ground between the valleys of the Girvan and the Stinchar rivers, together with the coast-line, and had extended the work northward about half-way to the town of Ayr. This survey included Ailsa Craig, Keats' "craggy ocean pyramid." The poet, in his well-known sonnet to this prominent feature of the Firth of Clyde, puts many questions about its history, chiefly involving length of time, which even the boldest geologist would shrink from attempting to answer. There was no evidence on the islet itself, so far as I could observe, which would decide whether this huge mass of rock belongs to the oldest or the youngest of the eruptive rocks of the Clyde basin. It may be connected with the Lower Silurian volcanic series, so well exposed about Knockdolian in the

Stinchar valley, and thus be among the most ancient rocks of the region. On the other hand it may belong to the same group as the granite of Goatfell in Arran, and thus be younger than the soft clay on which London stands. Popular legend has solved the problem in characteristic fashion, recording that once upon a time a noted witch, whose doings elsewhere have been alluded to by Burns, took it into her head to lay down some stepping stones between Carrick and the Irish coast. Selecting a hill in the vale of the Stinchar she took it up in her apron, but a few miles from the starting-place the apron-strings broke, and the huge hill fell into the Firth, where it now forms Ailsa Craig. The hollow from which it was taken is still pointed out.

The southern portion of my Ayrshire work having been completed from the centres of Dailly, Girvan, and Barr, I was able in the spring of 1865 to move northwards to the town of Ayr. From that centre, owing to the railway and its branch lines, a wide stretch of ground could easily be reached, so that changes of station would not be so frequent. Better accommodation could also be had there than in the remoter villages, and I had the pleasure of inducing two of my sisters to share my bachelor quarters in a pleasant villa close to the sea.

It was at Ayr, in the summer of 1865, that I finished the last pages of *The Scenery of Scotland.* The volume was dedicated to Murchison, to whose friendship I owed so much. Yet its main thesis was in such direct antagonism to his well-known cataclysmic opinions that I was prepared for some protest on his part, though he knew what my opinions were. But he accepted the compliment without demur. The further I advanced in the practical pursuit of geological studies I had become the more convinced of the paramount share taken by denudation in the production of the present surface of Scotland. Movements of the terrestrial crust had undoubtedly played their part. There had been extensive upheaval and subsidence, and enormous fractures had ruptured the crust all over the country. But the denuding agencies had been at work all the time, until no recognisable trace has been left of the aspect of the ground at the close of any subterranean movements. The present contours of the surface are unmistakeably the result of the operation of the various denuding agents on material irregularly raised into land, and presenting variable capacities of resistance to the action of these agents. The forms of the present surface are thus the result of a long process of sculpture which has been in progress ever since the dry land appeared above the sea. Like a mass of marble in the hands of a sculptor who gradually effaces all trace of its appearance as it was taken out of the quarry, each landscape owes its first origin to the operation of the forces which raised it above the sea. What was then its aspect we can only guess, but its present contours are the result of the combined action of all the eroding agents, which have been ceaselessly at work upon it since it was upraised from below.

The later pages of the *Scenery of Scotland* were passing through the press when the announcement came from the Royal Society of London which authorised the insertion of F.R.S. on the title-page. This was an honour which I deeply appreciated. I had little doubt that it was in no small measure due to strong backing from Murchison. Many years afterwards it was interesting to see the original certificate in my favour, and to find that the signatures attached to it were Charles Lyell, W. E. Logan, A. C. Ramsay, J. B. Jukes, J. Prestwich, W. J. Hamilton, J. Tyndall, J. Percy, D. T. Ansted, H. W. Bristow, E. W. Binney, R. Hunt, and T. Sopwith. The name of R. I. Murchison is not in this list; probably he was then on the Council of the Society and in that capacity was disqualified to sign a certificate. At the same election, William Huggins and Alfred Tennyson were made Fellows of the Society.

When the last sheets of the *Scenery of Scotland* were sent to press, I was free to make an excursion to Arctic Norway, which I had long planned, with the view of studying the glaciation of that region where the ice still remains on the heights and creeps down the valleys in glaciers, some of which even descend to the edge of the sea. The Ice-Age not being yet extinct in that region, there was every probability that from this visit light would be thrown upon the records of that geological period in Scotland, where the actual ice has long since disappeared. I was accompanied by two of my Survey colleagues, my brother James and William Whitaker.

Without halting by the way we made as quickly as possible for the Holands Fjord, a little to the north of the Arctic Circle, and spent a week there among the glaciers, ice-worn rocks, moraines, perched blocks, and raised beaches. Thence we sailed northwards through the intricate sounds and fjords, past the Lofodden Islands to the island of Skjaervö, where we left the coasting steam-boat, and boated across to see the glacier of the Jokuls Fjord, which was said to descend into the sea. We also visited the fjords on the north side of the great snow-field of the Jokuls Fjeld and reached Hammerfest, whence we began the return journey.

Opinion among geologists was still divided as to whether the glaciation of Britain, during what is known as the Glacial Period, was the work of floating ice at a time when most of the land was submerged beneath the sea, or of great sheets of land-ice that covered the country and, moving out to sea, broke off there into icebergs. Probably most geologists still clung to the former explanation. As already stated, the detailed investigation involved in the work of the Geological Survey in Scotland had convinced me that the views of the great Swiss glacialist, Louis Agassiz, provided the true solution of the problem. This conviction was amply confirmed by the exploration of northern Norway. A brief narrative of this trip and of the results obtained from it was communicated to the Royal Society of Edinburgh.[1]

[1] "Notes for a Comparison of the Glaciation of Scotland with that of Arctic Norway." *Proc. Roy. Soc. Edin.* 15th January, 1866. This paper was reprinted in my *Geological Sketches at Home and Abroad*, p. 127.

A few years later, when taking part in Sir Henry Roscoe's series of popular "Manchester Lectures," I gave one on "The Ice-Age," which contained some personal impressions of this excursion into Scandinavia. Landing upon the coast of Arctic Norway one sees precisely the scenery of the western coast of Scotland. I found that on the outer islands of the Norwegian seaboard the same type of smoothed and striated rock-surfaces, so familiar on the shores of Invernesshire and Argyll, could everywhere be traced; that the striae left by the stones under the slowly moving ice had been graven into and over all the little inequalities on the surface of the solid rock; that the striation went up over the hills, and diverged from the high grounds exactly as it does at home. I followed these rock-markings up one of the fjords which promised to yield the best results. They were found to wind in and out with all the windings of the valley exactly as they may be seen to do on the sides of the sea-lochs and glens on the western coast of Scotland. Tracking them step by step, I found them at last passing beneath one of the glaciers which comes down to the sea-margin from the snow-field. I crept under the blue ice and found bits of stone between the ice and the underlying rock. It was easy to see how the sharp points of these stones would score the rock as they were borne onward by the slowly gliding ice, and how in the lapse of time even the hardest materials would be ground down by this constant attrition. I caught the ice, as it were, in the very act of doing the work of which I had hitherto only seen the ancient results. The pleasure of actually looking with one's own eyes on the proofs which had been confidently expected was a memorable experience.

On this journey the Scots pronunciation of Latin was found to be of service, in default of acquaintance with the language of the country. When our little party disembarked from the coasting-steamer at a calling-place from which we should be able to reach the farm on the Holands Fjord, whither we were bound, we took our places in a row-boat which proceeded for a long way without revealing any village, or indeed any trace of human occupation. Beginning to wonder whether or not we had left the steamboat at the wrong place, I turned to a sedate Lutheran pastor who, with other natives, had got into the boat with us, and asked him in English whether the rowers were making for the hamlet for which we were bound. Most educated Norwegians know at least a few words of English; but he smiled and shook his head. I then tried him in French, with no better result, and lastly in German, but received only the same bland and benevolent smile. During the voyage from Hull to Norway, and in the coasting steamer from Trondheim, I had been working diligently with a grammar and dictionary of the Norwegian language, but I had not yet acquired enough of a vocabulary or of confidence to venture on a conversation in that tongue. It was a novel situation to be a foreigner in a land of whose language one had not knowledge enough even to ask one's way. With Ovid among the boorish Getae, who laughed at his Latin, I could exclaim

> Barbarus hic ego sum, qui non intellegor ulli.

But singularly, it was Latin that now came to my help. Our reverend fellow-traveller at last broke the awkward silence by asking a question in that tongue, pronounced of course in the broad continental way in which I was taught at the High School, and the rest of our talk was carried on, though somewhat limpingly, in the same language.

This experience quickened my zeal in studying the language of the country more assiduously than ever. The grammar and dictionary, however, received powerful practical aid from the conditions of travel in those northern regions. The absence of hotels, and the close touch with the native population involved in the system whereby a traveller is passed on by boat or carriole from one farmer's house to another, gave me many opportunities of getting into touch with the children. Making full use of the material supplied by the books, and asking the young folk endless questions, I rapidly increased my vocabulary, while at the same time affording infinite amusement to the children by my mispronunciation of words. At one of the fjords where we halted I found the little girls playing with shells of the northern clam (*Pecten islandicus*), an organism which is common in the Arctic shell-beds of the Clyde, though now extinct in our British seas. It was still evidently abundant as a living denizen of these northern waters, and with its elegant shape and bright colour it was appreciated as a plaything by the children who called the separate valves "Red Horses." Cultivating acquaintance with the young, I thereby pleased their mothers, who sometimes joined them in their lessons in Norse to this wandering Englishman. It was my good fortune to meet again, at one of the halting-places of the coasting steamboat on the return journey, the same native pastor. Coming on board he at once recognised me and gave a Latin salutation; but to his evident surprise I was now able to reply in his own language.

The stay at Ayr as the centre of my geological work extended over two pleasant years. The mapping from that centre proved wonderfully varied and full of novelty. It enlarged my acquaintance with the volcanic history of Scotland by revealing some fresh areas of eruption in the time of the Old Red Sandstone and the Carboniferous periods. It brought to light also, for the first time, proofs of the existence of a number of little volcanic vents later in age than the Coal-measures, together with a group of lavas and tuffs which had been discharged from these vents at the time of the deposition of the red (Permian) sandstones that overlie the Coal-measures. The discovery of a hitherto unknown epoch of volcanic energy in Scotland gave me much satisfaction. It was made in what is commonly known as the "Burns' Country," the poet's farm of Mossgiel lying in the midst of it. Certainly with more than geological thoughts in my head, I traversed the very fields which Burns toilfully ploughed, and ascended the little "gowany glens" and "hazelly shaws" where he used

> To stray and pensive ponder
> A heartfelt sang.

Many an hour was delightfully passed in the deep and picturesque ravines wherein

> Ayr, gurgling kiss'd his pebbled shore
> O'erhung with wild woods, thick'ning green;

and while maps and note-book received their due geological tribute, the poet's musical pictures of the scenery were often on my lips.

During the stay at Ayr one of the most interesting persons, of whom I saw a good deal, was Mr. Elias Cathcart of Auchendrane. He was a Scottish laird of a bygone type, full of humour and oddity and a most diverting talker. He had parted with a good deal of his land to Mr. James Baird, one of the great ironmasters of the West, who had built the mansion of Cambusdoon near "Alloway's auld haunted kirk." At a dinner party there, the laird of Auchendrane was alluding somewhat mournfully to the changes which, since his boyhood, had been made in the district, and he concluded by saying, "Well, if I were to come down fifty years after this, I daresay I would see just as much change again." Whereupon the host immediately ejaculated "What's that ye say, Cathcart, 'come down'! Take care ye hae na to come up."

Auchendrane Castle is a singularly romantic old stronghold overhanging the ravine of the River Doon, where the laird lived with his wife and two daughters. Among the attractions of the place was a row of noble silver-firs that towered above the surrounding woodland. Once or twice a member of this group had been struck by lightning, and the gentle and genial Mrs. Cathcart took me aside one day to ask whether by chance there could be a vein of iron-ore below these trees, whereby the thunderbolts were attracted to them. I was able confidently to affirm that her fear was quite groundless. She replied that she was thankful to hear it, adding that " if Cathcart took it into his head that there was a vein of iron below, he would have the whole place dug up to find it."

When Professor Ramsay came down to Ayr on inspection duty, he also was invited both to Dalquharran and Auchendrane. With the latter place and its inmates he was so enchanted that he made them the subject of a delightful paper in the *Saturday Review.* I have elsewhere given a few sketches of other Ayrshire worthies whom it was my good fortune to meet during my residence in their county.[1]

From Ayr, Dalquharran was within easy reach, and my " week-ends " were often spent there. There were also steamboat facilities for crossing to Arran, and from time to time I passed a quiet Sunday in that incomparable island. On one of these excursions Professors John Nichol and Edward Caird were my companions, and in the course of long walks we had much pleasant converse.

During the residence in Ayr I was occasionally called to London on Survey duty, and on the business of the Examinership in the London University, but only for a day or two at a time. On one of these occasions in the spring of 1866 I was invited to a pleasant dinner in the Garrick Club given by Alexander Macmillan. The party included Millais the artist, Samuel Baker the African traveller and his brother Valentine, W. H. Russell, the *Times* war-correspondent, A. V. Dicey, R. H. Hutton, editor of the *Spectator*, and one or two more. There was a great deal of excellent talk, in which I especially enjoyed Millais' contributions. Later in the same year, on the occasion of another brief visit to London, I dined, by Murchison's invitation, at the Geographical Club and met some of the leading geographers of the day. I was specially interested in John Campbell of Islay who afterwards published *Fire and Frost*, and of whom I was destined to see much in later years.

The winter of 1866-67 was mostly passed in the office at Edinburgh over the indoor-work of the Survey. On the

[1] *Scottish Reminiscences*, pp. 190-200.

4th February I read to the Royal Society there my first paper on the "Tertiary Volcanic Rocks of Scotland," embodying the results of the explorations in Mull and Skye. A few days thereafter I was summoned to London by the Director-General for a fortnight, in order to discuss with him certain official rearrangements which were then in contemplation. It had become obvious that if the Survey in Scotland was to make more rapid progress, as was eminently desirable, the number of its staff must be increased. But such an augmentation would throw much additional labour on Professor Ramsay, who as matters then stood found himself unable to overtake all the inspecting work in Scotland, and had been under the necessity of delegating a part of it to myself. The proposal ultimately took the shape of a separation of the Scottish staff into a distinct branch of the Survey, with an increase of its numbers and a separate Director, under the control of the Director-General. I arrived at headquarters in time for the annual dinner of the "Royal Hammerers," as the staff of the Survey and Museum at Jermyn Street had styled themselves, far back in the days of De La Beche and Edward Forbes. At these festivals Murchison for some years took the chair, and he would now and then follow the custom of versifiers in the staff, by writing a ditty for the occasion to the tune of some popular song. I remember one of his effusions about the "Royal Hammerers"[1] which in most unmusical fashion he sang or chanted to the air of the "British Grenadiers." A book, in which these doggerel productions were entered, is or was kept at the Jermyn Street Museum. But the young men who had spent most of the year in solitary country quarters were in no mood to be critical. They came to enjoy themselves, and generally succeeded. They met at these gatherings colleagues whom they had not seen for perhaps twelve months, and the cordiality and comradeship of the service were certainly thereby promoted.

[1] See *Life of Edward Forbes*, pp. 412, 498; *Life of Andrew C. Ramsay*, pp. 142, 175.

From these and other gaieties, including a pleasant Sunday with John Evans and his family at Nash Mills, I returned to the north to resume field-work in Ayrshire. As much ground having been mapped as could conveniently be reached from Ayr as a centre, I now pitched my camp at Ardrossan, which was then a far less busy seaport than it has now become. From the windows of my quarters I looked across the firth to Arran, and greatly enjoyed the sunsets over its noble mountain range. This island was now brought within easier reach than ever by a daily steamboat service, and I was still within hail of Dalquharran. Shortly after the change to Ardrossan, and at the end of many weeks of official correspondence between Murchison and the Science and Art Department, the following note reached me.

12th April, 1867.

MY DEAR GEIKIE,

I this day received the official authority appointing you Director of the Scottish Survey from 1st April (never mind the day !). . . .

Yours very truly,

ROD. I. MURCHISON.

# CHAPTER V

## START OF THE GEOLOGICAL SURVEY OF SCOTLAND (1867–1871)

THE staff of the newly constituted Scottish branch of the Geological Survey was organised as follows: One Director: one District Surveyor; two Geologists; six Assistant Geologists; two Fossil Collectors; one Porter. Nearly all of these officers had to be found, and when found had to pass through the ordeal of examination by the Civil Service Commission. The task of mustering the recruits proved a good deal more difficult than could have been anticipated. The most memorable incident in its progress was the rejection of an able algebraist, who had been warmly recommended by Sir William Thomson (afterwards Lord Kelvin) but failed in arithmetic. In his examination, as I heard, he had been required by the Commissioners to add up a long table of pounds, shillings, and pence within a limited number of minutes. Never having been accustomed to sums in compound addition, he not unnaturally failed to complete the summation within the prescribed time. The early doings of this Commission showed that little or no consideration was given to the kind of work for which the candidates were desired. The treatment of our algebraist might have been appropriate had he been seeking the post of a grocer's shopman, though even then, accuracy rather than rapidity of summation would have been the better test. In the end, however, we succeeded in adding the mathematician to the staff, of which he proved a valuable member.

It was my determination not to give up my own work

in mapping, so long as this did not interfere with the duties of superintending that of the staff, in the field and in the office. But in the course of a few years this effort was found to be no longer possible, and I reluctantly relinquished what had for so many years been to me one of the chief pleasures of life. But the surrender was in large measure compensated by the great interest of entering into the work of each member of the staff, and gaining thus a broader grasp of the geology of the country as a whole, as well as a detailed knowledge of each district, as it came within the area of survey.

Complete liberty having now been obtained to arrange the work and choose the methods to be employed, steps could be taken to meet a serious defect in our organisation—the absence of any satisfactory means of petrographical determination, in cases where the appliances of the field-geologist failed to suffice for the accurate discrimination of rocks, especially those of compact, close-grained texture. Application was first made for the assistance of chemical analysis, which led to an arrangement whereby such analysis should be conducted by the Professor of Chemistry in the School of Mines, who was an official in the same department with the Geological Survey, and who, in the days of De La Beche, was understood to undertake this kind of work for the Survey, as part of the duties of his office. But the result proved unsatisfactory. The work was put into the hands of some of the more advanced pupils of the school, and was not detailed enough to be of much service. After some years one of the assistant geologists, a capable chemist, was supplied with an improvised laboratory in Edinburgh, where he carried out chemical analysis as part of his winter indoor-work. Eventually, when the Director-Generalship passed into my hands, this branch of research was fully organised at the Jermyn Street Museum, where an excellent laboratory already existed, and where an accomplished chemist was installed as one of the regular staff, whose sole duty was to analyse the rocks and minerals met with in the course of the mapping.

Even more important than chemical analysis for the purposes of the field-geologist, was the application of the method of microscopical investigation, inaugurated for petrography by Sorby, who, as already stated, had pointed out to geologists in 1858 the remarkable results that could be obtained by the investigation of thin transparent sections of rocks. But though on the Continent a number of observers had adopted this method with increasing ardour, the geologists of this country were slow to perceive its value. My experience with Sorby when he was in Edinburgh working over the Nicol collection of thin sections of minerals and fossil woods (p. 28) prepared me for the adoption of this method of study, not merely by full-blooded petrographers, but by geologists in the field. During the mapping of Ayrshire I began to make thin slices of the igneous rocks which I met with in the field. My esteemed colleague A. C. Ramsay, however, had no sympathy with this "new-fangled geology," and, in his own words, did not believe in "studying a mountain through a microscope." But it was recognised in the end by some members of the staff in England, that for the precise determination of many close-grained igneous rocks, and for their correct insertion on the maps, the microscope must be made an essential instrument in the work of the Geological Survey. The preparation of one's own thin sections, however, was a laborious process, and for the most part only partially successful. Samples of the rocks were eventually sent to a London firm, which prepared microscopic slides from them. In later years two of the subordinate members of the staff in Scotland became proficient experts in cutting the sections, and in the end a trained petrographer was appointed as one of the permanent staff at the head office in London.

The British Association met this year (1867) at Dundee, under the presidency of the Duke of Buccleuch. At the Geological Section, of which I had been elected President, Lyell, Murchison, John Phillips, and Robert Chambers, besides a few other veterans, were present, and there was a good contingent of younger geologists. Since the Aberdeen meeting of the Association, eight years before, my studies of volcanic geology had been considerably extended. They led me to choose as the subject of the Presidential Address, the "History of Volcanic Action in the British Islands," and to make what was the first attempt to group in chronological order the sequence of eruptions in this westernmost part of Europe. Having also been asked to give one of the evening lectures, I discoursed on the origin of the present scenery of Scotland, dwelling especially on the paramount influence of prolonged denudation on the details of our landscapes. The views then put forward were only an application of the doctrines of Hutton and Playfair, but to many of the audience they were novel. Charles Kingsley, who was there, told me next evening that "the lecture had been a kind of revelation to him of what he had for years been groping and grubbing for."

The Saturday excursion, which, at the request of the local committee, the President of the Geological Section planned and conducted, was taken along the shore to the east of St. Andrews for the purpose of examining the singularly clear and instructive sections of the Lower Carboniferous strata and the volcanic vents, which have there been laid bare by the sea. The success of this excursion was enhanced by brilliant weather. There followed a banquet given by the University of St. Andrews in honour of the Association. Principal Tulloch, in the full vigour of life, took an active share in mustering and entertaining the guests. Dean Stanley, who happened to be at St. Andrews, had a place of honour at the table. Principal Forbes, though full of interest in the success of the meeting, was unable to attend the dinner; but I had the opportunity for a short talk with him at his house. It was the last time that I saw that great and good man. Robert Chambers as one of the croupiers at the dinner, made a genial speech; him also I never met again.

At the close of the Association meeting, Lord Kinnaird

carried off Murchison and me to Rossie Priory, where we remained a few days. One of the most memorable of these days was that on which we were taken to Fingask, to call on the Misses Murray-Threipland, whose old castellated mansion was a veritable museum of Jacobite relics. The ladies were in great glee over the visit that had just been paid to them by Professor Charles Martin, of Montpellier, who had been attending the Geological Section at Dundee. His enthusiasm for Scotland and everything Scottish captivated them. They told us how they bade him kneel, and how, taking a sword that had belonged to Prince Charlie, they laid it on his shoulder, and, as if the blade still possessed some royal virtue, dubbed him knight. Six years afterwards I chanced to meet this French professor in a river steamboat on the Tiber bound for Ostia, with a party of students led by their Professor from the University of Rome. I addressed him as "Sir Charles," which greatly pleased him, and he recalled with much vivacity the charms of Fingask and the distinguished ladies who had entertained him there. My acquaintance with Lord and Lady Kinnaird, begun at this time, ripened into a friendship which brought me from year to year to Rossie Priory, as long as he lived.

Though, as already mentioned, it was soon evident that the Director of the Survey would have so much to do in inspecting the field-work of his staff as to leave little or no time for any mapping by himself, he naturally wished to complete the survey of western Ayrshire and Renfrewshire up to the Firth of Clyde at Greenock. With this view, in the summer of 1868, quarters were shifted from Ardrossan to Largs, whence most of the volcanic plateau was mapped which rises so steeply from the coast, and stretches eastward to the coal-fields of Beith and Dalry. It was while at Largs that I made the acquaintance of Ferdinand Zirkel, who came thither with a note of introduction from Professor Ramsay. He accompanied me over some parts of the volcanic ground, and was much interested in the evidence there visible of two widely separated periods of igneous activity. He was also surprised and pleased to find the microscope, and thin slices of the rocks of the district, among my household goods. He was at this time Professor at Lemberg, and on the threshold of his career as one of the great petrographers of his day. Tall, broad-shouldered, and somewhat short-sighted, so that he always wore glasses, he was brimming over with geological enthusiasm and eager interest in what he saw. This brief visit from him laid the foundation of a firm friendship between us, and of a correspondence which continued for nearly half a century and only ceased with his lamented death in 1912. He will often be referred to in subsequent chapters of this narrative.

Later in the same summer, with four of my staff, I made an excursion to the Continent, partly geological and partly tourist, up the Rhine and through a portion of Switzerland. At Bonn we made the acquaintance of the venerable H. von Dechen, the Nestor of Rhineland geologists, and met Ferdinand Zirkel, who at every recess from his professorial duties, repaired to Bonn to be with his mother, to whom he was devotedly attached. To our surprise, we also stumbled upon the Director-General of the Survey, who had informed me, a week or two before, that owing to Lady Murchison's state of health, he thought of taking a short trip to the Continent, not for baths, but merely for amusement and relaxation. The younger members of our little party had thus an opportunity of shaking hands with their Chief, now in the seventy-seventh year of his age.

We lingered at some of the more picturesque little towns on the Rhine, and in our boating excursions on the river were much interested to find that by placing our ears close to the bottom of the boat we could distinctly hear the rattling of the pebbles on the bed of the river as they were kept rolling over each other by the onward current. We afterwards made the same observation at different points on the Moselle between Cochem and Coblentz. As students of denudation, we recognised that as the pebbles must have travelled many leagues from their source, they could not fail to have undergone a measurable loss of bulk by the prolonged abrasion, and that this loss must add to the muddiness of the rivers.

Furnished with letters of introduction from Zirkel to the geological members of the teaching staff of the Jesuit establishment at the Laacher See, we spent some profitable days exploring the volcanic riches of the Eifel, climbing its cinder-cones, following the tracks of its lava streams, finding dead birds at its mephitic gas-springs, and collecting the carbonic acid gas in an inverted tumbler of water at some of the more powerful emanations. We traversed the region to Daun in a clumsy *leiterwagen*, saw the famous *Maare*, and halting at Bertrich, struck the Moselle at the little village of Alf. As the water in the river was too low for the steamboats to ply, we had leisure to follow, on foot for some distance, the remarkable windings of the river in the sinuous ravine or cañon which it has cut for itself out of the tableland of plicated Devonian strata. The gorges that have been excavated by this river and the Rhine are a constant and ever-increasing wonderment to geological eyes, which at the same time are by no means insensible to the charms of their scenery, with its picturesquely-perched castles and quaintly antique little towns. Taking a boat at Cochem we floated and rowed down the forty miles to Coblentz, encountering a brilliant thunderstorm on the way, and arriving at our destination after the gates of the town were closed for the night. With some little difficulty we obtained admission and a night's quarters.

The ascent of the great gorge of the Rhine, with its memorable demonstration of river-erosion, led us to Bâle, whence we made straight for Grindelwald, in order to do just a little bit of mountaineering and see the glaciers. Securing Peter Michel, as guide, we made for a chalet some 1500 feet higher, so as to begin the ascent early next morning. The Strahlegg Pass, between 10,000 and 11,000 feet above sea-level, was duly reached by the party roped together. It was our first glimpse of the interior of the great Alpine mountain world. The panorama embraced some of the giants of the Oberland—Finster Aarhorn, Münch, Eiger, Jungfrau, Schreckhorn and many more.

Descending to the Unter Aar glacier on the other side we found the surface of this glacier much more even than that on the Grindelwald side. Walking was accordingly more rapid; but with so much to be seen on the way, by a group of geologists to whom it was all beheld for the first time, the walk of fifteen miles down to the Grimsel took between four and five hours. Nowhere are the superficial features of the ice-world to be seen in greater perfection and abundance than in this famous and classic space—ice-tables, dirt-cones, crevasses, glacier streams, moraines, perched blocks, and huge walls of bare rock admirably smoothed, polished, and striated.

In descending the Rhone valley by diligence, we halted for a night at Brigue, and had there an opportunity of witnessing the effects produced by the rapid flooding of a tributary valley. A few days of warm south wind had thawed the snow on the mountains and swollen the river which, rushing through the village, had swept away a bridge and partially demolished some houses. In the half that was left standing of one of these dwellings, a fireplace was exposed on the first floor, with a kettle still in its place. From Brigue we made for Geneva, and thence hurried homeward.

The autumn of this year (1868) was chiefly spent in inspecting work in the field, varied by two brief visits to London in November connected with the Examinership in the London University. I was one day surprised to receive a letter in somewhat curious English, and in a minute penmanship that ran obliquely across four closely-written pages. Here are the first few sentences:

Vienna, the 18th August, 1868.

Dear Sir,

Notwithstanding I have not the pleasure of our acquaintance, I wish to express you personally with what delight I remarked that Scotland at last had got a geologist who followed in the steps not only of Faujas St. Fond (Hutton, Macculloch left aside) but also of mine. I mean the so neglected study of igneous rocks. How is it possible that Great Britain would have given rise to a so numerous set of palaeontological geologists, without producing in the mean time a set of well-teached mineralogists and lithologists, to be apt for the solution of all the intricated problems of volcanic intrusions?

The signature on the last page was Ami Boué. It came like a voice from the grave. I had published appreciative reference to the excellent work which this pioneer geologist achieved in the early decades of the century, but imagined that he was no longer alive. The letter showed that he had retained a copious and accurate remembrance of the time which he had spent in Scotland. He correctly named a number of places which especially dwelt in his memory, adding that "if more than half a *seculum* had not elapsed since I had the good fortune of being able to hammer in Scotland, I would have return to your Scottish Mist, and endeavour to try there my geological experience."

Boué had a singular history. He came of a French family that could trace its annals back for several centuries. One of his ancestors was sent by his father out of France, concealed in an empty sugar-cask, and despatched to Amsterdam, in order to escape from the religious persecution under Louis XIV. Eventually this exile established himself in business at Hamburg in the year 1705, and became a rich man. In the fifth generation from him my correspondent was born in Hamburg in 1794. In his boyhood, while Napoleon was overrunning Europe, Boué pursued his education at Geneva, but the political horizon being so unsettled, and the possibility of the loss of the family property in Hamburg having to be considered, he was pressed by his guardians to choose a profession that would at least enable him to gain his livelihood, if the worst should befall. He resolved to take to medicine, and in order to avoid the turmoils of the Continent, he chose to place himself at the University of Edinburgh. As he remarked to me when I paid him a visit in 1869, " I really went to Scotland to escape from Napoleon."

Furnished with good introductions, he made friends and acquaintances in the best society of the Scottish capital. He continued his studies there until he took the degree of Doctor of Medicine in 1817. At the same time, while attending Jameson's lectures, he developed a strong love of mineralogy and geognosy, and appears to have spent much of his spare time in making geological excursions over a large part of Scotland. In these journeys he took copious notes of what he saw, which were eventually embodied by him in a volume written in French, which was published at Paris in 1820 with the title *Essai Géologique sur l'Écosse*—a most valuable treatise, which was far in advance of its day. The correspondence which he began with me at this time continued as long as he lived, and our friendship was further cemented by my visit to him at Vienna in the summer of 1869.

My early friend, Principal Forbes, died on the last day of 1868. As already mentioned his health had been slowly failing for some years. He gently passed away in the midst of his family, holding the hand of his eldest boy, and with the calm courage and trust of a true Christian. While my admiration of his scientific prowess was profound, not less was my affectionate regard for him personally. These feelings found expression in an obituary memoir, communicated as a presidential address to the Edinburgh Geological Society. The brief report of this address which appeared in the newspapers led his widow to ask to be allowed to publish the complete memoir in pamphlet form. This was accordingly done by her, and the account of Forbes's geological work appended to the address was included in his *Life* published in 1873.

The Royal Society of Edinburgh continued to show

its interest in the progress of the Geological Survey of Scotland, by from time to time inviting me to give an account of its progress at one of the evening meetings. The mention of this Society recalls its pleasant dining-club, to which reference has already been made, and of which I was now a member.[1] It included some choice spirits, who in ordinary society might be considered staid and demure, but who in the genial atmosphere of the club "let themselves go," and showed a capacity for humour and merriment that would have astonished those who only saw them at other times. With such men as Lord Neaves, Sir Douglas Maclagan, Professor Blackie, and Sir Daniel Macnee as performers, it may be imagined how delightful these evenings were. Some of the members, endowed with the faculty of verse-making, wrote topical songs which they sang on these occasions. It was said of Lord Neaves that when he was seen on the bench, apparently taking notes in some tedious case before him, he might really be inditing some humorous effusion to be sung by him at the club dinner the same evening.[2] The ditties by Maclagan were the gems of our Muse, and fortunately most of them were gathered together by him into a little volume published in his lifetime.

The membership of the Royal Society Club at Edinburgh was restricted to Fellows of the Society, though guests could be invited. Numerous convivial clubs which flourished in that town down even into the earlier decades of last century had died out. About this time (1869-70) the Edinburgh Evening Club was started as an attempt to revive the evening sociality of an older time. David Masson, one of its chief founders, constantly attended while his health allowed. As one of the evenings was that on which the Royal Society held a meeting, it was a common practice to stroll along Princes

[1] *Scottish Reminiscences*, p. 355.

[2] A selection of his metrical contributions was published with the title *Songs and Verses, Social and Scientific*, by an old Contributor to *Maga*, of which a third enlarged edition appeared in 1869.

Street to this Club after the meeting. Many of the more eminent men in law, literature, science, and general culture in Edinburgh joined this Club, the membership of which was limited to 150.

It now at last became evident that the duties of the Director of the Survey in Scotland could not be fully discharged if he continued to devote part of his time to personal mapping in the field. As already stated, I had been unwilling to relinquish the ground in Ayrshire and Renfrewshire until the mapping had been carried up to the shores of the Clyde. Since this work was begun by me in 1863, a strip of country nearly fifty miles in length and some three to ten miles in breadth had been mapped along the coast from Lendalfoot, south of Girvan, up to the northern boundary of Ayrshire. To complete the area up to the Clyde I took quarters at Gourock, and in three continuous months of the summer of 1869, had the satisfaction of finishing the work, thus completing the survey of a total area of between 200 and 300 square miles. Much to my regret it was not possible, on account of the pressure of other duties, to include in this work the higher part of the great volcanic plateau to the east. Never after this time, save as a rare and brief pleasure, could I resume the joyous work of independent mapping.

Correspondence with Ami Boué had now become so continuous and interesting that his invitation to pay him a visit could not be refused. The "Naturförscher und Aerzte," the German equivalent of our British Association for the Advancement of Science, were to meet at Innspruck in September 1869, and I arranged to attend their gathering, after having been with Boué. Leaving Scotland at the end of August, and travelling by Cologne and Nuremberg, I reached Vienna late on 4th September. My venerable correspondent met me at his town house in Vienna, whence next day we went about an hour's journey by railway to Vöslau, where he had a pleasant country retreat, on the last spur of the Alps as they dip down into the plain of the Leitha. Here he spent each summer amidst his vineyards, almonds, peaches, quinces, apples, and other fruits, some of which were then in full bearing. He was a keen maker of wine, contriving, by judicious use of the pump-well, to get more than one yield from the same grapes.

We had long talks, in which he gave, with much dry humour, reminiscences of his stay in Scotland. It was astonishing to find that his memory retained so many minute details of the places he had visited. Some of the recollections were droll enough, as when he recalled that as far as he could discover, there was only one washing-tub in Edinburgh, and it was in the hospital! We paid several visits to Vienna, and met there a number of the Austrian geologists, including Suess, Hauer, Tschermak, and some of the younger men. When we called on the venerable Wilhelm Haidinger, famous for his studies in mineralogy and optics, we found him sitting in his study. Like Faust in Gounod's opera, before he is transformed into a young man, he was wrapped in a long dressing-gown, his white beard flowing down his breast, his head covered with long lint-white hair that hung down his neck (as Boué told me, it was a wig), and his feet immersed in enormous woollen shoes. He spoke of his early time in Scotland, between 1823 and 1826, when he lived in Edinburgh with Mr. Allan, the owner of one of the best private mineralogical cabinets in this country, and when Haidinger's knowledge of minerals was of much service. His wife and daughter also received us with great courtesy.

Edouard Suess, with whom from this time a correspondence began which lasted as long as he lived, gave me full directions how to reach the Dachstein glacier, which I wished to visit on the way to Innspruck. He and Dr. Wolff drove me to Döppling to see the extended view across the vast plain of the Danube to the distant Carpathian mountains, and also to visit some geological sections in this neighbourhood.

From Vienna I went to Ischl and thence to Halstatt, where, through the attention of one of the officials, I was enabled to descend the salt mines and see the processes of working. With a good guide and a sufficient stock of provisions for us both, I started for the Wies Alp, a small hamlet, high up among the mountains, consisting of no more than four wooden huts, each tenanted by a young woman who attended to the cattle that were pastured there in summer. The hut in which we passed the night consisted of the ground-floor which formed the byre, and an upper storey that served at once as kitchen, food-boiling place for the cattle, and bedroom. It was about ten feet square, with the fireplace in the centre, and no chimney. The smoke, however, had hundreds of openings for escape, since one could see the stars between the rafters and the walls. The bitterly cold mountain wind blew in through these gaps. I slept in a boarded-off corner on a litter of fodder, with some sacking and an old garment as covering. The object of this excursion—to examine the end of the little Dachstein glacier and its terminal lake—was successfully accomplished.

At Salzburg a halt was made in order to pay due homage to the manes of Mozart, who from my boyhood has been to me the greatest master of melodious music. There was a pleasant little concert given in the Mozarteum, and one of his pieces was performed on the spinet on which he himself used to play. Innspruck was reached in good time for the opening of the forty-third meeting of the German Naturalists and Physicians. Here, among a number of acquaintances, it was a joy to find Zirkel, also Wolff and Charles Martin; pleasant likewise to come into personal touch with two geologists whose names were household words all over the scientific world—the venerable leaders of Swiss geology, Studer and Escher von der Linth; and to shake hands for the first time with Sandberger, Fuchs, Zittel, and others. My display of some sheets of our English Geological Survey, accompanied with a brief explanation of them and of our methods of mapping, formed the subject of discussion at one of the meetings, when the maps were scrutinised with some minuteness and received due commendation.

In the course of conversations with the members of

different Sections of the Association, and with professional men in Vienna, it was a surprise to find what a hold Darwin's views, as expressed in his *Origin of Species*, had gained in Austria. Again and again it was affirmed that the men of science in that country had gone much further in their acceptance of these views than was yet the case in the British Isles. One acquaintance, a physician of note in Vienna, remarked to me, " You are still discussing in England whether or not the theory of Darwin can be true. We have got a long way beyond that stage here. His theory is now our common starting-point." This was the case not only in the domain of science, but even in politics. It was remarked that three years ago, after the disastrous termination of the war with Prussia, when the Austrian Parliament assembled to deliberate on the reconsolidation of the empire, Professor Rokitansky, a distinguished member of the Upper Chamber, began a great speech with this sentence—"The question we have first to consider is whether Charles Darwin is right or no." A great empire lay in dire distress, and the form and method of its reconstruction were proposed to be decided after a consideration of the truth or error of the doctrine propounded by the English naturalist.[1] The decision arrived at by the politicians, even if based on Darwinian lines, has proved to be ineffectual in permanently cementing the irreconcilable component parts of the Austrian empire which, in little more than fifty years, has broken up into fragments under the stress of the Great War of 1914-1918.

From Innspruck by night diligence, twenty-six hours of continuous travelling brought me to Bregenz, where a short stay was made to allow of a good look at the delta of the Rhine in the Lake of Constance. The journey was continued by Zurich and Lucerne to Solothurn, where a few most instructive days were passed among the ridges, ravines, and crests of the Jura mountains, in the company of Professor Lang.

[1] A brief account of this meeting at Innspruck was given by me in the first number of *Nature*, published on 4th November, 1869.

In the course of the winter of 1869-70 I gave a short course of lectures on the "Principles of Geology" to the Industrial Classes at the Edinburgh Museum of Science and Art. A novel feature in this course was the introduction of excursions with the working men to some of the many places of geological interest in the immediate neighbourhood of Edinburgh. These artizans, like those in London, were always pleasant to meet and to instruct; they seemed so keen to learn and asked such intelligent questions. They mustered in good number at these lectures, and one could not but hope that some among them would be led to take up the study of the subject as a relaxation in such leisure as their daily labour allowed.

Up to the year 1870, when in Edinburgh, I continued to live with my parents. As this town had now become my permanent headquarters, I realised that it would be more convenient, both to my family and myself, were I to set up a separate establishment. The picturesque octagonal house, known as Ramsay Lodge, standing on the northern slope of the ridge which culminates, a little to the west, in the abrupt rock, crowned with Edinburgh Castle, had long taken my fancy as a most enviable abode. Rising out of an extensive area of public garden grounds in the very centre of the city, it not only overlooked the New Town but commanded a view of the whole range of hills from the distant peaks of the Trossachs on the west, along the chain of the Ochils and the uplands of Fife on the north, eastwards to the May Island in the North Sea. Besides the attraction of the site, the literary associations of the house appealed strongly to one's historic sense. It was built in the middle of the eighteenth century by Allan Ramsay the Scottish poet, as a suburban residence, wherein he spent the later years of his life, and where he died in 1757. The quaint octagonal shape of the main part of the building may have reminded some of the poet's friends of Swift's sarcastic account of Sir John Vanbrugh's house at Whitehall as

> "A thing resembling a goose-pie,"

for they appear to have given it this appellation. Its oddity of form, however, was much diminished by the poet's son, the eminent portrait painter, who added a new front and a wing, and in other ways made the dwelling more commodious.[1] He died in 1784. The "goose-pie" was afterwards the residence of Dr. Baird, Principal of Edinburgh University from 1793 to 1840. In the spring of 1870, hearing that this attractive house was soon to be vacant, I secured the tenancy.

Mr. George Julius Poulett-Scrope, already mentioned, was well known to the public as a writer of political pamphlets, but to a smaller circle he stood among the geologists of his day, as one of the highest authorities on volcanic problems, who also surpassed even Lyell in his recognition of the efficacy of denudation in carving out the present surface of the land. He sent me at this time a letter in which he wrote—"I have long wished it were possible to induce some thoroughly competent geologist, with clear ideas on the character of volcanic action, and a sound and quick judgment as a field-observer, with the power of reporting clearly what he sees, to undertake an examination of the volcanic districts of Southern Italy, but especially the Lipari Islands. I am too old myself, and disabled by loss of eyesight creeping upon me, or I should like nothing so much as to undertake this job, which circumstances have for many years hindered me from doing, sorely against my will." After stating some of the questions to be investigated and, if possible, solved, he continued—"I have so much at heart the obtaining a sound and impartial opinion on these points that I should be happy to contribute, say two hundred pounds, towards the expense of their exploration by a competent geologist, such as I have described above. Can you help in the matter?" The letter ended with an

[1] Both father and son were commemorated, without sarcasm, by Churchill:

> Thence came the Ramsays, names of worthy note,
> Of whom one paints, as well as t'other wrote.
>
> *Prophecy of Famine*, 125.

apology for his appeal that I should myself undertake the proposed investigation.

This tempting proposition, so entirely consonant with one's geological tastes, could not be accepted without much reflection. Besides my own personal affairs, having just taken a house and being in the middle of putting order and furniture into it, the official chiefs had to be consulted. In the end, however, all difficulties were removed. My eldest sister undertook to look after the furnishing of the house. Leave of absence was granted by the authorities from the middle of March until 3rd June, so that April and May could be spent in the Mediterranean area, after the spring gales were over, and before the heat of summer had set in.

Among the few days spent in London on the journey, one remains deeply impressed on my memory. With Alexander Macmillan I called one afternoon on Thomas Carlyle at his house in Chelsea. When the door of No. 5 Cheyne Row was opened by a little housemaid, the philosopher himself was seen standing at the far end of the narrow lobby, looking as if he would rather receive no callers. As soon as he recognised my friend, however, he came forward to meet us. He was wearing a black velvet cap, a grey coat or dressing-gown that reached to his feet, and had a long clay-pipe in his mouth. He brought into the room pipes and tobacco, but finding that we did not wish to smoke, he led us upstairs to his drawing-room—a quaint little chamber, with a number of antique pictures on the walls, some of them, perhaps, portraits of Mrs. Carlyle's forbears, and with old-fashioned nick-nacks dotted over the room. The grate with Dutch tiles up each side of the fireplace, the antique chairs and the various stools were all in excellent harmony, and one could see that some woman's hand, guided by good taste, had attended to the furnishing of the room. Mrs. Carlyle had died four years before this time.

But the man himself was the chief centre of attraction. He sat between us right in front of the fire. Fortunately he was in the mood to talk, and soon found himself dealing with Daniel Wilson's recently published biography of Chatterton, which in his opinion gave too flattering a portraiture of the poet. He looked on Chatterton as one who never seemed to get into the essence and truth of things, but was more taken up with the roll, and the marching, and the trumpet-clang. He passed on to David Gray, the poor Scottish poet, who, he said, had something in him, but lost himself from the foolish desire to be famous. No one, he thought, has any business to wish to be famous, but should be content to do his own work honestly and well. He enlarged on the "jackassery" of the world in its craving for novels.

Incidentally, too, he gave us some graphic glimpses of scenes in his own life. He remembered having climbed the Lomond Hills of Fife with Edward Irving, and found the Ordnance Survey sappers at work there, one of them a surly fellow, who at last was fairly won over by Irving's imperturbable good nature. His college days at Edinburgh came into his mind as he chatted. He laughed when he gave us some of his reminiscences of the lectures of Robert Jameson, the Professor of Natural History at the University. He said that Jameson had never done him any good, had no power of interesting him in the subject, and rambled so widely from one topic to another that it was hardly possible to follow him. Thus the Professor would suddenly stop in the middle of some details, and exclaim, "We shall now consider the order *Glires*," whereat, as he recalled the incident, Carlyle roared with laughter, adding how the Professor "dwelt upon the four chisel teeth." Jameson was probably somewhat dry as a lecturer. There can be no doubt that he had the reputation of being rigidly methodical in his prelections; but he certainly had the merit of inspiring some of his pupils (Edward Forbes and others) with a love of natural science. Yet we can well understand how he might repel the future author of *Sartor Resartus*.[1]

[1] In one of his early letters, written while he was attending Edinburgh University, Carlyle states that he had heard two of Jameson's lectures, but they made so bad an impression on him that he was

From Edinburgh and its University the talk branched off into the south-western Lowlands of Scotland, whence I was able to tell him some anecdotes of Ayrshire lairds and clergymen, and had the gratification of evoking from him such peals of laughter as I hardly ever heard before. He seemed sorry when we rose to leave, came down to the door with us, halting two or three times on the staircase to ask questions about the rocks of his native Dumfriesshire.

Alexander Macmillan at this time kept open house on the evening of Thursday at his warehouse in London, where he welcomed all friends and acquaintances who liked to talk and to smoke. These "tobacco parliaments" grew in popularity as, with the increase of the business of the firm, his circle of authors became ever wider. They began in Henrietta Street, Covent Garden, and were continued in the larger premises which were successively occupied in Bedford Street. The company was often a brilliant one, in which representatives of literature, science, art, politics, and the services mingled together. I enjoyed them when occasion took me to London, and before dense tobacco smoke painfully affected my eyes.

Travelling rapidly through France to Marseilles, I caught a steamboat for Italy and came in for a fierce "nor'-easter" on the way, which lashed the grey Mediterranean into a state of rage that seemed to a newcomer quite unworthy of its popular reputation for colour and calm. The vessel was some nine hours late in reaching Civita Vecchia, with its gingerbread fortifications in charge of French soldiers; for we were now in the Papal States, which France was protecting. The slow, slovenly ways of the papal régime became manifest even before we were allowed to land. Our passports were taken from us to be examined on shore. When at last we were permitted to follow them, the scene at the passport office gave us further experience of the ways of Pio Nono's

doubtful whether he ought to attend the class at all. *Early Letters*, vol. i, p. 192.

functionaries. My passport was not to be found, though my name was properly inserted in the list of arrivals. I had seen the tail-end of a piece of green silk projecting from the folds of the passport which one of the passengers was putting into his pocket. I asked him to open his document, and there was my passport enclosed within it. Had the precious piece of green silk, on which the passport was pasted, been lost, with its Foreign Office attestation, duly signed by Lord Stanley, my journey would doubtless have been seriously interrupted, if not stopped altogether.

It was the year of the great Oecumenical Council at Rome. The Eternal City literally swarmed with ecclesiastics who were arriving by every train, and disembarking at the small, ill-contrived, and unswept central station of the Papal Dominions. The streets were crowded with cardinals, archbishops, bishops, abbots,

> Eremites and friars,
> White, black and grey, with all their trumpery,

gathered together from every corner of the globe. Although much tempted to linger here a while, I had to content myself with a stay of only two or three days, during which, thanks to Mr. Pentland, at that time the great resident English authority on Rome (to whom Murchison had given me an introduction), I saw the chief pagan antiquities, and more especially the excavations then in progress upon the Palatine. It was an entrancing first experience to watch the workmen digging through the fifteen or twenty feet of rubbish which had accumulated in the course of centuries, and uncovering the walls of the imperial palaces with their frescoes still fresh upon them; to witness the revealing of ancient pavements worn smooth long centuries ago by the print of feet and the roll of chariot wheels, and to see from time to time pieces of pottery, rings, bone-pins, coins, lamps, or fragments of glass, and whole wine-jars, come up out of the earth. All these objects were being carefully collected and examined, and the best of them would doubtless find a resting-place in one or other of the museums of the city.

At Rome it was my good fortune to meet M. De Verneuil, Murchison's companion in Russia, and my friend Joseph Prestwich, President of the Geological Society. They invited me to accompany them in a geological excursion up the Anio to look at sections of the old alluvial deposits of that river. The blending in one's mind of geological facts with thoughts of Horace and his *praeceps Anio* gave to the excursion a singular mingling of attractions.

Hastening to Naples, I took the opportunity to make acquaintance with the general features of the extinct volcanic cones of the Campi Phlegraei, and could not resist the temptation to give a day to Pompeii. Mr. Prestwich and his charming wife, now on their wedding tour, also came to Naples, where he and I agreed to make the ascent of Vesuvius together. With an intelligent guide we climbed to the edge of the crater on 8th April, and on the way had a good look at some of the various lava streams that have successively descended the flanks of the mountain. We slept, or rather tried to sleep, at the Hermitage, not an attractive lodging, and were glad to resume the journey next morning. Descending from the Observatory ridge, we crossed the lava-field of the Fosse Veterano, and climbed the rugged interior front of Monte Somma, studying with keen enjoyment the succession of lava-sheets of which the precipice has been built up, and the innumerable vertical dykes of intruded lava which have filled the open fissures made at successive eruptions. Not less interesting was the descent on the northern side of the mountain, in one of the deep ravines which the rains of many centuries have carved out of the crumbling declivities. Rain began to fall in torrents ere we reached the village where we had arranged that a carriage from Naples should meet us. Before we had started on the little expedition, Mrs. Prestwich gave me many injunctions to keep her husband out of danger, and I had the satisfaction of presenting him to her at their hotel, not only safe and sound, but like myself delighted and instructed by our trip. There was perhaps some reason for a little anxiety on the part of strangers—we were aware that bandits infested this district only a few years before. Professor John Phillips, of Oxford, who came to Vesuvius in 1868 dared not venture to go where we went. We saw some suspicious-looking fellows at Ottajano, but were not in any way molested.

At Naples it was a great pleasure to meet the illustrious and venerable Mrs. Somerville, now nearly ninety years of age, and her two daughters. She was still in the full possession of her faculties, and being interested in volcanic matters, discussed with me the projected visit to the Lipari Islands. Professor Guiscardi of Naples, who placed his local knowledge at my service in the Phlegraean Fields, likewise gave me useful information for my guidance among the islands, furnishing me also with an introduction to one of the higher ecclesiastics at Lipari.

A pleasant but brief voyage brings the traveller from Naples to Messina, whence a vessel landed me at Lipari on 17th April. Unfortunately, either at Rome or at Naples, I had caught one of the forms of malarial fever, which showed itself only after I had arrived at my destination. Not knowing what ailed me, but anxious to get to work, I spent three days in a preliminary examination of the nearer parts of the island, and a fourth in an excursion to Volcano and Volcanello. On the evening of that last day I had some difficulty in walking back to the miserable *locanda*. Hardly able to stand, I went to bed, and the courier, whom I had taken with me from Naples, sent for a doctor. In a short while, various residents found their way into the bedroom, and sat discussing my condition and the object of my visit, two or three speaking at a time, and filling the place with their kindly-meant but trying noise. At last the doctor came, the crowd was dismissed, and after the usual examination, he pronounced the case to be one of what he called "rheumatismal fever."

The doctor was a young graduate from Naples who spoke French fairly well, and all our talk was in that language. He was remarkably attentive, usually visiting his patient twice a day, and prescribing "sweating-potions" on a gigantic scale, for they filled flagons like bedroom ewers. The food of the little inn being poor, he kindly supplemented it from time to time with a present of quails, which, being a keen sportsman, he would shoot in the leisure hours that seemed to form the greater part of his time. He did his best to keep up one's spirits by the assurance that he looked forward to the resumption of my researches in a few days. But week succeeded week with varying diminutions and accessions of the fever, and with a continuous loss of bodily strength. Had I come entirely on my own errand, I would have tried to get back to Messina and take a homeward-bound steamboat for Marseilles. But as long as there was the least hope of being able to fulfil the duty entrusted by Mr. Scrope, one felt bound to remain in Lipari. Towards the end of a month, however, most of which had been passed in bed, it had become impossible for me to make any bodily exertion. Only with difficulty could I walk up and down my room.

Eventually, the fever having somewhat abated, with my kind doctor's full consent, I embarked on a south-bound steamboat on 18th May and landed at Messina on the same day. A French boat was to call there next morning, and I arranged to start by it. But the end of the trial was not yet. The landlord of the hotel, seeing my feebleness, and hearing that I meant to leave so soon, unknown to me, asked his medical attendant to come and see me. I did not know who it was that introduced himself pleasantly, and began to enquire about my doings in Lipari. He explained that he was a medical man, and asked that he might feel my pulse. When he had done so, he said very gravely that it was quite out of the question, if I valued my life, to start for France next morning, for I was in a high state of fever. When the morning came, instead of embarking for Marseilles, I was in bed, delirious and quite unconscious of my surroundings. I have no recollection of that day, or of how long time may have intervened before my senses were regained. I learnt afterwards that the landlord had brought to my room such English people as he could find, to see if they knew me and could communicate with my friends. I still remember with a shudder some of the visions that haunted me in the delirium. The first incident that remains in my memory of what occurred after consciousness returned, was the visit of a kind Italian, who came and sponged my face and hands, and spoke to me in English with a strong American accent. He was a Sicilian, named Victor Calorese, who had spent some time in the United States, where he had naturalised himself as an American citizen. He had come back to his native island, and was now acting as a courier at the hotel. From his cottage on the hills above Messina he came down each morning, with his pockets filled with oranges from his little garden, which he peeled for me, and he took much pains to discover in the hotel any kind of food that I could take.

The Messina doctor continued to pay his daily visit, but could give no assurance of when the homeward journey could be resumed. The complaint seemed to have settled down into what Juvenal calls a "domesticated fever." At last my faithful Sicilian came to me one day and pressed me to leave Messina, otherwise recovery was impossible. He assured me that if I would go, he would come with me to London. In spite of the doctor's opposition we embarked on 31st May in a Messageries Impériales steamboat for Marseilles.

The handwriting of my letters had been growing increasingly unsteady, and had become so shaky in the last brief note, written on the day of my arrival at Messina, that the cessation of further communications gave rise to much anxiety at home. Sir Roderick Murchison got Lord Clarendon at the Foreign Office to telegraph to the British Consul at Messina for information. The reply by telegram came that I had left that place.

Landing at Marseilles on 3rd June we took train next day for Paris, which was reached in the midst of the political excitement that preceded the outbreak of the Franco-Prussian war. Without lingering there for a day, we pushed on for London, and Calorese delivered me safely at the suburban residence of my friend Alexander Macmillan. The kind-hearted and capable Sicilian went straight back to his native island, assured of my lifelong gratitude. I kept in touch with him for some years, and then lost trace of him.

It was a week or two before the journey to Scotland could be undertaken. One of the earliest friends to pay me a visit was Sir Roderick Murchison, who, not wholly reassured by the sight of me, asked Dr. Charles Murchison, the chief authority on fevers, to overhaul me and report to him. The middle of June, however, found me established in my new home in Ramsay Lodge. While regaining strength there, I was able to form one of a quiet dinner party at Professor Masson's house, to meet Thomas Carlyle who was his guest at the time. The sage was in excellent form, except when the scream of the railway whistle from the valley below reached his ear. He would then launch forth a torrent of vituperation on the noises that made life unendurable in a city. But he soon became vivacious again, and laughed, as only he could, at various Scots stories that were told to him.

But the poison of the Lipari fever had not yet been eradicated from the system. Before long a severe recrudescence of it brought a renewed prostration. The best authority on fevers in Edinburgh, Dr. Warburton Begbie, was called in, who took the case thoroughly in hand. By careful treatment, the virulence of the relapse was gradually reduced, though the fever germs were even then not wholly eradicated, for they asserted themselves at intervals afterwards for no less than seventeen years. There now came yet a final trial in the form of abscesses on the bone which required surgical treatment. Thus month after month slipped away, during which I lay for the most part on the sofa, prisoner in my own house.

Fortunately, in spite of this disablement, it was now possible to get all the threads of the Survey work once more in hand, except the inspection in the field. By degrees also one could take up again a task which had been entrusted to me by my lamented friend, J. B. Jukes, some time before his death in 1869. When his health began seriously to fail, he had asked me to relieve him of the labour of revising the text of his *Student's Manual of Geology* for the third edition of the work. When the revision, which had been begun during the previous winter, was now resumed, much more change was found to be requisite than had been anticipated, some of the chapters having to be largely altered, and more than one to be practically rewritten. Hence when the volume was published in October 1871 it was in many respects a new book.

The autumn of 1870 was overshadowed by the great Franco-German War, which we all watched with feverish anxiety. At first, in ignorance of the true cause of the outbreak, as afterwards revealed, many were inclined to blame the French for having declared war, but as the contest turned against them my sympathies were increasingly on their side. The correspondence with Ami Boué revealed how a man could be affected who had relatives on both sides of the combatants. A passage may be given here from the letter he wrote on 21st November: "The dreadful war pre-occupation did take me all time for thinking at scientific matter, and now perhaps that distress will approach till nearer our abode! When you will know that I have very good and near Parents in both armies, and you perceive the possibility of parents killing themselves without recognising themselves, nor having the opportunity to do so, you will understand that I have often headach, when I ride the newspapers or hear from the quite useless slaughters which have been prosecuted only by those men at the head of the human society. I have parents in Paris; other exiled in Spain, in England, in Switzerland. The country-houses of some by Paris are German hospitals or barracks."

At Edinburgh during the winter of 1870-71 there arose a little domestic excitement in educational circles on the subject of the erection of a Chair of Geology at the University. Dr. Allman, on 26th August, had intimated his desire to resign the Professorship of Natural History, which included geology within its scope. The wish had long been expressed, more strongly since the death of Edward Forbes, that advantage should be taken of the first vacancy to separate the department of Geology and Mineralogy and to make it the business of a new Chair, leaving Zoology as the main subject of the Natural History professorship. It was a matter of common talk that Sir Roderick Murchison intended to found and endow such a Chair, and he himself often spoke to me on the subject, saying that he would like to do this in his native Scotland, and to see me the first Professor. Preliminary steps had now been taken to carry out his design, when further progress was suddenly arrested on 21st November by an attack of paralysis which struck the old Chief down. For a while it was doubtful if he would recover. But in an enfeebled condition he continued to live for nearly a year longer. He had provided in his will for the endowment of the proposed Chair, but he now desired to have it established in his lifetime. He applied to the Government to divide the Natural History Professorship at Edinburgh into two, and to consider Geology and Mineralogy as the subjects of a new Professorship, towards the endowment of which he offered to contribute £6000, on the understanding that he should nominate the first Professor. Owing to the state of his health and certain official difficulties, the negotiations were protracted for some months. The Home Office objected to the donor of a large part of the endowment having the nomination of the first Professor, on the ground that this would infringe the prerogative of the Crown. The Science and Art Department objected to the holder of the Professorship retaining his post as Director of the Geological Survey of Scotland. With such a Scylla and Charybdis in front, it seemed for a time that the Chair would not be created. When its establishment was eventually secured, the official objections to my appointment appeared too formidable to be overcome. As long as he was able, Murchison fought stoutly against the contentions of the two Government Departments, and when he could himself no longer fight, his views were strongly supported by several of his influential friends, and especially by Dr. Lyon Playfair, Member for Edinburgh University. After many weeks of correspondence, interviews and negotiations, the following note came to me:

THE ATHENAEUM,
15 February, 1871.

MY DEAR PROFESSOR,

I have the pleasure of being the first person to call you by that title, as I have just returned from the Home Office, with the assurance that your name will be immediately submitted to the Queen for formal approval. I wish you all joy in your new appointment. Tell Prof. Wilson.

Yours sincerely,
LYON PLAYFAIR.

The creation of a Chair of Geology at Edinburgh could not but awaken memories of the time, nearly a century by-past, when, in this city, a brilliant band of geologists, including James Hutton, John Playfair and James Hall, placed this science on a firm basis of observation and experiment, and founded what has been known as the Huttonian, Plutonist or Scottish School; when the antagonistic Neptunist doctrines of the Saxon Professor, Abraham Gottlob Werner, were introduced into the Edinburgh University by his devoted pupil Robert Jameson, and when, as Brewster has remarked, "the rival theories of Fire and Water were discussed with all the warmth or even bitterness of political or theological controversy." With the ultimate discomfiture of the Neptunists and the triumph of the Plutonists, geological zeal gradually died down in Scotland. Principal J. D.

Forbes, who in his youth may have shared in the controversy, used in later life to contemplate with much regret the decay of geological enthusiasm in his native country. In an impressive address given to the Royal Society of Edinburgh on 1st December, 1862, he found full expression for this regret. " It is a fact," he said, " that the Scottish Geological School, which once made Edinburgh famous, especially when the Vulcanist and Neptunian War raged simultaneously in the hall of this Society and in the class-rooms of the University, may now almost be said to have been transported bodily to Burlington House. Roderick Murchison, Charles Lyell, Leonard Horner, are Scottish names, and the bearers are Scottish in everything, save residence. Our younger men are drafted off as soon as their aquirements become known. Professor Ramsay was early called from his voluntary labours in Arran to English soil; and we only retain the services which our townsman, Mr. Geikie, volunteers for our instruction, so long as the central forces of Jermyn Street suffer him to linger within the Scottish borders. Others, who still reside in Scotland, not unnaturally seek a larger audience, and a more rapid publicity for their memoirs, by transmitting them to London. This is reasonable and inevitable. . . . Of all the changes which have befallen Scottish science during the last half century, that which I most deeply deplore, and at the same time wonder at, is the progressive decay of our once illustrious Geological School. Centralisation may account for it in part, but not entirely."[1]

Had the revered Principal lived three years longer he would have rejoiced over the introduction of Geology as a distinct department of study in his old University of Edinburgh, with the prospect that the science which he had himself prosecuted in his earlier years, and in the promotion of which he remained keenly interested up to the end, would now have a better chance of being domesticated once more in Scotland. My earnest intention certainly was to do what in me lay towards

[1] *Proc. Roy. Soc. Edin.*, vol. v. (1862), p. 17.

recovering the ground which he thought that we had lost in Scotland. "The central forces of Jermyn Street," which he dreaded, suffered me "to linger within the Scottish borders" for ten years after the time at which he wrote. They were years of hard work mainly devoted to the investigation of the geological structure of Scotland, and, if that might be, to the restoration of the fame of the Scottish School.

The Royal Warrant creating the Chair of Geology in the University of Edinburgh, and appointing the first Professor was issued on 10th March, 1871, with the bold signature of Queen Victoria at its head. I was duly inducted at the meeting of Senatus on the 25th of that month. Thus the early premonition that I would return to the University as Professor was at last fulfilled. How it was to come about remained always dark, and I would sometimes ask myself Horace's question, "An me ludit amabilis insania?" The fond illusion, however, had become a reality, and looking backwards one could now see how one incident after another, though unsuspectedly at the time, had brought about its accomplishment.

In the first half of the year 1871 I was thrice summoned to London on urgent business, partly in connection with a contested Parliamentary Bill for bringing a fresh water-supply to Edinburgh. I took advantage of one of these visits to make an offer of marriage to a young lady whom I had met the year before and of whom I this year saw a good deal more. Alice Gabrielle Pignatel lived with her mother and two older sisters not far from the residence of my old friend, Alexander Macmillan, at whose house I made her acquaintance. Her mother was English: her father, who had recently died, belonged to a branch of the Pignatelli family of Southern Italy, which had migrated into France some generations before and had settled in Lyons as merchants. Her cousin, Victor Pignatel, was one of the Directors of the Crédit Lyonnais Bank in Lyons and Paris. I was first captivated by her beauty and the charm of her voice as she sang, but soon still more by the gentleness and gaiety of her nature. She accepted my hand, and the wedding was arranged to take place in the following August.

After all the vexations arising from the unfortunate Italian journey there was a delightful sense of restored liberty and renewed strength in resuming inspection work in the field, and more especially in the wide region of Galloway. Several members of the staff were now at work in this part of the country, where much ruggedness and wildness, almost equal to that of the Highlands, is combined with broad strips of lowland moor, well-cultivated valleys, interesting little towns and villages and a wonderfully varied coast-line whence, in clear weather, the Isle of Man and the hills of Antrim can be seen.

The British Association for the Advancement of Science met for the third time in Edinburgh on 10th August this year, with Sir William Thomson (afterwards Lord Kelvin) as President. The Council had again elected me President of the Geological Section, and the Local Committee entrusted me with the selection, arrangement and conduct of the geological excursions. Arthur's Seat was chosen for an afternoon walk, while for the scene of our more distant ramble on foot, it was arranged to take the picturesque coast-line from the Siccar Point to Fast Castle in Berwickshire—ground classic in the history of geology. The starting point of this walk was easily reached by railway to Cockburnspath, but the distance to Fast Castle, the furthest point, was too great for the average pedestrian. The long line of sea cliff would have been best viewed from small row-boats, but as this mode of locomotion was impracticable, we lingered for a little at the cliffs about the Siccar Point, and then the less athletic members of the company found seats on country carts which had been procured from farms in the neighbourhood. The party numbered about 120 and included a good many ladies. There was sufficient accommodation in the carts for all who did not care to walk, and the day turned out wonderfully fine. The crisp sea air, the striking range of cliff scenery, with its plicated rocks, the lonely ruin of Fast Castle and its association with the *Bride of Lammermuir*, the unwonted ride on straw in lumbering carts along a rough road, the plain but ample luncheon, and the general hilarity of everybody combined to make the excursion eminently successful. It was at a gathering of the "Red Lions" during this meeting of the British Association that Alexander Nicolson, already mentioned (p. 38), produced and sang to the tune of "The British Grenadiers" his verses on the British Association, which have often enlivened subsequent dinners of the "Lions."

There was just time, after the close of this meeting, to reach London before the 17th August, the date fixed for my marriage. Our wedding tour began by a journey to Bonn in order to attend the Beethoven Festival there, for which my wife, being ardently musical, had secured tickets some time before. We lingered for a while among the little picturesque towns on the Rhine, diverging to Heidelberg and halting at Strasburg, now in the possession of the Germans. We drove through parts of that town which had been most battered by the recent bombardment—a melancholy spectacle, whole acres of streets, houses, shops, and public buildings involved in one common ruin. On the houses still standing hundreds of bullets had left their pittings, while here and there a shell had riven a large hole in a solid wall of masonry. But perhaps the most striking feature of the place to our eyes was the swagger of the Prussian officers on the pavements, and still more within the hotel. They let it be plainly seen that they were the lords and masters of the place, and that their pleasure and convenience must first be considered.

After a short and quiet tour through Switzerland we crossed into France in order to pay a visit to my wife's relatives near Lyons. Her cousin, Victor Pignatel, met us at the railway station and drove us to the old family home, the Château de St. Didier, on the upland a few miles from the city. There we were overwhelmed with kindness, and I made my first acquaintance with domestic life in France. Victor's mother, now an elderly, but still active lady, was the central figure of the whole household, and around her in separate suites of rooms, but under the same capacious roof, covering three sides of a square, her son and his family, and her daughters and their families were all accommodated beneath her beneficent sway. It was the French custom, she said, for a newly married couple to be placed next each other at table for a year and a day, after which they were rigorously separated.

On our way home we again came upon the track of the German invaders ; at Dijon there was a guard of Prussian soldiers at the station, and once more, as we approached Paris, the spiked helmets were to be seen garrisoning Fort Charenton. On reaching Paris I found a telegram urging my speedy return to London, as Sir Roderick Murchison was much worse, and had repeatedly expressed his wish to see me. We at once hurried back. The lapse of the few weeks since I was with him in August had produced a marked and sad change. His speech had become so indistinct that even his nephew, who assiduously watched him, sometimes could not make out what he said. His face brightened with the old smile as I sat down beside him for the last time. There was something that he wished to say, but he tried in vain to express it. He then had recourse to the pencil, with which for a week or two he had been able to make his wants known to those about him. But the fingers could no longer form any intelligible writing. His eyes filled with tears, and he sank back into his chair. Three weeks later, on 22nd October, he quietly and almost imperceptibly passed away.[1]

His death deprived me of the greatest benefactor and one of the warmest friends I ever had. Having known him in almost every variety of intercourse, I had ample opportunity of forming a just judgment of his character, which I have tried to express in the *Life* of him. Underneath a somewhat pompous manner he carried a most kindly heart, and many were the friendly acts by which, to my knowledge, he assisted others. To myself

[1] *Life of Murchison*, vol. ii. p. 343.

from the very first interview he was uniformly sympathetic and helpful, treating me more like a son than a subordinate member of his staff.

He was buried in Brompton Cemetery beside his wife. I returned to London for the funeral. In the carriage in which I was placed there sat Mr. Gladstone, Prime Minister (who talked most of the time, partly about an East Anglian poet of whom I had never heard and whose name I have entirely forgotten, partly on other matters of literary interest, with stray questions on geology and Murchison), Mr. John Murray, Murchison's publisher, and one of his Executors, and Trenham Reeks, Registrar of the School of Mines, another of the Executors. I was asked to return to the house in Belgrave Square where the will was read. I then learnt from it that I was nominated Literary Executor with a legacy of £1000.

# CHAPTER VI

## CREATION OF A CHAIR OF GEOLOGY IN THE UNIVERSITY OF EDINBURGH

(1871–1882)

On 6th November, 1871, the newly-created Department of Geology and Mineralogy in the University of Edinburgh was opened with an inaugural discourse on "The Scottish School of Geology." To most of the audience it was probably unknown that geological science had once been strenuously pursued in Edinburgh, and that several of its Scottish cultivators of that time are now, all over the world, reckoned among the founders of the science. The subject was chosen for the discourse with the view of setting in clear light the work accomplished by these early observers, as a stimulus to follow their example. The same hills and valleys, the same crags and rocky shores which inspired them, were still making their mute appeal to the present generation, and were teaching the same lessons which these great men were the first to understand. The establishment of a Chair of Geology in Edinburgh by a distinguished Scotsman, Sir Roderick Murchison, afforded an opportunity for combined and sedulous work, such as had not been previously obtainable in Scotland. It was the Professor's earnest hope that with diligence and enthusiasm on his part, and responsive activity on the part of the students, the renown of the old Scottish Geological School might yet be revived.

It so happened that there was still living a survivor of those who studied geology in Scotland when the

Scottish School was at the height of its renown—Ami Boué. I sent him a copy of the inaugural lecture, in which he was himself alluded to. From his characteristic reply dated 3rd December, 1871, I extract a few sentences:

"You must think I am already sleeping in my grave; receive my excuses for my negligence. When I did not write, I did not forget you. I advice in time Hauer of your new Professorship by Murchison munificence. I wrote about it a letter of sincere thank to Murchison for himself. Now I sended to Hauer your good-styled Inaugural lecture and your well-combined Syllabus. Your Geological Map of Edinburgh[1] recalled me all my 58 year ago made excursions; but what a progress since that time! I had not the idea of the possibility to arrive at such a perfect map for such a complicated part of Land. My best congratulations! What a guide for students! Middle Scotland is the country in whole world which has the most been burst and dislocated by plutonic-volcanic action, yet South and North Scotland remained united!

"Your Inaugural Speech recall me that I was contemporaneous of Hall, Playfair, Macculloch, Dugald Stewart, Leslie, Neill, Ellis, Brewster, Wilson. Sir James Hall turn soon during my stay at Edinburgh, to become half-mad or very excentric at least. But Playfair house was my very friendly abode; what a clear mathematical head he was! With Macculloch I travel in the Hebrides; I sailed in his cutter around Arran. I think that the scientific world generally did not pay him the attention due to his knowledge; his particular intercourse, his sarcastic elocutions may have injured his reputation.[2] Then he was to little attracted by the newer progress of science. His work made the effect of having already attained the limits of possible petrological and geological knowledge. Old Dugald Stewart I visited in his mansion in Linlithgowshire and spent some

[1] This map accompanied my Presidential Address to the Geological Section of the British Association.

[2] See pp. 54, 127.

instructive days in that scientific house with his wife and daughter."

The new Professorship of Geology was placed in the Faculty of Arts, no Faculty of Science having yet been established in the University. Geology, being practically a new subject in the curriculum of the studies of the University, was not as yet required for any examination or degree. Consequently the class in which it was taught might be expected to be attended only by students who had both a liking for the study, and time to give to it; perhaps also by a few leisured outsiders with the same taste. It was therefore gratifying that the class numbered upwards of forty in the first year of its history.

The Professor started under difficulties which are probably always more or less unavoidable at the introduction of a novel branch of tuition into an old educational establishment which is only sparingly endowed. He had no lecture-room for himself, but shared one with two other professors. This room was quite unsuited for the teaching of his subject. Moreover, it could be entered only half-an-hour before his hour of lecture, the diagrams and specimens of the previous occupant having first to be removed. With some difficulty he obtained the use of a kind of cellar in the south-west part of the College buildings, in which to keep his diagrams, specimens, and apparatus; but from deficiency of light, distance from the class-room and difficulty of access, this receptacle was only fit to be used as a store-room. He made his diagrams with his own hands, the lantern with transparent slides not having yet been made available for the teaching of geology. His long practice with pencil and brush now stood him in good stead; and in the end he succeeded in making a collection of diagrams and pictorial representations sufficient for his needs. He had no available specimens of minerals, rocks or fossils. These for a time he had to collect and purchase. The old University Museum which, under Robert Jameson's fostering care, was rich in this department of natural history, had been handed over some years before to the

Government, and now formed part of the "Scottish Museum of Science and Art." There were difficulties in the way of making free use of its materials. Eventually, however, from the duplicates in that Museum, a large series of specimens was obtained, and the foundation of a Class-Museum was laid, whence the students could be furnished with what was needful for practical examination. In the course of a few years this subsidiary museum grew in bulk and educational value, until it required a separate room or hall for its accommodation.

Further, there was no apartment anywhere available for the practical teaching of petrography, so essential a branch of geological science, nor any apparatus for the purposes of instruction. In the end, however, I was able to start a class for blowpipe work, and to introduce, for the first time, I believe, in the Universities of this country, practical instruction in modern petrographical methods, with the aid of the microscope and thin slices of minerals and rocks. This practical work was taken on Tuesday and Thursday, the lectures being on Monday, Wednesday and Friday, while Saturday was reserved for an occasional excursion into the field.

It was arranged that the Geology course, like the other courses in the Faculty of Arts, should be held during the winter half of the year. Its duties thus kept the Professor in Edinburgh from the end of October until the beginning of April, that is, five months of fairly continuous work. There was no long interruption of holiday at the end of the year, like the Christmas vacation south of the Tweed. But I introduced the practice of taking a "long excursion" at the end of the winter session, lasting a week or ten days, for the purpose of visiting some district of geological interest in Scotland. This practice I continued during the whole of my tenure of the Professorship.

It was the obvious duty of the Professor to make full use of the altogether exceptional advantages of Edinburgh, as a centre for the practical demonstration of many of the fundamental principles of geology, by a detailed study of the crags and ravines within and around the city, and of the shores of the Firth. From the outset, this field-instruction was made a prominent part of the work of the Class. During the session the afternoon of Saturday was often spent in visiting the nearer exposures of rock, and the whole day was occasionally given up to a more distant locality. Not less practically useful was the "long excursion" above referred to. At the end of the session, teacher and students spent a week or ten days in a visit to some ground of geological interest in the Highlands or Lowlands. They lived together in little country inns or cottages, and long days were spent in the open air, actively prosecuting geological observation. Of these gatherings, which were eminently fruitful, some account will be given in the later pages of this chapter. The young men were brought into closer touch with each other than could otherwise have been attained. By learning how to use their eyes, their faculty of observation was exercised, while in not a few of them, the love of nature and especially of geology, now stimulated, became a lifelong source of pleasure. The Professor, too, was strengthened by this more intimate converse with his pupils, and was often able to help them in the course of the free and friendly chat by the wayside or among the hills. In the course of years the "long excursion" was frequently attended by students of former years, and their presence added to the good fellowship of the party (see *postea*, pp. 165, 166).

The Geological Class was not the only one in the University which made long excursions into the field. The custom was followed both in the Natural History Class, and in that of Botany. Now and again two of these excursion parties would combine. Thus when Professor Carus took the place of Professor Wyville Thomson, who had started in the *Challenger* expedition, the Natural History and the Geology classes joined forces in Arran. There during the day, while the zoologists were busy afloat, dredging the prolific waters of the Firth of Clyde, the geologists were not less active on the shores and among the glens and peaks of the island, the two companies spending a merry evening at the hotel. On another occasion, when Professor Balfour took his class to the wild Grampian district of Clova, a favourite botanical field, I joined him there with my little army of geologists. The two companies worked together over the same ground, the botanists learning a little geology by the way, while the geologists were practically instructed on the spot in the Alpine flora of those high grounds.

While the excursions allowed the Professor to become personally acquainted with his students, there was still another old University custom which tends in the same direction, and which I resolved to keep up—the practice of inviting each of the pupils to breakfast at least once during the session. They came to Edinburgh University, not only from all parts of Scotland, but from all quarters of the globe. Most of them had no friends or acquaintances in the town, and would often be lonely enough in their lodgings. To be introduced into a Professor's house and family circle is to many students an important event, which gratifies them at the time, and remains as a pleasant memory Sometimes I made the breakfast the prelude to a Saturday excursion, especially after my removal to the southern suburbs, within easy reach of the geological attractions of the Braid and Pentland Hills.

For several years there had been an active movement in Edinburgh to establish University training for women. Vigorous efforts were made to open the medical school to them, but the opposition of the majority of the medical teaching staff was so strong that in my time these efforts came to nought. My sympathies were with the advocates for the admission of women. It seemed to me that there was ample room in the world for female medical practitioners. They might never become numerous, and, as a rule, would be only those women who were thoroughly devoted to the subject, and who would doubtless make admirable physicians, especially for their own sex. But as the agitation went on I could see that, if the matter were pressed to the bitter end, it would lead to the breaking up of the great Edinburgh Medical School.

The object of the women would no doubt be gained in the end, but the time for it, at least here, was obviously not yet, and could not be expected to arrive until the generation of professional opponents had passed away.

On the other hand, the endeavour to obtain for women University training in Arts was not only unopposed, but met with encouragement among the Professors. A Ladies' Educational Association was formed, and two or three Professors undertook to teach classes of women. A suite of rooms was engaged at the west end of the town, where courses of instruction were given similar to those in the University. Mixed classes of the two sexes not being yet tolerated, the Professors who undertook this task had double duty to perform. In my second year as Professor (1872-73) I introduced science into the Association by giving a course of lectures on geology, and continued to do so as long as I held the Chair in the University. The ladies mustered in good number at these lectures, and acquitted themselves well in the examinations, which were held twice or thrice during the session. They also attended excursions to places of geological interest in the immediate neighbourhood of Edinburgh, of which there is so large a choice. It was an entirely new start in Scottish educational procedure when on Saturday, 14th December, 1872, I conducted a party of thirty ladies over Arthur Seat. This innovation was so much enjoyed by the company that these Saturday afternoon excursions became a regular part of the curriculum of the Ladies' Association, as they were in that of the University. Occasionally the whole Saturday would be devoted to a more distant expedition. One of the first of these longer trips dwells in my memory. Its object was to see the evidence of great denudation, and of fluviatile deposition in the Moffat valley, and to visit the wonderfully perfect glacier moraines that still dam back the waters of Loch Skene, nestling in its corrie among the bare Southern Uplands. A large omnibus drawn by four horses carried us up Moffatdale, and, dropping us at the foot of the

Grey Mare's Tail, went on to the pass at Birkhill. After a climb and scramble among the moraines we crossed the moorland, with its rough heather and bogs, and descended on Birkhill. The cottage there was only a shepherd's shieling, but I had often lodged in it when inspecting the progress of the Survey, and was on friendly terms with old Jennie, the shepherd's widow, who tenanted it. My young wife had come with the party as "matron," and when I presented her to my old acquaintance the widow, Jennie looked at her for a few moments and then said slowly to her, "Weel now, ye'll never anger him, and ye'll aye see that he has a dry pair of stockings to put on when he comes hame frae the hill."

The social life of Edinburgh at this time was remarkably varied and attractive. In a city of such moderate size, calling and dining-out involved no vehicular difficulties, and everybody was, in some way or other, known to everybody else. The legal circle, the University circle, the mercantile circle, and a numerous body of residents attracted to the town as a pleasant place of abode, mingled freely together, each class supplying its own distinctive flavour to the general converse. The spirit of conviviality being well developed in the community, many were the dinner parties to which the new Professor and his young wife were invited. As it was the custom at that time to have music in the drawing-room, and as she was discovered to possess a good voice which had been well-trained, she was often asked to sing. There were more especially two houses in which she was often to be heard, those of Professor Lister and of Thomas Stevenson, the well-known harbour and lighthouse engineer. The future Lord Lister and his admirable wife, being fond of music, often asked her to come and sing to them. I can recall an evening at Mr. Stevenson's house in Heriot Row when his son, the future Robert Louis, of literary fame, then a young advocate, stood at the piano, turning over the leaves of music for her, as she played her accompaniment, his long hair hanging over his pale cheeks, and his eyes intent on the printed score.

This allusion to the Stevenson family brings back to my recollection many talks with the parents of Robert Louis about the career of their son. The father had taken for granted that the youth would follow the engineering profession wherein the family had so long been eminent. One day he unburdened his mind to me about his son. " I don't know what to make of Louis," he said, and he told how he had induced the young man to study engineering, and how the young student had shown great promise for a time; but that eventually it became evident that he had no great inclination to join the family calling. He was next induced to take to a career in the Law and advanced so far as to be called to the Bar. His mother fondly hoped that he would rise to eminence as an advocate, and she wore attached to her watch-chain a gold coin which she greatly prized, being her son's first fee as a barrister. But here again he could not prevail on himself to take up the law as a serious profession. Meanwhile, his literary fame was rapidly growing, and probably no one was so much astonished by it as his worthy father.

At this time there was not a little histrionic talent in the University circle at Edinburgh, and it not infrequently enlivened the evenings after dinner. On rare occasions Sir Robert Christison and Professor Douglas Maclagan were persuaded to sing a most humorous duet in the most doleful tones, acting the parts of two distressed seamen, begging on the street. It was indescribably comical, and all the more as the two performers were among the most dignified members of the medical profession.

Of a more serious caste were the recitations of Mrs. Lushington, sister of Alfred Tennyson, and wife of the Professor of Greek in the University of Glasgow. Known to the world from the graceful allusion to her marriage in the Epilogue to " In Memoriam," she sometimes recited one of her brother's poems. This she did standing outside the open door of the drawing-room, so that her strong deep voice filled the whole staircase and reached to the furthest part of the room. Her repetition of " The Northern Farmer," with her full command of the local dialect, might almost have been taken for the voice of the poet himself. Professor and Mrs. Fleeming Jenkyn, who were themselves accomplished actors, from time to time gave most excellent private theatricals at their parties.

At our picturesque aerie on the Castle Hill my wife made the most of our surroundings. Her gardening instinct was at once roused to improve the little domain. Carts of fresh soil were brought in, flowering plants were introduced, decayed shrubs and trees were removed, the greenhouse was restocked, and by a judicious arrangement of walks and vegetation, the small and irregularly-shaped piece of ground was made to look its best and biggest. Thus readjusted, the place became more than ever a pleasant *rus in urbe*. On a fine morning we would sometimes mount to the top of the slope, stretch a stout carriage rug on the grass, and with a stool, a cushion and a few books, a screen of trees behind us, birds singing and hopping near at hand, and in front, beyond the whole sweep of the New Town, glimpses of the blue firth and the hills of Fife ; it might almost have been imagined that instead of being in the very heart of a populous city, we were in a secluded spot of open country, far from the busy haunts of men.

In due time it became our duty to return " the sweet civilities of life " which had been so plenteously bestowed on us. The quaint octagonal rooms of Allan Ramsay's " goose-pie " became the scene of many a pleasant gathering. Among our guests we greatly enjoyed the company of the genial physician Dr. John Brown, author of " Rab and his Friends," and other delightful essays. His thoughtful talk, ever and anon rippling with humour, gave him a special place in the Edinburgh society of his day. He discovered one evening that his hostess was well acquainted with Mr. Ruskin, who used to visit the Winnington school where she was educated, and where he was almost worshipped. Dr. Brown loved to discuss with her the ways and opinions of the brilliant author of *Modern Painters*. In one of his published letters to

Ruskin (27th December, 1873) he playfully alludes to these talks—"we have one of your old disciples here, Mrs. Geikie; she and I like to take you to pieces." Another interesting guest, who dined with us when he visited Edinburgh, was Dr. John A. Carlyle, brother of the more famous Thomas, and an able Dante scholar. His talk, in the broad Annandale accent, abounded in graphic pictures of his travels abroad, sometimes with reminiscences connected with his brother's career.

Our quiet evenings at home were unfortunately somewhat hampered by the amount of literary work which I had on hand, and for which I had no other leisure time. I usually succeeded, however, in reading aloud for a while, and thereafter, while my wife sat plying her needle or reading, I went on with my pen. The *Life of Murchison* was the chief occupation of these leisure hours during the years from 1871 to 1875. The quantity of manuscript material left by the old Chief was so enormous and bewildering that a considerable proportion of the time taken in preparing the biography was consumed in examining and sifting the vast mass. It almost seemed as if he never destroyed any piece of manuscript that came to him. Before the end of the year 1873 the first four or five chapters were ready for the printers. At the desire of the executors the book was to be published by Mr. John Murray, who had been Murchison's publisher, and he agreed to my request that it should be printed in Edinburgh by Messrs. Constable. Proofs of the earlier chapters were sent to the executors, whose approval was an encouragement in what was proving to be a somewhat heavy task.

Besides this *pièce de résistance*, my friend Norman Lockyer had engaged me to write certain books on elementary science, which were to be published by Messrs. Macmillan & Co., and which were eventually included in the series of *Science Primers*, under the general supervision of Professors Huxley, Roscoe, and Balfour Stewart. My contribution to this series was to write the *Primer of Physical Geography* and the *Primer of Geology*. The composition of those two little books was one of the most difficult tasks I ever undertook. How often the manuscript of the first of them was torn up and thrown into the fire or the waste-paper basket, cannot now be recalled.[1] Simplicity of treatment, brightness of style, and clearness of exposition were the chief qualities aimed at; but it proved far harder than might be supposed to reach one's own ideal of what the little books should be.

An occasional lecture or address outside the University was at this time a duty which could not always be declined. Thus in 1872 a course of lectures to artizans on the "Elements of Physiography" was given in Edinburgh, two lectures were delivered at Leeds, and a discourse on "The Ice Age in Britain" (already alluded to, p. 108) was given to one of the large audiences which Professor Roscoe assembled for his series of "Manchester Lectures." Before long, however, extraneous lecturing was found to be too great a tax on time and strength, and the invitations to undertake it had usually to be declined, not always without regret. In the winter of 1875-76 there came a temptation which I could not resist—to address two audiences of a kind entirely new to me. The first company, consisting of the assembled inmates of the Morningside Lunatic Asylum, was quite orderly and attentive, though now and then one of the audience would stand up, gaze around in an enquiring manner and sit down again, without saying a word. A number of ladies in the front row came forward at the close to ask me some intelligent questions, the subject of the discourse having been "Scots Pebbles." As they were retiring, one of them pressed me to promise to come to her reception next Thursday, adding, "You know, I am Mary, Queen of Scots." I then discovered that each one of those ladies was a patient of the excellent institution, suffering from some derangement or hallucination which required medical observation and treatment.

[1] This experience recalled two expressive lines of Ovid:

Scribimus et scriptos absumimus igne libellos,
Exitus est studii parva favilla mei.

The second audience was composed of the inmates of the Deaf and Dumb Institution in Edinburgh. With those of our fellow-men who have no power of hearing or speech I had a hereditary sympathy. My uncle Walter, who, I have been told, was the best draughtsman of his day among the Royal Scottish Academicians, was one of these unfortunate sufferers. He died before the end of my second year and I have no certain remembrance of him. But his memory was fresh in the family when I was old enough to have impressions indelibly fixed on my mind. He was so fond of me as an infant that he would sometimes insist on having me on his knee as he painted or etched. In my early days I not only heard a great deal about him, but some of the canvases, sketch-books, brushes, pencils, and other artists' materials which he left were still to be seen in our house, where he lived and worked. As my father had become thoroughly efficient in the finger alphabet of the deaf mutes, and as I used often to watch him conversing rapidly on his hands with some of these unfortunates, I even learnt this manual alphabet, though never able to use it fluently in conversation.

It was with all these distant memories reviving in my mind that I discoursed to the inmates of the Deaf and Dumb Institution on "A Piece of Coal." One of the officials of the Institution sat on the platform beside me, and at the end of every few minutes I paused, while he reproduced, in visible speech, the gist of what I had been saying. I much regretted then that I had allowed my knowledge of the finger language to "rust disused." The audience though so strangely silent was most attentive. Some of them had possibly acquired the power, which has since those days been so greatly developed, of actually making out what is said by watching the movement of the lips. But generally they kept their eyes rivetted on the hands of my interpreter, and pleasant it was to see the smile of intelligent appreciation which passed over their faces as they followed his report. At any playful allusion that I had ventured to make the smile would broaden into a grin, or even be answered by some audible but inarticulate response.

By the winter of 1875 all the literary work on my hands was cleared off. The two *Science Primers* were published in 1873. Their immediate and extraordinary success was a complete surprise to me. The first large edition of the *Physical Geography* was exhausted in six months. From the beginning up till the end of December 1922 no fewer than 513,000 copies of this little volume were printed in England. To these must be added successive editions printed in America. The annual sales, even after nearly half a century, still continue to be large. The book has been translated into most European languages and is used in Continental schools. It has likewise been translated into one or two of the languages of India.

In one of his "autobiographies," Gibbon could affirm that in his day "the conquests of our language and literature are not confined to Europe alone; and the writer who succeeds in London, is speedily read on the banks of the Delaware and the Ganges." How greatly had the passing of a century since his time widened and quickened the spread of English books all over the world! I have often been much interested in the course of my journeys at home and abroad to meet with natives and foreigners, who, on hearing my name, have claimed acquaintance with me as one of the instructors of their youth.

In the spring of 1875 the *Life of Sir Roderick I. Murchison* made its public appearance in two octavo volumes, with notices and a number of portraits of scientific celebrities who were his contemporaries, together with a sketch of the rise and growth of Palæozoic Geology in Britain. It met with a favourable reception from the press, and was the occasion of a number of gratifying appreciations from my Chief's old friends who were best qualified to judge whether my portrait of him was a truthful one. I tried to be quite fair in my estimate of his character and of his work, and they assured me that they were satisfied. Even some who in my opinion had in his lifetime misjudged him, confessed to me that they now could see him in another light. One of these was George P. Scrope, who wrote, "I have been reading through (or rather hearing read) your two volumes, and I am anxious to express to you my thanks for the kindly and flattering mention you have made of me therein, and also my admiration of your work. You have indeed raised a monument to your hero, of enduring reputation, and made the very most of your subject, interweaving into your memoir so many sparkling and eloquent passages as cannot fail to captivate even the general unscientific reader. You have put his character into a more favourable light than I could have thought possible, and even exalted the old geologist in my eyes."

In the same year in which the *Life of Murchison* was published, further biographical work of a much less onerous kind was gratefully undertaken. My friends Sir William E. Logan and Sir Charles Lyell had now passed away. Sir William, born in 1797, and Sir Charles in 1798, had each done noble service to the cause of geology in the course of a long and devoted life, and from both of them I had received kindness and encouragement in youth, which ripened in later years into warm friendship. To Logan, geology was indebted for the earliest researches into the stratigraphy of the older rocks of Canada, and for the organisation and development of the excellent Survey staff by which the subsequent investigation of the subject had been carried on. Lyell was looked up to in all English-speaking countries as the great leader of modern geology, in whose hands the principles of the science were enlarged and perfected by a luminous review of the geological operations now in visible progress. His influence was profound even on men who would not call themselves geologists. Charles Darwin was never tired of admitting his great indebtedness to him.[1]

[1] My obituary appreciations of these two distinguished men appeared in *Nature*, vol. xii, pp. 161, 325, and I also contributed an obituary of Lyell to the *Proceedings of the Royal Society*.

It had long been my intention to compile a text-book of geology on a plan which I had fully considered. This intention received a stimulus from a proposal made to me in the autumn of 1876 by Messrs. A. & C. Black, the Edinburgh firm of publishers. I was asked to contribute the article "Geology" to the new edition of their *Encyclopaedia Britannica*, then in preparation. It was to be a fully detailed exposition of the science, and it eventually proved to be the longest article in the whole *Encyclopaedia*. I stipulated that I should be at liberty eventually to reprint the article in book form. Having agreed with Messrs. Macmillan to prepare an advanced manual or text-book of geology, I intended that the article should be the ground-work of the volume. This firm would thus publish for me three distinct educational works on the science—the primer already issued, the class-book now almost ready, and the text-book yet to be written.

In an earlier part of this chapter brief reference was made to the "long excursion" with which I brought each session of the Geological Class to a close at Edinburgh. As an example of this form of instruction, I may refer to the excursion which took place in 1878. The object on this occasion was to follow the course of a river from its rise among the mountains, and to note the characteristic features of each part of its journey down to the sea. We assembled at the little inn of Dalwhinnie, in the heart of the Grampian Hills, and rambled over a wide area of the wonderful group of moraine-mounds which have been left by the glaciers that once filled all the high corries and valleys of these mountains. From the heights we descended into the valley of the River Spey, examining its fine series of fluviatile terraces, passing thence down the gorge of the Findhorn to Forres and the Old Red Sandstone at Nairn, completing the programme by a visit to Elgin and the Triassic sandstones so well exposed on the coast at Lossiemouth.

It was always pleasant to welcome former students to these excursions. On this occasion we were favoured with the presence of Henry Drummond, who was the first to enrol in my first session as Professor, and to whom frequent reference will be made in a later chapter. One of the students of the year, W. A. Herdman (now Sir William and Emeritus Professor of Natural History in the University of Liverpool), took part in this excursion, and has given some account of it in his recent *Memoir of Sir John Murray*, from which the following sentences may be quoted :

" One very pleasant and not the least instructive part of the geological course [at Edinburgh University] was the series of geological walks conducted by the Professor, not merely Saturday walks in the neighbourhood of Edinburgh, but also longer expeditions of a week or ten days at the end of the session, to localities of special geological interest farther afield, such as the Highlands or the Island of Arran. I well remember one such long excursion to the Grampian and the Cairngorm Mountains and Speyside, when we had, as somewhat senior members of the party—in addition to Professor Geikie—Dr. Benjamin Peach, and Dr. John Horne of the Geological Survey ; Dr. Aitken, of the University Chemical Department, Joseph Thomson, the African explorer, and John Murray, of the *Challenger*. The rest of us were ordinary students of science, and you can realise how we enjoyed and profited by the conversation of these senior men, how we dogged their steps and hung upon their every word." [1]

These students' rambles and the love of geology which they fostered have dwelt ever since those days in Sir William Herdman's memory, and though he has become eminent in another field of natural science, he has assured me that it was the remembrance of his experience in the Geology Class at Edinburgh and its excursions, which led him in recent years to found and endow a Professorship of Geology in the University of Liverpool. It is not always that a teacher lives to see the fruition of his labours. Certainly no incident connected with my professorial

[1] *Thirty-first Annual Report of the Liverpool Marine Biology Committee*, 1917, p. 26.

career has given me keener pleasure than this generous liberality of a former student.[1]

Life was now fuller to me of varied and interesting work than it had ever yet been. In the first place, the Geological Survey, which I always considered to be my most important charge, generally kept me engaged in its duties during the most part of the summer. These duties were often of the most absorbing kind. They brought me face to face with the various problems presented by the detailed mapping of the geology of the Highlands; they enabled me to maintain the most intimate and cordial relations with my colleagues, and they thus cemented some of the closest friendships of life. The winter months, when the staff was assembled at the office in Edinburgh, afforded opportunity to compare notes and discuss the difficulties which arose in the field-work, as well as to prepare the maps for the engraver and to write the descriptive text for these maps.

The mapping of Scotland was advancing across the southern counties from the Tweed on the one side to the shores of Kirkcudbright and Wigton on the other. It was likewise extending northwards to the edge of the Highlands in Perthshire and Dumbartonshire. I was as yet unwilling to plunge into the complicated structure of the Central Highlands, being of opinion that it would probably be wiser to begin at the base of the whole series of Highland crystalline rocks as exposed in the west of Sutherland and Ross. In the meantime, without scattering the staff too much, it was convenient to carry the field-work from the Old Red Sandstone of Kincardineshire round the eastern end of the Highlands into the area of the Moray Firth, and thence northward, until the time had come for taking up the Highland problem as a whole. As the field-work in the northern half of Scotland would thus lie to a large extent on the Old Red Sandstone, it was desirable to make some preliminary traverses of the

[1] Since this passage was written, Sir William Herdman has given a further sum of £20,000 to the same University, for the creation of a Geological Department.

northern counties, and perhaps to extend them into the Orkney and Shetland Isles, where the formation is well developed. I have elsewhere given some account of my first experiences among these northern isles, with which a wider and more intimate acquaintance was gained in the yachting cruises of later years.[1]

While these northern traverses were in progress I was amused to receive from the proprietor of one of the Orkney islands the announcement that there had been found on his estate in the strata of the Old Red Sandstone, "the footprints of men, women, children, and animals," and he offered to send me samples of them. A human community living in the north of Scotland during the deposition of the Old Red Sandstone! Soon afterwards a hamper, weighing several hundredweight, arrived at the College from the northern isle. The contents, as I fully anticipated, proved to be slabs of flagstone from which variously shaped calcareous concretions had decayed or fallen out, leaving sharply defined hollows, some of which had a rude resemblance to footprints on a surface of half-dried mud. One or two of the oddest specimens were added as curiosities to the Class museum.

Having been engaged for some years in the study of the Old Red Sandstone in all parts of Scotland, I spent part of the winter of 1877-78 in writing a memoir on the subject which was read to the Royal Society of Edinburgh.[2] It gave a general account of this section of the Geological Record as developed in the British Isles, and a more detailed description of the most northerly of the areas over which it spreads, and which, for shortness of reference, I called Lake Orcadie. This memoir was intended to be the first part of a monograph which would ultimately include all the other areas, but as a few years later I removed from Scotland, this design could never be completed. In the *Ancient Volcanoes of Great Britain*, however, I have given a succinct description of the other

[1] *Scottish Reminiscences*, pp. 275, 280.

[2] Read before the Society, 1st April, 1878, and printed in vol. xxviii. of the *Transactions*, pp. 347-452.

Old Red Sandstone tracts, with more special reference to the ample records of volcanic eruptions which they have preserved.

During my tenure of the Professorship some interesting instances were observed in Edinburgh, of the rate at which certain geological phenomena may be carried on. The North Bridge, which spans the valley between the Old and New Town, was proving too narrow for the increasing traffic upon it. To gain additional breadth it was determined to insert stout iron girders into the mass of masonry so as to provide on either side a broad platform on which a pavement could be laid. While the work was in progress Mr. David Stevenson, of the well-known firm of engineers, who had it in charge, asked me to come and see a curious disclosure laid open by the operations. I found that a series of vaults had been met with in the spaces between the arches of the bridge, and that the roof of each vault was hung with stalactites in the form of long pencil-like white stalks which in some cases reached the floor. In the course of a century, the rain filtering down from the causeway of the bridge through the masonry, had gradually extracted from the mortar some of the lime which, as the moisture evaporated in the vault, was re-deposited in these pendent forms as well as on the floor. I went into one of the vaults for a short distance. It was like walking through a close-set thicket of reeds, with this great difference that instead of bending aside, as the reeds would have done, the brittle stalactites, growing downwards, snapped through at once and fell with a kind of metallic ring on the floor. It was an interesting example of the rate at which this geological process may advance, for all this transference of substance from the mortar had been accomplished since the year 1772 when the bridge was opened. I wrote an account of the observations in a short article entitled "The Spar-caves of the North Bridge, Edinburgh."[1]

In the course of my journeys all over Scotland there was always an interest in examining the tombstones in the

[1] *Nature*, x. p. 8, 1874.

graveyards, first for the light which they so often throw upon the families that have lived for generations in a district, their names, occupations and the characteristic mortuary language in which their lives are commemorated; and next for the valuable and unexpected information which they frequently afford as to the manner and rate of decay which various kinds of stone, employed for monumental purposes, suffer from the disintegrating effects of the atmosphere. In the spring of 1880 I communicated to the Royal Society of Edinburgh [1] a paper on this subject, in which it was pointed out that while the decay is generally more marked in the open air of large towns where coal is extensively burnt, and where therefore various gases are given off that are absorbed by falling rain, yet that even in remote country places, where few or no such deleterious additions are poured into the atmosphere, traces of disintegration seldom fail to be observable. The least durable material used for our open-air monuments is undoubtedly white statuary marble. As the result of my observations, the conclusion was reached that in the open air of a large town the inscription on a white marble tombstone freely exposed to the weather may be entirely effaced in less than a hundred years. There was one monument of special interest to which I called attention, that of the illustrious Joseph Black, in the churchyard of the Greyfriars at Edinburgh, who died in 1799. This sumptuous tombstone consisted of a solid framework of hard siliceous sandstone into which a large upright slab of white marble had been firmly fastened, recording in Latin, with pious reverence, the genius and achievements of the discoverer of Carbonic Acid, and Latent Heat, and adding that his friends wished to mark his resting-place by the marble while it should last. Less than eighty years, however, had sufficed to make the inscription in part illegible when I examined it in 1879. But there were ominous signs that besides illegibility of the letters, the whole substance of the

[1] *Proc. Roy. Soc. Edin.*, x. (1880), p. 518. Reprinted in *Geological Sketches at Home and Abroad*, p. 182.

marble was in a state of decay and chemical transformation. The slab, still firmly held in place by the metal fastenings all round its margin, had bulged out considerably in the centre, forming a large blister-like expansion, which had been rent by numerous cracks. Twenty years afterwards I revisited the spot and found that in the interval the decay had advanced so far as to necessitate the entire removal of the crumbling marble, and in its place the insertion of a slab of far more durable sandstone on which the original epitaph had been carefully copied. This reparation was carried out by the Town Council of Edinburgh, as is stated below the inscription, and the date "A.D. 1894" is added. Thus this costly marble monument utterly disappeared in less than a hundred years. At the time of my later visit to Joseph Black's tomb I re-examined the other graveyards in Edinburgh where I had formerly noted the condition of the marble monuments. The further progress of decay in twenty years was found to be remarkable. No trace was to be seen of a number of the marble slabs which were in a crumbling and illegible state in 1879, and others erected since that date were already showing the initial stages of decay.[1]

Sheltered from rain, as in the interior of a building, white marble will last for an indefinite time. A cynic may say that, in the vast majority of cases, it will be no great matter if, at the end of a hundred years, a marble monument has become illegible, or fallen to pieces, and that people may be allowed to put up memorials in this perishable material which in most cases is likely to last at least as long as the memory of the deceased. But in the case of open-air monuments that are fine works of art and are meant to last for many generations to come, it is fatuous, in such a climate as ours, and especially in large towns, to make choice of such a stone as white marble. It was

[1] In a letter in the *Times*, published on 10th June, 1919, I called attention to the desirability of avoiding the use of statuary marble in the erection of the many Memorials of the Great War which were about to be erected all over the country.

accordingly a surprise and grief to me when the noble monument to Queen Victoria, erected in front of Buckingham Palace, was wrought in white marble. The consummate beauty of the design and of the workmanship are now seen at their highest perfection, which will slowly but inevitably fade until a later generation will be called on to rebuild the structure in some less exquisite but more lasting material.[1]

During the early years of my married life and tenure of the professorship, the most important interruption of the regular routine of life was an occasional excursion abroad, either in spring or summer. Thus, the winter of 1872-73 having been one of continuous and somewhat exacting duties, when spring arrived my wife insisted on carrying me off for a complete change of scene and employment by spending the latter half of April and all May in a visit to Italy. We went as far south as Naples, and lingered at many halting-places on the way back. Traversing the Maremma we rested at Leghorn for the purpose of visiting a little property among the adjacent hills, which had belonged to my mother-in-law and where my wife's brothers and sisters had been born. On returning to Leghorn I found myself apparently in the first stages of malarial fever, and unable to appear at dinner. It was rather late in the season, and my wife was alone at the dinner-table until another guest was placed opposite to her. He began conversation with her in French, but as she saw that English would be much easier for him she eventually spoke in his own tongue. In the end she said she must return upstairs to her husband, who seemed to be suffering from a return of the fever that had nearly cost him his life some years before. The stranger sympathised, and remarked that a friend of his had nearly been carried off by a malarious fever when geologising in the Lipari Islands. He turned out to be no other than Dr. War-

[1] The statue of Queen Anne, erected in 1712 in front of St. Paul's Cathedral, London, in course of time became so decayed and unsightly that in 1886 it had to be removed. It was replaced by another statue similar in design and materials, which before the end of the present century will have become another wreck.

burton Begbie already referred to (p. 140)—the one man in the world who had watched and understood my case. He was then on his way to Amalfi to see a relative who was seriously ill. His steamboat had touched at Leghorn, and instead of dining on board, he had come ashore to have his dinner at a hotel. He went upstairs with my wife, sat for an hour in talk with me, prescribed a remedy, and in a day or two I was able to resume our northward journey. This " coincidence " is by far the most remarkable that has ever occurred in my experience.

My mother-in-law and one of her daughters having now made their home in Picardy there was always the temptation to spend a holiday with them. And to this temptation we occasionally yielded during the years of my Professorship at Edinburgh. Finding that part of France to furnish ample material for water-colour sketching, I spent there many a pleasant hour in this pursuit. There was a picturesque coast-line, with cliffs of brightly tinted rocks, ruined Martello towers, reminiscent of the old French wars, fishing villages with all their characteristic life and features, while inland, quaint farms and windmills, antique chateaux, dark woods and rolling downs offered a wide range of subjects. Next to geological exploration I know of no occupation so delightful as that of the sketcher, whether he be a true artist or merely what Dryden calls a " sign-post dauber." He sits face to face with nature, and his soul, if he has one, communes with her. He looks at every detail of her features and sees on every side manifold variety and beauty of form and of colour that escape most eyes. A landscape means more to him than to most people, especially if he be also a geologist. As he tries, with more or less success, to paint it, its lineaments are printed on his memory and remain there as a possession for ever.

Had I remained a bachelor, with Edinburgh as my residence, it would have been hardly possible that any inducement would ever lead me to leave Ramsay Lodge. It was an ideal home for a single man,—in the very heart of the town, within a walk of a few minutes to everything and everybody, yet open to all the air of heaven, with a panorama to gaze upon which was a perennial joy, and with a breadth of distance that allowed all the varied moods of sky and cloud to be watched, as from a commanding hill-top. But drawbacks began to appear. My wife, as I have said, was fond of gardening, and lavished her attentions on the little bit of slope that stretched from the front of the house up to the wall and high railing of the Castle esplanade. She had stocked it with flowers which, under her fostering hands, grew as well as could be hoped for in a poor soil and in the midst of a town atmosphere. But the bright blossoms of her plants set the " pugging tooth on edge " of some loafer, who early one summer morning carried off nearly every plant in bloom. In the little greenhouse, built into a corner near the entrance-gate to our small domain, she tended a few more delicate things that contrived to grow, though sadly disfigured by the descent of black smuts so plentifully supplied by the hundreds of chimneys in the neighbourhood.

The horticultural difficulties might perhaps have been endured with patience, but a little daughter, Lucy, had been born to us, and we were not long in perceiving that Ramsay Lodge, in spite of the many attractions for us, was not an ideal place for a child. The steepness of the slopes and the scarcity of flat ground on which a child could run about, were drawbacks which we never realised until a little pair of feet came to point them out. Eventually I surrendered my lease of the house, and purchased a recently built house on the Colinton Road, to the south of Edinburgh, standing in about an acre and a half of ground. It commanded a view of the range of the Pentland Hills on the south, while to the north, from the upper windows, we looked across the Firth to the Ochil chain and the Trossach mountains beyond. We had only entered into possession of this new home a few days before our son Roderick was born.

The ground on which the house stood was still only a bare field, on which, on one of our first visits of inspection, we scared a lark from her nest. I well remember the vivid sense of ownership which the uprise of that bird gave me. I felt that I had acquired a bit of land where the wild life of Nature could still be found.[1] As the whole garden had to be created, it provided a delightful domestic occupation for the next few years. We laid out all the walks, planted every bush, shrub, and tree, and had the satisfaction before long to have our own flowers and vegetables, and even some screen and shade from our own trees.

On 19th February, 1879, there came to me from Mr. Augustus Lowell in Boston, Mass., an invitation to give six lectures at the Lowell Institute during the autumn of that year. It had long been my earnest desire to visit some parts of Western America, where the phenomena of denudation and of volcanic action are displayed on a scale and in a manner to which no equal can be found in Europe. The remarkable discoveries of Powell, Gilbert, King, Dutton, Hayden, and others, so well described and illustrated in scores of reports and memoirs, were almost as familiar to me as if I had myself beheld the scenes depicted in them. To combine a journey to the Far West with a couple of weeks in Boston was an enterprise well worthy of being carried out, if it were practicable. It would require an absence from Scotland of about three months. Such a holiday could easily be intercalated between two winter sessions at the University; hence the duties of the Geological Chair need not be in any way disturbed. But the invitation could not be accepted without the sanction of the authorities of the Science and Art Department, my official superiors on the Geological Survey. Application was accordingly made to that governing body for six weeks of special leave, which, together with the usual six weeks of ordinary annual

[1] It was the same feeling as that expressed by Juvenal on becoming the possessor of a lizard :

> Est aliquid, quocumque loco, quocumque recessu
> Unius sese dominum fecisse lacertae.
>
> iii. 230.

leave, would cover the time required. The application was granted.

The preparation of the Lowell Institute lectures was promptly begun. At the request of the Royal Geographical Society I had recently read to that body an address on "Geographical Evolution";[1] and I chose an amplification of the same theme as the subject of the lectures at Boston. Meanwhile it was necessary to "clear the slate" in regard both to the work of the Survey in the field and at the office, and to my non-official writings which were in progress. Of these last the most important was a memoir descriptive of the volcanic history of the Carboniferous Period in the basin of the Firth of Forth, and containing in condensed form the results of my researches in that region during nearly a quarter of a century of more or less continuous application. Determining in the field the chronological succession of the different vents and outflows of material, I had studied the petrography of the rocks by means of thin slices of them viewed under the microscope. After many vain efforts to have these thin sections prepared with the required transparency, two members of the Scottish Survey staff were successfully trained for this work, and a large and important collection of slides was ultimately formed at the Edinburgh office, illustrative of the petrography of Scotland In the memoir on the basin of the Firth of Forth, the chronology and petrography of the eruptive rocks of the region were for the first time described, and their minute structure was illustrated with coloured plates. In accordance with the advice given by Sir Robert Christison (*ante*, p. 65) the paper was read before the Royal Society of Edinburgh at two successive meetings, and was published in the Society's *Transactions*.[2] It was peculiarly gratifying to receive from the Society this year its Makdougall-Brisbane medal, the first honour of the kind that had come to me.

[1] Reported in *Geological Essays at Home and Abroad.*

[2] Vol. xxix. pp. 437-518.

Everything at the College having been cleared off, and the Survey inspection work in the field completed, the 31st of July, 1879, found me ready to start for the New World. I was fortunate in securing, as my companion on the expedition, Henry Drummond, already mentioned (p. 166). Having seen a good deal of him since he attended the Geology Class in 1871, I had formed a high opinion of his mind and character. He proved in every respect an admirable fellow-traveller. His subsequent career—too soon cut short by death—showed him to be an able naturalist, an accomplished writer, and one of the most earnest and spiritually-minded men of his generation. He undertook the arrangements for our passage in the *Campania* of the Anchor Line, and his acquaintance with the heads of the Company led to our receiving every care and comfort on the outward and return voyage.

We spent not more than a day at New York, where we were met by Dr. F. V. Hayden, the geologist who first revealed in detail the wonders of the Yellowstone Park, and with whom I had for some years been in active correspondence. He had kindly procured for me letters of recommendation to the resident military authorities in the region over which the excursion was meant to extend. One of these letters was from the Secretary of War to commanding officers of posts in Utah, Wyoming, and adjacent territories, requesting them to show me due courtesy and afford such assistance as could be given. Another letter from the Quartermaster-General commended me to the courteous attention of the Quartermaster's Department throughout the sphere of my explorations. Within twenty-four hours after landing, my companion and I were looking out from the windows of a Pullman car that rapidly swept past the blue reaches of the Hudson River.

An account of the more personal experiences in the journey was sent by me in 1881 to *Macmillan's Magazine*.[1] A few of the incidents that stand out above the rest may

[1] In two articles which were reprinted in *Geological Sketches at Home and Abroad* (1882), pp. 205-273.

be briefly referred to here. One of these happened at the beginning of our wanderings in the Far West. We had halted at the station of Fort Bridger, a disused military post, where we were hospitably entertained by Judge Carter, the patriarch of the district, who, at the request of Dr. Hayden and also of Dr. Joseph Leidy, the Nestor of American comparative anatomy, met us at the railway and drove us to his home. We wished to ascend the Uintah Mountains, which rise at no great distance from the Fort, that we might get above the timber line, and from the bare summit of the chain be able to obtain an extended view of the whole surrounding country. Provided with horses, tent, provisions and guide we made our way through the thick forest as far as the animals could go. Leaving them in charge of the old trapper who accompanied us, we climbed the craggy slopes till, after emerging above the limit of trees, we found ourselves upon a bare broad plateau between 11,000 and 12,000 feet above the level of the sea. The prospect from this elevation was a magnificent and most instructive panorama of peaks, buttresses, amphitheatres, pinnacles, columns, and all manner of picturesque architectural forms carved by denudation out of a vast flat arch of dull-red sandstone. It was the most impressive display of the results of age-long erosion which my eyes had ever beheld, far surpassing in extent and grandeur the instructive landscapes of Le Puy. Having lingered over this marvellous scene as long as time permitted, we rejoined our guide and the horses, and struck down to regain our tent. Before long I became convinced that we were being led too far to the right hand. Unwilling at first to admit that he had lost his bearings, our guide at last consented to turn round at a right angle to the course he had been following. But we had lost so much time by this deviation that before we had gone far through the dense woodland, darkness began to descend, and there was nothing to be done but to spend the night among the pines. We selected a small open space in the forest, where the guide was to kindle a fire, while my friend and I set out to find water, in which search we succeeded not by sight, but in the gloom by stepping into a pool. On our return we were not sorry to find a fire blazing, on which we proceeded to cook a few trout which Drummond, who was a keen angler, had caught in the afternoon. This was all the food available, our supplies being in the tent, while our only coverings were the saddle-cloths of the horses, under which we stretched ourselves on the hard and rough ground. Before long, the fire, which our trapper had carelessly lighted near an old pine tree, spread to that tree, which was soon a sheet of coruscating fireworks. Some of the adjacent trees next caught fire, the flames creeping along the branches and rushing up the stems to the feathery crests of the tall pines, whence showers of sparks flew out and fell in long lines through the profoundly still air. As tree after tree joined the conflagration the loud reports from the exploding branches, the hiss of the leaping flames, and the crash of the falling firebrands, together with the ghastly glare that now died down and anon shot up into renewed splendour, would have made sleep impossible, even had there been no sense of peril in the situation.

Had a wind arisen the fire might have spread far and wide. But fortunately the night was absolutely calm, and the conflagration did not extend beyond the trees at one side of the open space on which we bivouacked.

As soon as daylight began to return, the horses, which had been labouring all night to find a meal among the brushwood, were harnessed, and we resumed the march in a most glorious morning. So tranquil had the air been all night, that long wreaths or wisps of blue smoke lay at rest among the pines, like streaks of cloud or mist on a mountain side. Following the same altered direction which we had been pursuing the evening before, we had gone scarcely half-a-mile when we emerged from the forest at the edge of an open valley, and there in front stood our tent gleaming white in the sunlight. We came upon several serious forest fires some weeks later, but could skirt them on open ground, and were never again surrounded in what might easily have been a fatal position. Henceforth I made it a point of duty to see, before we left a camping ground, that our fire was extinguished and that its site was well drenched with water.

Our journey through the Yellowstone region has been described in the second of the two articles already referred to. Apart from the interesting geology, my memory specially dwells on the genial hospitality of the officers stationed at Fort Ellis, especially of Lieutenant Alison, the quartermaster, by whose attentive care we were supplied with a tent, horses, mules, provisions, and an escort of two men for our ride of some 300 miles. I recall the contrast of temperature between day and night in the dry air at the elevations which we reached. While in the sunshine we felt as warm as on an ordinary summer day at home, the thermometer at night sometimes fell to 19°, my sponge in its bag inside the tent being solidly frozen, so that I could have broken it with my hammer. Another striking meteorological feature was the remarkable clearness of the atmosphere at night. During the day there might be a delicate haze, and now and then we saw good examples of mirage; but the nocturnal sky was undimmed by a cloud, and every visible planet and star shone with all its lustre. Wrapping myself in the buffalo-cloak, with which Quartermaster Alison had so kindly provided me, I used to lie apart in the open for a while, gazing up at the deep sky, so clear, so sparkling, so utterly and almost incredibly different from the bleared cloudy expanse with which we must usually be content in Britain. Every familiar constellation, with marvellous brilliance, seemed to swim overhead, while behind and beyond it the unaided eye could penetrate into space, and see the heavens aglow with stars hardly ever discernible at home.

Our most exciting experience was on the last day of our journey through the Yellowstone region, when we came upon some Red Indians in full war panoply among their native wilds. We had reached the head of the last valley that had to be traversed before we debouched into the great lava-plain of Idaho, across which ran a branch railway to the main Union Pacific line. While we paused at the watershed our scout's sharp eyes detected a little cloud of dust at the far end of the valley, and he soon made out that it was a small party on horseback and, from their way of riding, that they were Indians. As they came nearer the same unaided eyes reported that there were four mounted men with four led horses. Jack the scout dismounted and got his rifle ready; the soldier who had charge of the mules did the same. Neither of them spoke a word, but they both covered with their rifles the foremost Indian, who now spurred on rapidly in front of the rest, gesticulating to us with a whip or rod in his hand. "They are friendly," said Jack, and down went the rifles. The first rider when he came up to us was cross-questioned by our scout in some Indian tongue, but with here and there a word of broken English. We were told that they were bound for a council of Indians up in Montana.

Four more picturesque savages could not have been desired to complete our reminiscences of the Far West. Every bright colour was to be found somewhere on their costumes. One wore a light blue coat faced with scarlet; another had chosen his coat of the tawniest orange. Their straw hats were each encircled with a band of down, and surmounted with feathers. Scarlet braid embroidered with beads wound in and out all over their dress. Their rifles (for every one of them was fully armed) were cased in richly-broidered canvas covers, and were slung across the front of their saddles, ready for any emergency. One of them, the son of a chief whom Jack had known, carried hanging at his saddle-bow a twopenny looking-glass with which, during our interview, he engaged in pulling out some hairs on his face that had escaped notice at his last toilette. The absence of hair on their faces added to their somewhat feminine aspect.

Knowing the familiar Indian greeting, we saluted them with "How!" whereupon each of them gravely rode round and shook hands with Drummond and me. We were glad to have seen the noble barbarian in his war-paint, and amid the scenery with which his forefathers had been familiar for many centuries before the white man appeared to dispute its possession. Our satisfaction, however, would have been less had we known then what we only learnt when we arrived in Utah, that the neighbouring tribe of the Utes was in revolt, had murdered the Government agent and his people, and had killed a United States officer and a number of soldiers who had been sent to put down the rising, and further, that rumour represented the active disaffection to be spreading into other tribes.

After this interview with the Indians our way lay for many miles across the vast lava-plain of Idaho, which has been deeply trenched by the Snake River and its tributaries. The walls of many deep cañons, that have been excavated in this plain, display fine sections of the successive sheets of basalt which have been poured out over each other throughout an area of many hundreds of square miles. We had seen in various parts of our journey that valleys, cut through and between mountains of trachyte, have in comparatively late time been flooded with basalt, but we everywhere failed to observe any great central cones whence these younger lavas could have come. It was in this last part of our expedition that we found the solution of the problem presented by these floors of basalt in the valleys, and by the wide basalt-plain traversed by the Snake River. They were not emitted from central volcanic cones of the type of Etna or Vesuvius, but from numerous longitudinal fissures in the crust of the earth, many of which are now revealed as dykes of basalt running for miles through all the other rocks. The region was a magnificent example of the colossal volcanic type of massive or fissure eruptions, so well diagnosed by my old friend Baron F. von Richthofen. It was in this ride over the vast Snake River volcanic plain that the mists fell from my eyes as to the origin of the Tertiary basalt-plateaux of Scotland, Ireland, the Faroe Islands, and

Iceland. The problem with which they had puzzled me for many years was here solved. They were now recognised to be an older and more wasted example of the same type of fissure eruptions. It may be imagined with what satisfaction my volcanic studies in Western America came to a close, and how I longed to be able to return to the plateaux of Antrim and the Inner Hebrides, in order to apply to those familiar areas the lesson which had now been learnt.

Having reached the branch railway-line, we there waited in the midst of the arid wilderness of lava till a train appeared, which at our signal took us in and conveyed us to the main Union Pacific line at Ogden Junction. With the first train available we went on to Salt Lake City, where a few days were agreeably spent, partly in geological excursions into the Wahsatch Mountains and along the ancient terraces of the Great Salt Lake, and partly in bathing experiments in the lake itself. After being a fortnight in the saddle, and consequently with somewhat chafed skins, we first realised the high salinity of the water by the sharp pain inflicted on each wound. It was said to be easy to float in a sitting posture in this brine, with the hands clasped round the knees; but our experiments in this posture led to uncomfortable submersion of heels over head. The large proportion of salt in the water and the great dryness of the air were well shown by the stalactitic drips of salt under the wooden steps that led from the dressing-chambers down into the lake, and by the extent to which we were each of us crusted with salt, like Belfast hams, before we reached the rooms, where a douche of warm water was ready to remove the incrustation.

At our hotel I was surprised and greatly pleased to be accosted by Captain Dutton, the distinguished explorer of the High Plateau of Utah, with whom I had been for some time in correspondence, though we had never met. He told me that another geologist, on his way to San Francisco, would pass that evening through Ogden Junction. This was none other than Major

J. W. Powell, the famous pioneer who first descended in a boat the Grand Cañon of the Colorado. Captain Dutton went up to Ogden, intercepted the one-armed veteran and brought him down to Salt Lake City, thus giving us the great pleasure of a day or two in the company of these two excellent men and enthusiastic geologists. They took us up to Fort Douglas, the military post established by the Government to keep the Mormons in order, and the officers in charge there were unwearied in their attention. Major Powell had lost an arm in the Civil War of 1861, and it was after this partial disablement that he made his exploration of the Colorado gorge. One of his most outstanding qualities was his extreme modesty. This struck me now on this first meeting, and the impression was amply confirmed when I was again in the United States eighteen years later, and became more intimately acquainted with him. Not only was he one of the foremost geologists of the United States: he was also an able antiquary, and did signal service to the investigation of American ethnography by presiding over the Government Department created for the purpose.

At Salt Lake City I had an introduction to one of the leading Mormon elders. He was good enough to take us over the chief sights of the place. It turned out that he had once been Secretary at Edinburgh to the Forth and Clyde Canal Company, before the days of railways, and had been induced to join the Mormons early in the career of Joseph Smith.

The return journey to the east was tolerably rapid and continuous until we reached Niagara, where we took time to see the Falls well from both sides of the river, and to watch the rapids above, which are almost as striking as the cascade. Diverging thence into Canada and crossing Lake Ontario, we halted at Toronto, where I found my cousin Walter, one of the medical professors there, and his two sons, also doctors. The evening express brought us to Kingston by daybreak, and we there caught the steamboat down the St. Lawrence. At Montreal we passed a brief but agreeable time with Dr. J. W. Dawson, the eminent Principal of M'Gill University, and his wife. Mr. A. Selwyn, who had been one of the staff of the Geological Survey of the United Kingdom under De La Beche, and was now the Director of the Geological Survey of Canada, showed us the rich collection of specimens which Sir William Logan and his associates had gathered together in the Museum. On our way down the Hudson we halted at Albany to see the venerable James Hall, and his valuable assemblage of Palæozoic fossils. Taking advantage of our proximity to the Catskill Mountains, we spent a day or two among them, and from the almost empty hotel on the summit had the good fortune to be in time for the "turning of the leaf." The panorama visible from that elevation left a vision of wide and quiet beauty on the memory, while the rapidity of the change of colour on the rich tree-foliage of the nearer landscape filled us with wonder and admiration. Almost from hour to hour the eye could mark the increase in the depth of the tints, and the air was vocal with the insect shouts and responses of the "katydids" and "katydidints." It was an unforgettable experience of the glory of an American autumn.

We reached Boston in good time before the date of my first lecture there. The fortnight passed in this "hub of the universe" was made memorably pleasant by the kindness and cordiality with which we were everywhere received. Mr. Lowell, the life and soul of the Institute, was untiring in his efforts to make my stay agreeable to me. He invited me to his charming residence at Brookline, and gathered there a group of the most notable men of Boston and Cambridge. At dinner he placed me between Longfellow and Oliver W. Holmes. There were present also Charles W. Eliot, the distinguished President of Harvard University, and William Rogers, one of the oldest and most honoured of the geologists of the United States. The poet was exceedingly quiet at dinner, and I found continuous conversation with him to be difficult. But his face always beamed with a kindly smile. Mr. Holmes, on the other side, was full of talk which ranged, like his writings, over a wide sweep of subjects, always racy and original. I remember that we had a long discussion on Border ballads, with which his memory was well stored. Two years afterwards I reminded him of our meeting, and sent him prints of the two articles in *Macmillan's Magazine*, already referred to. He made the following reply:

Beverley Farms,
Mass., Aug. 12th, 1881.

MY DEAR PROFESSOR GEIKIE,

I am much pleased to find that you have not forgotten me, for you may be assured that I remember well the very pleasant evening I passed in your company.

I have read your adventures and your scientific experiences in Wyoming which you have kindly sent me. I am all ready to become an "Erosionist" under your agreeable guidance, and if I could throw off half a century or so of time's *impedimenta*, I should like to cross the continent with you. Let me congratulate you and everybody that the bullet which might have stopped you on your journey contented itself with giving a job to some worthy glazier.[1]

Believe me, dear Professor,
Very truly yours,
O. W. HOLMES.

Another pleasant family circle was frankly opened to me in the house of the venerable botanist Asa Gray, with whom and his wife I spent several quiet but eminently enjoyable evenings. Before leaving America I went for a day or two to New Haven, and saw as much as time permitted of the Nestor of American geologists, Professor James Dwight Dana, with whom I had corresponded for several years. He kindly arranged an

[1] Alluding to a pistol-shot through the railway carriage window as the train was about to enter Chicago Station. *Geological Sketches at Home and Abroad*, p. 211.

excursion for me to some places of geological interest, and thus gave me the gratification of seeing the rocks around his home, and hearing them pointed out and expounded by the great veteran himself. At New Haven, Professor Marsh was also infinitely kind and helpful, and now began with me a friendship which increased in cordial intimacy as long as he lived. At Philadelphia, where I spent a few days under the hospitable roof of Dr. Hayden, it was with peculiar pleasure that I made acquaintance with the revered comparative anatomist, Joseph Leidy, who had led the way in the description of the huge fossil reptiles and mammals which have been so abundantly exhumed from the Secondary and Tertiary rocks of the western regions of the United States. He remarked to me that the remains of these animals used always to be sent to him, but that his younger successors in the same field of research, Marsh and Cope, being both rich men, had been able to acquire all the new material in vertebrate palæontology. And thus, finding his occupation in that department gone, he had returned to an earlier love, and was now busy with his microscope over the Rhizopods.

After a journey by sea and land of some 15,500 miles I was back in Edinburgh before the date for the opening of the session at the University. The three months of absence were time profitably spent. I had come into personal contact with scores of new acquaintances, especially among the scientific men in the United States, and with many of them had laid the foundation of sincere friendship and helpful correspondence. It was thenceforth a pleasure to me to make more known on this side of the Atlantic some of the remarkable geological results obtained by our brethren on the other side, and with this view I not infrequently contributed papers to the columns of *Nature*. Besides this great gain of new friends and acquaintances, I had considerably extended my geological knowledge by the experience obtained in the Far West, especially the finding of the clue to the origin of the Tertiary volcanic plateaux of western Europe, and the fresh and impressive examples of denudation which it had been my good fortune to see.

I was enabled also to add some new items in elucidation of the geology of the United States. Thus, among the Wasatch Mountains I found that what had been described as an Archæan granite, that is, one of the oldest rock-masses in the country, was actually younger than the Carboniferous Limestone which it had invaded and metamorphosed.[1] The existence of traces of ancient glaciers in the Yellowstone district, which appear to have been missed by the earlier explorers, were found by me to be abundant, and I obtained evidence of the great thickness and extent of the ice.[2]

In the spring of 1880 there came from Vienna the last communication made to me by Ami Boué. It was a long letter, dated 26th March, written with a shaky hand, and here and there hardly legible. It contained much information as to his health and his bodily symptoms, combined with many references to what was going on in Vienna, and allusions to foreign politics, thus showing no abatement of his interest in the movements of the great world outside, nor any slackening of his outlook on the progress of geology at home and abroad. Some of its autobiographical details and allusions may here be quoted :

"M. Mojsisovics goes to day to Paris to the 50 year Jubilaeum of the Geol. Soc. of France, which was created in my library-room the 1st April 1830; present were Alex. Brongniart, Cordier, Ferussac, Blainville, Constant Prevost, Jobert, all dead. They wish I should preside this solemn meeting, but at 86 y. of age (the 16 March) with my infirmities it was impossible. I leav only by a very severe sanatory diet. . . . I must congratulate you for your fine journey to North America—countries which I had visited also if I had now 25 years of existence. I was a young man 60 or 70 year ago; all the facilities for travelling of our time were not yet discovered and

[1] *American Journal of Science*, xix. p. 363.

[2] *American Naturalist*, xv. No. 1.

used. . . . Your Gladstone think always in Metternich's Politik,—a short sight, only able to remain a short time. . . . It is quite wright that our enlighten Emperor did only laugh at Gladstone ridiculous expectorations, for they will furnish once to a dramatic poet the matter of theater stages, where Gladstone will excite only the laughter of the public, when he will, for instance, compare our Emperor to a Nero." After in nearly illegible lines justifying the political action of Austria in reference to the surrounding countries the letter ends thus :—" Now enough for you to laugh at | I remain, Dear Sir, Yours most sincerely, Ami Boué."

One day late in November 1881 there came from Vienna a packet addressed to me in Boué's well-known handwriting. It contained a small volume, giving in French a sketch of his own life. On the title-page he had written my name, designation, and address; possibly he may even have affixed the postage stamps. But he left instructions that the packets of which mine was one should be put into the post after his decease. He died on 22nd November, 1881. The title-page runs as follows :—Autobiographie | du | Docteur médécin Ami Boué | Membre de l'Académie Impériale des Sciences de Vienne &c | né à Hambourg le 16 Mars 1794 | et[1] | mort comme Autrichien à Vienne | Le seul survivant quoique l'ainé de trois frères et d'une sœur | (La distribution de cet opuscule n'aura lieu qu'après sa mort) | Vienne, Novembre 1879 | Imprimé chez Ferd. Ulrich et Fils | The little volume contains an interesting account of his family, of his own life, and of his contemporaries, and his writings. He married in 1826, and towards the close of his long career he could say that, looking back on their happy union, he would leave this world loving his wife as he loved her at the first. The narrative of his wanderings over Europe, and his allusions to the men of science whom he knew throw a good deal of light on the state of geology in the first half of last century. Living until he had reached his eighty-eighth year he was

[1] This blank left for the insertion of the date of his death.

a venerable living relic of that interesting time. Up to the last his memory retained a marvellously detailed impression of scenes and persons he had known, and he still kept in touch with the progress of science. Aided by his accomplished wife and a succession of adopted daughters, his genial domestic nature found scope for its activities in his picturesque country home at Vöslau, and in the modest apartments in Vienna, where he shared in the life of the Austrian capital. It is to me a delightful recollection to have seen him at both places, in the scarcely impaired vigour of his faculties, and to have been for some thirteen years in correspondence with one of the last veterans of what has been called " the heroic age " of geology.

In the spring of 1881 I was called to London to attend the anniversary meeting of the Geological Society, when the Murchison medal was conferred on me—an award specially welcome not only as a mark of the Society's regard, but also as another link connecting me with the life and work of my Chief, the founder of the medal.

A pleasant surprise befell me at this time. I was not aware that Murchison had, many years before, proposed my name in the book of candidates at the Athenaeum Club, and I now had notice that my election would be coming on soon. I was duly elected; and, sooner than at the time seemed likely, had ample opportunity of enjoying the advantages of this admirable club.

A good deal of literary work was completed about this time, partly in the form of memoirs communicated to the Royal Society of Edinburgh and of papers and articles contributed to journals, but chiefly in the preparation of the *Text-book of Geology* which I had undertaken for Messrs. Macmillan. The volume of the *Encyclopædia Britannica* containing my long article had now been published, and I had begun to expand this article, with the incorporation of detail necessary for students of the science, and to prepare the numerous illustrations required.

The early portion was now sent to the printer; but the labour involved was so great, and the time that could be devoted to it so restricted, that the book was not completed and published until October 1882.

Most of the inspection-work of the Survey now lay along the margin of the Highlands and to the north of them. Rambles among the eastern Grampians have always had a peculiar charm for me. The wide areas of tolerably level moor on the summits of these mountains impress one as an unexpected feature of the landscape, and as a striking contrast to the narrow, jagged crests and peaks of the Western Highlands. At heights of 3800 feet one comes upon extensive tracts which, if they lay in the Lowlands, might be used as golf links or as racecourses. An invigorating influence seems to pervade these relics of an ancient table-land. One may walk there for miles without fatigue in the bracing air, often with impressive views of distant heights in the range, or of the plains of the low country. Then there is always the sight of the living creatures of the high mountain fauna—a herd of red deer may be startled, the blue Alpine hares watch the human trespasser for a time before they scamper off, and the ptarmigan rise in alarm from their nests, leaving their broods to run helplessly among the quivering bent. If one has also eyes for plant forms, there is the Alpine flora to add another source of interest, and another problem to the many geographical questions which the walk suggests.

In the course of the summer of 1881 the Geological Survey made a remarkable discovery of fossils in the lower Carboniferous rocks of Eskdale in the south of Scotland. A number of new species of fishes were obtained, but the most startling feature of the collection was the presence in it of abundant well-preserved specimens of an ancient type of scorpion. A brief account of this discovery which appeared in *Nature* brought me a letter from Charles Darwin, which is inserted here as an instance of his keen and watchful interest in the progress of palæontological research, and his generous impulse to further it by contributing to its financial support.

Down, Beckenham, Kent, Nov. 1881.

My dear Sir,

I have been much interested by your account in *Nature* of the great " find " in the Lower Carboniferous strata. As so many scorpions were found, one might hope for other terrestrial animals and plants, if some new places were searched by blasting away the overlying rocks. But I daresay you would not think yourself justified in employing the officers of the Survey in such work. This leads me to make an offer,—and I hope and trust that you will not think that I am taking a liberty in doing so,—namely to subscribe £100 or £200 if you can find anyone whom you could trust to send, and if you think it worth while to make further search for the chance of fresh and greater palæontological treasures being discovered. If my offer seems to you superfluous or presumptuous pray forgive me and believe me.

My dear Sir,

Yours sincerely,

Ch. Darwin.

In thanking the great naturalist for this characteristically sympathetic offer, I was able to assure him that we should probably obtain from our own resources the means of completing the investigation. I added that his hope that other animals might be found, had already been partly realised, for we had unearthed two specimens of small amphibians. So well preserved were some of the scorpions that their chitinous tests were still in some degree elastic, and the poison gland was sometimes still recognisable.

At this time a detailed examination of the detrital deposits of the deep sea, collected during the voyage of the *Challenger*, was in progress in Edinburgh in the hands of Dr. John Murray and the Abbé Renard. As the Belgian geologist remained for some time in Scotland, we saw a good deal of him in my family circle, where his talking French, and his evident fondness for children and domestic life, made him a favourite visitor. He was in some respects a remarkable and interesting man. Trained as a Jesuit, he had refrained from making the last profession before entering the Order. In the many long conversations which I had with him, I could see that his opinions were broadly liberal in regard to ecclesiastical questions, but he still remained attached to his Church. He had a great desire to see some part of the Highlands, and as my duties led me to the north of Scotland, I invited him to accompany me in a visit to the extreme north-west of Sutherland. By the courtesy of the Engineers of the Northern Lights I had the privilege of the hospitality of the lighthouse at Cape Wrath. As we boated from Durness to the Cape, a good deal of swell was surging through the narrow passages at the foot of the cliffs, and some skill on the part of the boatmen is then needed to thread these channels safely. The worthy Belgian was visibly alarmed at times, and it was to his evident satisfaction that we passed finally into the little bay at the end of which is the landing-place for the lighthouse. Coming upon a boat that was here fishing lobsters, I made a purchase from the crew, pitched a coin into their boat and received into ours a fine lobster to increase our commissariat. Renard had never, I think, been before in a small boat upon the open ocean, and certainly had never been so far north. He had come prepared for all the hardships of an Arctic climate, and huge was his astonishment to find that his heavy greatcoat and other wraps were not needed, and that we could stand, even bare-headed, at the top of the lighthouse tower, and gaze at the waste of Atlantic waters beneath which the sun was sinking. It was indeed a glorious evening. The clouds had grouped themselves into the semblance of a varied terrestrial landscape, with hills, plains and lakes. On this new land, now discovered by us, we amused ourselves, *more geographico*, in naming the more prominent features in the vast celestial country after the leaders in geological science. As we turned to descend for our frugal dinner, the Abbé, who was at this time manfully striving to wrestle with the vagaries of the English language, exclaimed with much exuberance, " Vat a vedder we have, to be sure ! "

In August 1881 Professor A. C. Ramsay, Director-General of the Geological Survey, paid a visit to the staff in Scotland. Besides his official duties, he was desirous of spending a few days in seeing once more some of the places that he had known well in his youth. There had been an official enquiry into the progress of the Survey in Great Britain and Ireland, which involved the preparation of tables of figures, maps and reports. This work proved a heavy task to the Director-General, although we all did what we could to assist him. When he joined me, I was at once painfully struck with the great change which he had undergone since we last met in the previous year. In bodily strength he had aged considerably, and he had lost much of the merry vivacious manner that endeared him to us all. He said of himself, in the words of the old ballad, " I'm weary wi' hunting, and fain would lie down." So his visit proved really to be farewell to us in Scotland, for obviously his resignation of the burden of his office could not be long delayed. As he wished to see some of the Highland lakes once more, we made for Kenmore, sailed up Loch Tay, discussed the structure of Ben Lawers, and the relation of the lake to his famous theory of the origin of lake-basins. In further connection with that theory, we saw in succession Lochs Earn, Lubnaig, Vennachar and Achray, passed through the Trossachs, sailed up Loch Katrine, and from Inversnaid sailed down Loch Lomond —scenes that had been familiar to him in early years and which he knew he was now looking upon for the last time.

By the end of September, Professor Ramsay had definitely made up his mind to resign the office of Director-General. His retirement was marked by the bestowal of a simple knighthood upon him—a tardy and inadequate recognition of forty years of devoted work in the public service. The usual newspaper paragraphs duly appeared, and many guesses were hazarded as to his successor. If the appointment were to be made within the ranks of the Survey, the three Directors of England, Scotland and Ireland were eligible. Of these I was senior Director, though not senior in the service. But the question of seniority might not be considered, and to avoid internal friction, the post might again be given to some one entirely outside of the Survey, as was the case when Murchison succeeded De La Beche. On the evening of 9th December a somewhat mysterious telegram reached me from Colonel Donnelly, Secretary of the Science and Art Department, advising me not to be in a hurry about a house in London. It was capable of more interpretations than one, and my household was puzzled. Next day, being Sunday, when for an hour in the morning letters were obtainable at the General Post Office in Edinburgh, I walked into town to see if there were any communication that would throw light on the cryptic telegram. The following letter was then handed to me :

Privy Council Office,
9th Dec. —81.

Dear Professor Geikie,

The present Director-General of the Geological Survey retires towards the end of this year.

I have much pleasure in offering the Post to you and also the Directorship of the Museum in Jermyn Street.

It is right to mention that some changes may be made in the Museum, and although I cannot specify them at the present time, I beg that you will note that they are in contemplation, and I do not wish you to be unacquainted with this intention when I offer the post to you.

I am sure that your appointment will be of much public utility, and that you will be able to press forward the completion of the Surveys in the three kingdoms.

Yours truly,
Spencer.

My first feeling on reading this letter was naturally one of satisfaction. The prospect was opened out to me of a larger life in London, with the probability of greater usefulness, and with the honour of filling the highest post in my profession. It would have been a disappointment to me, after more than a quarter of a century of hard work in the service, to be placed under the rule of an outsider. At the same time I could not overlook the fact that there were two or three members of the staff older in standing than myself, who might not unnaturally resent that a younger colleague was advanced over them. This was most to be anticipated in the case of my esteemed friend and colleague, H. W. Bristow, who though well over sixty years of age, might hope, as Director for England and Wales, to obtain the promotion. I therefore wrote at once to him to announce my acceptance of the post, and to assure him that my one keen regret came from sympathy with him in having been passed over, that I would gladly do all in my power to be of service to him, and that I sincerely hoped our old brotherly relations would continue unbroken. His generous reply assured me that we would remain as fast friends as ever.

The excitement was great in the family circle, when on the Sunday morning I returned with Earl Spencer's letter. My wife's early associations were with the south, and though she had become deeply attached to her adopted home, and to the many close friends she had made in Scotland, she could not but be pleased to return to her oldest friends and her nearest relatives. The children, now numbering three girls and one boy, were too young to be able to form an opinion, but there was a prospect of change to London; their father was to be promoted; their mother was happy; and they were delighted.

And yet when, after long suspense, the matter was thus finally settled in my favour, the thought of leaving the scene of all my work, of severing the many associations that bound me so firmly to Edinburgh, of losing the continued intercourse with the students and with the varied interests of University life—would sometimes fill me with a pang of regret. And this feeling grew stronger as the months slipped past. But there never was, nor could there be, any real hesitation in accepting the preferment.

It was important that the new appointment should have the approbation, not only of the Geological Survey, but also of the officers of the Jermyn Street Museum. I was consequently much gratified to receive from Professor Huxley, the most outstanding of their number, a letter of welcome in which he wrote: " I sincerely congratulate Lord Spencer and you on the appointment. There is not the smallest doubt in my mind that it was the right thing to do; and I need not say how glad we shall all be to have you among us in London."

Among the letters received at this time from friends in this country one from Henry Drummond may be quoted here. Since our travels in the Far West he had been appointed Lecturer on Natural Science at the Free Church College, Glasgow, was widely known for his evangelical work among students, and was yet further to distinguish himself by travel in Africa and by his remarkable contributions to scientific literature. I had written to him on other matters a few days before; his reply contains the following sentences:

" Your very kind letter duly reached and cheered me. I am still alive, still unmarried, unengaged, and hopeless.

" But I am too full of another subject to speak of myself. The news about you has just come from London; and of the many whom it will gladden, I am sure there is none whose heart more sincerely rejoices than mine. I remember one day in the Rocky Mountains, when the vast solitude set us talking about destiny, you said you had a strange hankering to spend part of your life in the great metropolis. And now your thoughts have been prophetic, and the dream comes true. I heard a whisper of the change a few days ago, but I had no idea it was so nearly realised."

Of the congratulatory letters from abroad none touched me more than one from Major Powell, who was now

Director of the Geological Survey of the United States. He said that he remembered with pleasure the hours we had passed together at Great Salt Lake, and assured me that he was deeply gratified with my appointment. Not less welcome were the greetings of another eminent geologist of the United States, G. K. Gilbert, whom at that time I had never met, but with whom I had in later years friendly relations, both in his own country and in mine. As an illustration of the kindly regard that links together men of science on the two sides of the Atlantic who have never met, the letter is here given :

Washington, D.C.,
March 21st, 1882.

Dear Sir,

Once before I have taken up my pen to express my satisfaction at your promotion, but it occurred to me that I had no assurance that you were aware of my existence, and I laid it down. Since then, however, I have met with some mention of my work in a publication of yours, and now comes your Circular of March 8th.[1]

It was a subject of great regret on my part that I was unable to meet you when you were here, for there is no other geologist who holds quite so exalted a position in my estimation, and it was with corresponding pleasure that I learned of your appointment as Director-General. It is fitting that the first post of honor be filled by the one who stands first in attainment.

In tendering you my congratulations permit me to express the hope that executive duties will not absorb your time to the exclusion of scientific work.

Very truly yours,

G. K. Gilbert.

It was arranged that Sir Andrew Ramsay should demit his office at the end of the year. As the change had come about in the very middle of my course of lectures at the University, which I felt bound to complete, the first duty

[1] This was a formal card sent to correspondents, announcing my appointment and change of address.

that called for action was to apply to my official superiors for permission to fulfil my engagement at the University which would end in April. As a question of finance was involved, the application was referred to the Treasury, which sent the characteristic reply that "My Lords offer no opinion as to the compatibility of the two offices, but they cannot agree to his receiving salaries for both offices at the same time." Of course, there never was any question of a double salary. It was settled that I should retain the Professorship and its salary until 15th May, but begin to discharge some of the duties of Director-General without salary on 1st January.

Sir Andrew Ramsay retired at the end of December, and I took his place on the Monday morning following, 1st January, 1882. A week in London sufficed to give me a general view of the condition of the office, and the state of the English branch of the Survey, and I was able to resume my duties in Edinburgh by the end of the Christmas vacation, though it was necessary to pay repeated brief visits to London during the winter and spring

With an undercurrent of sincere regret, I continued and finished my last courses of lectures at the University and at the Ladies' Association. Teaching had become so great a pleasure that the approach of its complete cessation could not but sadden me. The close contact with young and eager natures, which the open-air work of geology so plentifully supplies, was a constant source of satisfaction, and the enthusiasm shown by the students as each fresh branch of enquiry was opened out to them acted as a continuous spur to one's own spirit of investigation. In this my final session it seemed to me that the consciousness that we were soon to part, and that the personal intercourse which had counted for so much would never be resumed in after years, reacted both on teacher and pupils. Certainly the students worked well; the class-work was excellent, and the excursions were never better attended. It so happened that this, my last class at Edinburgh, was not only one of the largest in the whole of my tenure of the professorship, but contained a specially marked group of men who have since made their mark in the world. Among them were Robert Chalmers, G.C.B., Joint-Secretary to the Treasury, now Baron Chalmers, George L. Gulland, Professor of Medicine in the University of Edinburgh, R. T. Harvey-Gibson, Professor of Botany in the University of Liverpool, and Robert Francis Scharff, Acting Director and Keeper of the Natural History collections in the National Museum, Dublin.

The class at the Ladies' Educational Association was also large, and the quality of its members, as tested by examination, remarkably high. In their case, also, the influence of the feeling that teacher and taught were working together for the last time could be recognised. There never had been greater industry in the class-work, nor fuller attendance at the excursions. At the last Saturday ramble forty-four ladies in five waggonettes drove with me from Edinburgh to Carlops in order to examine the Upper Silurian deposits of the Pentland Hills, alluded to in a former chapter (p. 55).

As the scene of the last "long excursion" with my students I chose the island of Arran. This classic spot in the history of Scottish geology had been the scene of my own earliest geological exploration, it was the first distant place to which I led my class at the start of the Chair of Geology, and now, with all its inexhaustible attractions, it seemed the fittest place in which my active career as Professor should come to a close.

Mixed classes of both sexes were then impracticable, and of course mixed excursions were absolutely impossible. I arranged that my young men should come first for a week or more, and that after they had left the island the ladies should arrive. The northern half of the island was selected as the area to be examined, with the village of Corrie as headquarters. We arrived there on 24th April, filling the little inn and overflowing into the cottages on either side. The weather continued generally fine throughout the whole time. Nearly all of the students had never been on the island before, and with bright skies overhead, and a succession of admirably varied and instructive rock-sections, alike among the mountains and along the shores, the long days of climbing and scrambling passed so quickly and pleasantly that the last of them came too soon, and much to the regret of both the teacher and his pupils. When the young men left, Mrs. Morrison, the worthy innkeeper, came to me with a grave face to express her opinion that I had "made a mistake in sending away the young gentlemen; for it would have been such a merry company with them and the ladies too!"

The second or ladies' contingent duly arrived in the afternoon to the number of twenty-six, more than half of the whole class. My wife and Mrs. Fraser, wife of one of the Judges of the Court of Session, accompanied the party, in order that it might not be said that they strayed forth unchaperoned. The ladies distributed themselves, as the students had done, in the inn and among the cottages. The utmost good humour prevailed in circumstances which were so novel to them, and the united company appeared in the inn each day for dinner, and to spend there a merry evening. The good weather continued. We were consequently able to profit by every day and to have long rambles on the shores where so many striking geological features are laid bare, and in the glens and corries, climbing to the summit of Goatfell, and scrambling along the jagged granite crests before descending to our starting-point. For the more distant excursions the company filled five waggonettes, exploring the western and northern coast, and picnicking by the shore.

One feature of these days with the ladies may be mentioned here. When we returned each evening I wrote before dinner a minute of the day's proceedings in the field, mentioning what we had seen and recounting any personal incidents that might have occurred. In this narrative numerous blank spaces were left for the insertion of adjectives, which were afterwards supplied to me as they happened to suggest themselves promiscuously to the company, who, not having yet heard the minute read, had no idea of the appropriateness or otherwise of the epithets which they dictated. Read aloud each evening after dinner, while the experiences of the day were fresh in the minds of the actors, these minutes gave rise to inextinguishable laughter. They formed a novel and pleasant variation of the supposed serious solemnity of workers in science. It was resolved that they should be printed, and that a copy should be sent as a souvenir to each lady who took part in this last long excursion.

On 9th May my University students entertained me at a farewell dinner. I was hardly prepared for the warmth of kindly feeling which they showed me, and which they assured me was shared by my former students in all parts of the world. They presented me with a beautifully-illuminated Address, designed by themselves, with appropriate geological vignettes, and signed by some 150 of present and former students. At the same time they handed me a handsome silver casket embossed with Scottish stones collected by themselves.

The ladies were not less sympathetic and kindly in their farewell letter, and my colleagues in the University made a formal entry on their Minutes in the usual stately but friendly language—"The Senatus Academicus have received with regret the intimation by Professor Geikie of his resignation (on promotion to a higher appointment) of the Murchison Chair of Geology, of which he has been the first incumbent. In accepting this resignation, which they will forward to the University Court, the Senatus resolve to record their sense of Professor Geikie's distinguished merits as a Professor and of his agreeable qualities as a colleague."

# CHAPTER VII

## FIRST YEARS OF THE DIRECTOR-GENERALSHIP OF THE GEOLOGICAL SURVEY

(1882–1891)

THE duties of the Director-General of the Geological Survey comprised the supervision and co-ordination of the operations of the three branches of the service in England, Scotland and Ireland, and the direction of the Museum of Practical Geology in Jermyn Street, London. These duties being mainly administrative, the holder of the office was naturally obliged to spend much of his time in London, though, if he could also keep himself in personal touch with the mapping in the field and with the officers by whom it was carried on, he would obviously make his supervision of the maps and memoirs more thorough, and at the same time be able to give encouragement, and perhaps occasionally even assistance, to his colleagues in the difficulties of their field work, as De La Beche is recorded to have done.

Being still in the prime of life and in full vigour of body, I resolved from the outset that, as far as possible the administrative duties at headquarters should not prevent me from keeping in personal touch with the progress of the field work in each of the three kingdoms. Of the details of the mapping in Scotland I was, of course, already fully cognisant. The authorities decided not to appoint a successor to the Directorate of the Survey of Scotland, until such time as Mr. H. H. Howell could be transferred to that post from his work in the north of England. Consequently for two years and a half the duties of the Scottish office were superadded to those of my new post.

It was one of my first objects to make the personal acquaintance of each officer on the staff in England and in Ireland, visiting him, if possible, at his country quarters, seeing his maps, and learning on his ground the kind of work on which he was engaged. This pleasant duty was in great measure accomplished in the course of my first year of office. So helpful to myself in the superintendence of the Survey was this personal relationship with the members of the staff and their mapping, and so generally did it appear to be appreciated by them, that I kept it up as far as practicable during the whole of my tenure of office, which extended over nineteen years. During this period the work of the Geological Survey formed the main occupation of my life. In this and the following chapters, as in those which have preceded, allusion will occasionally be made to the progress of the Survey in the three kingdoms, sufficient to indicate its general nature, without entering too far into technical details.

At the outset of my headship of the Survey there were urgent reasons why no delay should arise in coming into personal touch with the Irish staff. The mapping of Ireland was now far advanced, being concentrated mainly in the northern counties, though some revision was also in progress in the south. It was common knowledge throughout the Survey that things did not always move smoothly in the Irish branch, chiefly owing, as was reported, to the insubordination of one of its members. Rumours, perhaps more or less exaggerated, of this man's peculiarities used to find their way to our ears across St. George's Channel, even as far back as the days when J. B. Jukes had charge of the Irish Survey. And fresh reports of his doings had recently come to me. Being particularly anxious to have an interview with this rebellious colleague, I arranged that my first visit should be to him. After a few days spent in Dublin with the Director and those of the staff who happened to be there,

I went down to the pleasant Vale of Ovoca, where the officer in question had for some time been engaged in the revision of the published maps. He met me at the railway station with cordiality—a tall, broad-shouldered, stoutly-built man, with handsome features, a dark bushy beard, a strong voice, and a full Irish brogue—evidently not the kind of antagonist one would care to encounter in a personal scuffle. Taking me to his house, he introduced me to his wife and daughters, who kept me to luncheon. He showed me his maps, which I carefully examined, and we had a long and amicable conversation over them. I found him entirely reasonable, ready to listen to any criticism I might make on the mapping, and evidently pleased that I had taken the trouble to come at once to him in his country quarters. The ladies were not less courteous and pleasant As I was leaving, they gave me an armful of beautiful flowers for the wife of the Director—a peace-offering which the same evening I handed in to its destination in Dublin.

The impression left on my mind by this first contact with the alleged author of so many disturbances was that if he were treated justly and firmly, with the fullest consideration of his point of view, he might be moulded into a peaceable and kindly comrade. In the course of my subsequent experience of the Irish temperament this first impression, though modified, was never quite effaced. Before the end of the summer, however, the Irish Director reported to me that this stalwart member of his staff was on the war-path again. So long as these two members remained in the service, the chief trouble of my reign was to compose their differences and prevent them from injuring the progress of the Survey. Each of them was an Irishman, with many excellent qualities, which, however, included neither a capacity for understanding each other's habit of mind nor a spirit of forbearance.

After this first interview with one of the Irish staff on his ground, it was with good hope that I moved into the north of the island and paid a visit to each man in the midst of his field of work in the wide tract from Donegal to Larne. The year 1882 was an anxious time in Ireland. The Phoenix Park murders had taken place on 6th May, and outrages had occurred in many parts of the country. In the course of a stroll through Dublin one Sunday afternoon, I witnessed no fewer than five street " rows."

In the course of the inspection in Scotland this year the month of October was marked by a succession of storms worthy of the grimmest winter. By fierce gales, heavy rains, and dense mists we were fairly driven out of the hills. Descending to Comrie, on the line of the great boundary-fault—a locality famous for its frequent earthquakes—we lingered a while in the gloaming, hoping that the underground Powers, in sympathy with the atmospheric tumult above, might possibly favour us with one of their performances. But after waiting for an hour or two with no answer to our hope, we drove off to our quarters at Crieff. It was tantalising to learn on coming down to breakfast next day that there had actually been an earthquake between three and four o'clock of the morning, when I was sound asleep and neither heard nor felt it, though the tremor affected a wide district and knocked down a few plates and glasses.

In the midst of the multifarious doings of this period I found time for a little literary work. It was a pleasant and easy task to collect various essays and lectures which had appeared during the previous twenty years. They were now published (April 1882) as a volume entitled *Geological Sketches at Home and Abroad*. Much more serious was the completion of the first edition of the *Text-book of Geology*, on which I had been engaged for five years, and which was at last published on 17th October, 1882. The sale of this volume both at home and in America proved unexpectedly rapid, for in little more than a year the whole of the large impression was sold, and a second edition was called for.

With the charge of the Scottish Survey still resting on me, in addition to the duties of my new post, it was desirable to defer, as long as might be, all engagements connected with the Councils and Committees of the various scientific Societies in London to which I belonged. But this abstention could not be for long. These Societies have a right to count upon the co-operation of their members in the work of administration. Elected into the Geological Society in 1859, I became a member of Council in 1883, and from that time to this, with the exception of a few years, I have continued to serve. Two years later I entered the Council of the Royal Society. To be asked by the managers of the Royal Institution to give one of the Friday evening discourses is an honour which can hardly be declined. In response to this invitation I gave an address on "The Cañons of the Far West." These marvellous chasms had greatly fascinated me in the journey in Western America, and had led me to make a study of river-erosion under different climatal conditions, on different forms of terrestrial surface, and in connection with other geological influences. The results of my reflections were comprised in an article on "Rivers and River-gorges in the Old World and in the New," which appeared in the *English Illustrated Magazine* (1883).

In the early months of the following year, I gave, at the Royal Institution, a course of five lectures on "The Origin of the Scenery of the British Isles." Tolerably full abstracts of these lectures appeared in *Nature*,[1] and it was my original intention to write them out fully, with many illustrative views and diagrams. But leisure for this performance never came.

The celebration of the three hundredth birthday of Edinburgh University took place successfully in April 1884, and as a former Professor it was my duty to attend it. The organisation of the festival was excellent, and the attendance large, from all parts of the world. As a retired Professor, according to custom I was included in the list of those who were to receive honorary degrees. We assembled in the fine old Parliament Hall of Scotland, where large cards, fixed on stands, marked the part of the

[1] These abstracts were reprinted in my *Landscape in History*, 1905, pp. 130-157.

hall where each group of visitors was to muster. Freeman, the historian, was wandering helplessly about, evidently never noticing the sign-posts. I got hold of him and led him to the proper place of assembly. The mention of the historian's name reminds me of many pleasant parties at Knapdale, Mr. Alexander Macmillan's suburban home at Tooting, where Freeman, John Richard Green, Canon Ainger, and William Aldis Wright were sometimes together. Freeman regarded Green as a disciple and protégé of his, and was evidently attached to him, but sometimes Green could not resist the temptation to poke fun at his literary father. The mention of Froude's name was enough to inflame the historian of the Norman Conquest. Before he could get his heavy artillery ready, Green was discharging another little pin-prick in the rear. And so the discussion would go on to the infinite amusement of the company, Ainger getting in his witty and appropriate hit, and Aldis Wright quietly watching and enjoying the whole performance. I remember when Froude's *Oceana* was published, this volume became the subject of talk at one of these Knapdale parties. Ainger wondered where and how Froude obtained his consummate mastery of style. Freeman would not admit this excellence, or any other. There were times when Freeman had a cold or other slight ailment, and then the whole household was upset. He was an utterly intractable patient, and although everything was done for him that kindness could suggest, he was always grumbling. The children of the house were afraid of him, and remained subdued and quiet as long as he was under the roof.

On the other hand Canon Ainger was devoted to children, and they were not less fond of him. My own young folk rejoiced when he paid us a visit. His fun and frolic on these occasions were endlessly delightful. He seemed for the while to be as juvenile as any of them, and as ready for any kind of amusement. He was an excellent mimic, and could imitate some sounds to perfection. A favourite performance of his, and dearly loved in our household, was his chase of a blue-bottle fly round the room. His imitation of the insect's buzz was remarkably good, and as he ran on after the supposed fly, the young eyes were strained to see the little object which they could so plainly hear. Then at last he would make a dash with his arm, and clasp the palms of his hands together, the buzz still going on apparently between them. He would end the play by cautiously opening the hands, and lo! nothing was to be seen.

As the children grew older, Ainger found other ways of amusing them and retaining their devotion to him. I remember one occasion in the country when he had to drive about three miles in a dogcart with my son and his cousin, both about ten years old, to catch a train. Before they started he said that the three of them must make fifty puns before they reached the station. Leading the chase himself with two or three before they were outside the gate, they kept up the fun and reached the fiftieth pun as they halted at the railway.

At this point in my narrative, allusion may be made to some domestic rearrangements, which affected my life in London and were of great importance to my family. The change from the bracing climate and the open-air life of their northern home to the atmosphere and more indoor life in London was not long in introducing a succession of children's ailments which our young folk had hitherto escaped. It had been planned that, on leaving the preparatory school at Hunstanton, where we had placed him, our son Roderick should enter Harrow School. Believing it to be a great advantage that, where possible, the earliest part of a boy's life at a great public school should be spent under the parental roof, and amid the kindly influences of home, we finally decided to remove for a few years to Harrow, where in 1886 we obtained a house and garden, with an extensive view over the vast plain of the Thames, from St. Paul's dome to the towers of Windsor. This change proved of much benefit to the health and happiness of the household. Our eldest daughter, Lucy, however, had not been able to throw off a delicacy for which winter in a warmer climate was recommended. She was accordingly sent to Cannes where her grandmother had taken a villa. There she was enabled not only to overcome her ailment, but to go through a course of training in drawing at the studio of Mademoiselle Mercier, one of the most accomplished of French painters in water-colours.

Before our migration to Harrow, we were in the habit of transporting our young folk from time to time to the rocky and breezy coast of Newquay in Cornwall. We were there in the summer of 1884 when I received a letter from Professor Zirkel who had come to England for a short time, one object being to see a little of the geology of Cornwall. He was easily prevailed upon to pay us a visit at Newquay, and as I had been studying the rocks far and wide from that centre, I was able to take him on a different excursion every day. Entering with spirit into the pastimes of the children on the picturesque coast-line, he became a great favourite in the family. The correspondence, which began soon after we became acquainted at Largs in 1868, had continued ever since that time, and now grew more frequent and intimate after this sojourn in Cornwall. He was one of the most gentle, courteous, sympathetic, and modest men I ever knew. Certainly I never encountered another German professor of the same charm. He was a devoted son who looked after his mother with the most sedulous care, escaping at every vacation from his professorial labours, first at Lemberg, then at Leipzig, in order to hurry back to Bonn, and be beside her. After her death he charged himself with the care of his married sister and her children, and when she became a widow he grew more solicitous than ever in his tender care of her, and the superintendence of the education of her young family. The most distinguished petrographer in Germany, he had none of the narrowness, vanity, and self-assertion, unhappily so prevalent in German professorial circles. With many admiring friends all over the globe he kept up cordial relations. His handwriting was a remarkable index of the character of the writer—a clear, round, simple, legible Italian hand, not the angular and often illegible German Handschrift. He not only would not use that national style of handwriting, but also insisted that all his publications should be in Roman type. He used to say that he never could understand why his fellow-countrymen persisted in using a handwriting and a form of printed type different from those of all civilised nations, and which placed their country with the backward Russians and Turks.

In the spring of the same year some weeks were spent by me in Ireland, partly at the Dublin office, but chiefly in the field. For some years before this time various revisions had been made of the published maps on the scale of one inch to a mile, and it was now desirable to consider whether and to what extent new editions of these sheets should be prepared. To be able to decide this matter, I had to go over some portions of the south and west of the island. As a large part of the revision had been done by Mr. Alexander M'Henry, one of the assistant geologists, he was the fittest member of the staff to conduct me over the various traverses that had to be made. He proved now and in subsequent years to be one of the most helpful and efficient travelling companions I ever had the good fortune to meet with. He undertook all the arrangements as to trains, cars, hotels, and communications; always had his wits about him, and being personally agreeable and conversable, he made even the longest drives and the poorest inns less irksome. We spent some time in the south-west of County Cork, crossing the parallel ridges and valleys to Killarney. In the course of these tramps he gave me graphic accounts of the experiences through which he and several of his colleagues had gone during the time of the Fenian troubles, when every Government official was an object of suspicion and dislike. On a quiet country road, after the end of one of his narratives, I remarked that it was well that he no longer needed to carry a revolver. He made no answer, whereupon I asked if he still went about armed in that way. "Indeed, I do," he replied, "for you never know what is going to happen," at the same time pulling out the weapon from behind his coat. As we drove to Bantry our jarvey was careful to point out to me the exact scene of every murder, or attempted murder, during the last few years—a melancholy list, in the length and completeness of which he evidently prided himself.

The inspection of the Survey work in progress in the north of Ireland brought one into a wholly different class of scenery, a more varied geology, and a distinct type of peasantry. In the course of one of my tours of inspection in this region, when our work was stopped by a heavy rain which prevented the use of the maps, one of the party suggested that, until the shower had passed, we should take shelter in a neighbouring farm which was tenanted by an acquaintance of his, an old Scotsman. The farmer, who received us hospitably, had so strong an Irish brogue that after we had left his house I asked how long ago he or his family had left Scotland. The answer came at once that they had for some generations been all born in Ulster, and that their progenitors came over in the "Plantations" of the seventeenth century. And yet the Celtic populace still looked upon this man as a Scot! We got into talk with him about "poteen," the small-still whisky illicitly made in the lonely recesses of the hills. In my walks over the district I more than once kicked up on the moors recognisable fragments of stills, and in passing a police barrack, near Letterkenny, I had seen a pyramid of rusting stills, more than six feet high, which had been seized in the surrounding hills and broken up by the constabulary. Our farmer knew all about poteen, and when he found that I had never seen or tasted it, he fetched a dusty bottle of the liquor. It proved to be the most fiery fluid I had ever met with.

The geological mapping of Scotland had now been carried westward from Aberdeenshire along the northern base of the Highland mountains, but it had not been pushed far into the uplands, pending the detailed investigation which was about to be made in the north-west counties of Sutherland and Ross. I retain, however, some pleasant recollections of visits to the northern flanks of the high grounds of Aberdeenshire and Banff. One excursion was specially memorable, when with my colleague, John Horne, I ascended the Tap o' Noth, a hill not more than 1851 feet in height, yet so placed as to command a remarkably wide area of the north of Scotland. From the vitrified fort on the summit, the weather being clear and bright, we looked across the Moray Firth to the Caithness hills, gleaming in tints of pearly blue, seventy miles to the north, while to the south, at a distance of thirty miles, rose the snowy crests and dark chasms of Lochnagar, and far behind them the white summits of the further Grampians. Not less exhilarating to the geological eye was the wonderful complex of schists, serpentines, and other crystalline masses which underlie this region. It was difficult to tear oneself away from these rocks, so varied and beautiful that one would fain have carried off a specimen of each of their endless varieties. As it was, I contrived to collect and send up to the Jermyn Street Museum a bulky box, containing a series of lithological treasures, such as the cases there did not yet contain. Another day was given to the examination of the range of sea-cliffs between Cullen and Banff. But the calm east winds were now succeeded by furious gales from the south-west, the full force of which caught us on the exposed coast. We had sometimes almost to lie down to avoid being swept over into the sea. Yet the sky remained clear, and in sheltered places the air was warm. From the top of the cliffs the Moray Firth presented a majestic sight—a vast expanse of angry surf, lit up by the sun and torn into clouds of spindrift by the gusts that swirled over its surface. Later in the afternoon the sky became completely overcast, and we tramped our last four miles of moor in rain, hail, and snow. After eleven hours in the field we were not sorry to return to the warm comfort of the hotel at Huntly.

The time had now come when the mapping of the

Highlands must be undertaken. As a prelude to this work it was necessary that the true structure of the region of Sutherland and Ross, which Murchison believed that he had settled, should be determined beyond all question. The validity of his interpretation of the ground having been disputed by several observers, it was obviously desirable that by detailed mapping the true order of succession of the rock-masses should be accurately ascertained before the attempt was made to unravel the structure of the Highlands as a whole. Accordingly, in the spring of 1884 I selected B. N. Peach and John Horne as the officers to whom this important, and probably difficult, task could most confidently be assigned. They were instructed to get at the truth, regardless of anything that had been published on the subject The structure of the ground proved to be extraordinarily complicated, but by the end of summer these careful observers had advanced so far as to be convinced that Murchison's interpretation of the structure of the ground could not be maintained. I had implicitly accepted that interpretation, and had shared in Murchison's attempt to apply it to the rest of the Highlands. Naturally I could not bring myself to abandon it until I had seen with my own eyes such evidence as would convince me of its error. Accordingly I made for the extreme north-west of Scotland before the middle of October, and joined my colleagues at Durness. We spent some days in going minutely over the sections in that now classic region, and tested the results by further traverses to the south. Unfortunately the weather became so tempestuous as greatly to hamper field-work. The wind came in such powerful gusts that I was blown down, and all of us had at times to hold on with hands as well as feet. In spite of mist, rain, and gale, however, we did what we set out to do. I was completely convinced by the evidence so fully worked out by my two colleagues, that the Murchisonian view of the order of sequence in the rocks of the north-west of Scotland must be abandoned. It should be remembered, however, that nature had so cunningly concealed the true structure that no one could have definitely proved it by a mere cursory examination of the ground. The demonstration of the actual structure required laborious mapping in minute detail. The correct explanation of this structure introduced to geologists a new type of displacement in the earth's crust.

I arranged that Messrs. Peach and Horne should prepare and send up to London a conjoint report, giving succinctly the results of their investigation, with the necessary illustrative diagrams. To this report I wrote a preface, with a frank confession that I had been mistaken, and that I entirely accepted the opinion of my colleagues. These documents appeared in *Nature* of 13th November, 1884. They mark a noteworthy incident in the history of Scottish geology.[1]

Five years had slipped away since my return from the Far West of the United States, during which time, amid the pressure of official work and domestic claims, it had never been possible to resume the study of the Tertiary volcanic rocks of the Inner Hebrides. I was now at last able to return to that interesting region. Revisiting ground that was familiar to me upwards of thirty years before, I could now look at it with fresh eyes, quickened by experience in many parts of Britain and in foreign lands. One of the first observations now made enabled me to correct an error into which Macculloch had fallen and in which I had accepted his teaching. I had since suspected that the white marble of Strath which was once worked and exported,[2] was not altered Lias limestone as Macculloch believed, but belonged to a much more venerable series of rocks. On coming afresh to it from the ancient limestones of the North-west Highlands, I was

[1] The history of the investigation of this complicated area, with full references to all the observers who have taken part in it, is given in the massive volume on the *The Geological Structure of the North-West Highlands of Scotland* in the *Memoirs of the Geological Survey of Great Britain* (published in 1907) which will be mentioned in a subsequent chapter.

[2] It is said to have been used in laying down the pavements of the Palace of Versailles.

at once struck by its resemblance to them and proceeded to look for fossils in parts that had escaped much metamorphism. My former student, Mr. H. M. Cadell, who accompanied me to Skye, found the first fossil, and on subsequent searches I obtained a number more which conclusively proved the rock to be the same as the Durness limestone, now known to form an important group in the Cambrian series of the North-west Highlands.

After an active and profitable time in Skye I went alone to the isle of Eigg for a brief visit to my colleague in the Edinburgh Senatus, Professor Norman Macpherson, owner of the island. Thence I crossed in an open rowing boat to the island of Rum, the proprietor of which, not being in residence there at the time, had been good enough to instruct his bailiff to make me comfortable in the "big house." This large mountainous and picturesque island had then few inhabitants and no roads. It was a piece of wild nature, seemingly still left untouched by man. The details of its geological structure proved extraordinarily interesting, and supplied further striking evidence of the order of succession of the volcanic phenomena which had been found to extend throughout the region of the Inner Hebrides. The scenery, too, was singularly impressive, and to be entirely alone in the midst of it added to its effect on the imagination.

The enthusiasm and delight of these weeks among the isles found vent in a letter to a friend: "Io Triumphe! Here I am back again from the West, after the most brilliant weather, and with entire success in reaching the objects of the journey. My good fortune in the way of weather is almost incredible. That I should have been for a whole month in that weeping climate and never lost one day, is a marvel. The last days in Rum were perhaps the finest of all—so splendid in scenery and so bewitching in geological interest that it seems all like a dream. I can hardly believe that I actually saw such noble scenery under such superlative conditions. In fact my work has been a kind of intoxication. I have seen almost too much of grandeur and beauty for a frail mortal to take in. Surely there is no scenery in the world to beat these Western Isles, when the weather is at its best." [1]

The third International Geological Congress met in Berlin towards the end of September 1885, under the Presidency of the venerable Nestor of Rhineland geologists, H. von Dechen. Though seventeen years had passed since we last met, I found him remarkably well and vigorous. Addressing the assembly in French, the official language of the Congress, and looking towards the only lady in the company, Mrs. Hughes, wife of the Cambridge professor, he began, "Madame et Messieurs." An account of the proceedings was given in one of my letters from which I extract a few sentences. "The Berlin Congress was a success so far as the seeing of German geologists went. After the meetings during the day, every night a "Kneipe" at which Zirkel, Lossen, Reusch, Lehmann, Neumayr, Mosjsovics, and a host of younger men took part. Such a noise amidst air blue with tobacco smoke! Capellini from Bologna (where the previous Congress was held) was conspicuous and active at all the meetings. I went to the Harz with Lossen, and have struck up quite a friendship with him. What a fine fellow he is! but so deaf. I also went to Stassfurt and, with the party which made that excursion, descended into the principal salt-mine, which had been elaborately prepared for our reception—illuminations, banquet, poetry, music, speechifying, etc. A gnome, from behind a pillar of salt, read a German Ode of welcome to us. The tables were of solid salt, and everything else in keeping. I proposed the health of the Bergleutee, which seemed to be appreciated by them. Leaving the excursionists I went on to Bonn in order to spend a day or two with Zirkel and his mother, and to see a little more of dear old von Dechen."

One of the most valuable acquaintances made by me at the Berlin meeting was with Albert Auguste de Lapparent, one of the most distinguished of French

[1] The results of this summer's exploration were given in the Memoir described on p. 223.

geologists. We were a good deal together at this time and laid the foundation of a cordial friendship and correspondence, which lasted with growing intimacy till his lamented death in 1908.

The brief halt at Bonn was memorably pleasant, I now made the acquaintance of Zirkel's mother, and could quite understand the strength of the attachment and devotion which he had for her. She was a gentle old lady, domestic and simple in her ways, full of sympathy with her son in his pursuits, and proud of the distinction which he had reached. I remember how pleased she was when he showed her a box of English lavender water, which, knowing his fondness for that perfume, I had brought with me from London for him. We had some delightful rambles; did not fail to climb the Drachenfels, to look once more over "the wide and winding Rhine," and in some long and earnest talks we cemented more closely than ever the friendship which united us.

At the anniversary of the Royal Society on 30th November 1885 I was elected into the Council, at the same time that Sir George G. Stokes was chosen President. Though there are only some eight or nine meetings of the Council in the course of a year, the attendance at them is usually so good that to miss even one of them may bring the absent member within the application of the rule which provides for the dropping out of those who have been present the smallest number of times. I appreciated the honour of being elected into this select company, though believing that my frequent absence from London would probably limit my tenure of office to one year. The new President appointed me one of his Vice-Presidents, and eventually I found myself in a position to retain my seat at the Council next year.

This year saw the completion and publication of the second edition of my *Text-book of Geology*; and the first edition of the *Class-book of Geology*. On the latter volume I had been for a long time engaged. The original intention was that it should be issued soon after the appearance of the *Class-book of Physical Geography*. But the pressure of many other calls had delayed it until now. The postponement, however, was not without some advantages, for it enabled me to devise and mature a treatment of the subject which I hoped would make a useful book not only for the purposes of the class-room, but for the unaided solitary student. The scheme thus adopted was original, and the success of the book fully answered my hopes. Successive editions of it have been printed, and it still continues in demand.

In the spring of 1887, on the invitation of the Junior Scientific Club, a brief visit was paid to Oxford, where a pleasant company assembled at the museum, under the genial auspices of the venerable Sir Henry Acland, and where I gave an address on the influence which the geological features of Britain have had on the races that have settled in these islands.[1] During this visit I was the guest of my old friends Sir Joseph and Lady Prestwich. This eminent veteran had now determined to retire from his professorship in the University. Though seventy-five years of age, he continued as active in mind as he had ever been, and was then hard at work on the proofs of the second volume of his treatise on geology. When the Chair of Geology thus became vacant, I was pressed to accept it by one of the University authorities, a member of the Committee of Electors, who assured me that if I would consent my appointment would take place without delay. By another Oxford man the ease and dignity of an Oxford Chair were extolled to me, in contrast with the constant work and worry of the Survey, which would obviously become more burdensome as years passed. Though frankly admitting the tempting nature of the proposal, I never seriously entertained it, but took every means in my power to promote the appointment of my friend and old colleague in the Geological Survey, Professor A. H. Green of Leeds, who soon after was elected to the vacant post.

[1] This address was afterwards published in *Macmillan's Magazine* and reprinted in the volume of collected essays published in 1908, to which it gave the title *Landscape in History*.

The last return of the malarial fever of 1870 disabled me for a little in the summer of 1887; but its effects were dissipated by an inspiriting round with my colleagues, first in the northern counties of Scotland, and thereafter in a traverse of some mountainous ground in the north-west of Ireland, where the surveyors had met with much difficulty in the mapping of the metamorphosed rocks. The crystalline schists, limestones, and quartzites of that region were naturally assumed to be a continuation of those of the Scottish Highlands. We had long walks and still longer car-drives; we examined many miles of rocky coast-line, we climbed mountain after mountain, each bleaker and rougher than the last, but in the end I was compelled to conclude that the Irish staff had not yet been more successful than their Scottish brethren in ascertaining the true sequence of this ancient series of rocks. At the same time there was every reason to hope that the experience which was accumulating in the north of Scotland would eventually furnish the key to the solution of this Irish stratigraphical problem.

One incident of this tour made a lasting impression on me. Having reached the most westerly headland of the region, we proceeded to climb the imposing mountain of Slieve League which rises precipitously from the northern shores of Donegal Bay to a height of nearly 2000 feet. This district of Ireland is sacred ground in the history of the Celtic Church and in the life of St. Columba, whose name and fame are perpetuated in the name of the valley—Glen Columbkille. In our ascent of the landward side of the mountain we came upon some of the rude rock-shelters which the Irish missionaries used as their "deserts," or places of retirement for meditation and prayer. Musing, partly on the historical associations of the place, and partly on the geological puzzle presented by the rocks of the rugged acclivities which we were climbing, I was unprepared for the surprise which awaited us at the summit. The mountain had been mapped as entirely made up of quartzite, but we had not traversed more than a few yards of the flat top when we came upon a block of sandstone enclosing a well-preserved fossil tree! A brief examination of the ground showed that this was no mere chance boulder, dropped during some time of submergence on the sunken hill-top, but formed part of a cake or platform of sandstone, a few acres in extent, which lay as a capping upon the main body of quartzite. This sandstone, with its stems of *Stigmaria*, was obviously an outlier of the group of strata which includes the Irish Coal-measures. What a picture was here presented of the stupendous denudation which the surface of the island has undergone! The Carboniferous system, including, perhaps, its coal-bearing portion, may once have spread over this north-west region of Ireland, but it has now been worn away and washed into the Atlantic Ocean, leaving only this little patch of its lowest platform on the top of Slieve League, as a memorial of its presence. Verily the woes of Ireland date from a remote past! Another memorable lesson in denudation was taught by the same mountain. Climbed on its landward side, it presented only the familiar craggy slopes of the Donegal hills, but when its top is reached the climber finds himself on the edge of a vast precipice that plunges steeply down into the waves that burst into foam along its base. It looks as though that side of the mountain had been cut away by the sea.[1]

The year 1888 was marked in the history of the Geological Survey by an important rearrangement of the staff. Mr. H. W. Bristow, the Director for England and Wales, now above seventy years of age, retired on his pension. His office and that of the Director for Scotland were conjoined, as they had been up to 1867. Advantage was taken of this abolition to appoint a Petrographer to the three branches of the Survey, who should have his headquarters and laboratory at the Jermyn Street Museum, but should visit the officers in the field when their work involved petrographical assistance. He would thus keep in touch with the mapping, and ensure correct discrimination of the rocks. Various minor improve-

[1] *Landscape in History*, p. 58.

ments in the staff were at the same time effected, with the result that the service was made more efficient, and the pay of the officers was slightly increased. Thus at last I had the satisfaction of seeing one important branch of the work of the Survey, which I had myself introduced and had striven in the face of many difficulties to maintain, put on an efficient footing, not only worthy of the service but likely to add to its reputation. We were fortunate in securing for our first petrographer J. J. H. Teall, who had shortly before completed his large and important volume on *British Petrography*, in virtue of which and of his published papers he now stood in the front rank of living petrographers. His capacity and geniality were soon felt and acknowledged on the staff, as well as in the scientific and official world outside, insomuch that when, thirteen years afterwards, I retired from the Survey, he was at once, with universal approval, appointed as my successor.

Two years later, in 1890, advantage was taken of the opportunity to reduce the staff of the Geological Survey in Ireland. The first mapping of that country being now not far from completion, and several of the senior members of the staff being qualified to claim their pensions, the authorities arranged that these members should retire on 30th September, and that two of the remaining staff should be placed in the Scottish branch. There were thus left in Ireland a Senior geologist in charge at the Dublin office, three geologists for the completion of the maps and memoirs, an assistant for petrographic work, and a collector.

During the work of inspection in Scotland in the summer of 1888 I attended a meeting of the Royal Society of Edinburgh, and then communicated in outline the results of my prolonged study of British Tertiary Volcanic history. To this Society, which had published my earliest papers on this subject, it seemed proper to offer the memoir which completed these investigations. This memoir, the bulkiest I had yet written, was accepted and published in the sumptuous manner in which the

Society issued its *Transactions* (Vol. XXXV., part 2, 1888, entitled "The History of Volcanic Action during the Tertiary Period in the British Isles"). It makes a quarto volume of 164 pages with two maps and fifty-three illustrations in the text. It was afterwards embodied, together with the results of subsequent explorations, in the second volume of *Ancient Volcanoes of Britain*.

The fourth International Geological Congress assembled in London on 17th September, 1888, under the presidency of the veteran Joseph Prestwich. His large circle of Continental friends sent a goodly contingent of their number to do him honour In spite of his years, he discharged the duties of the office with complete success. He gave his presidential address in French—a language which he spoke and wrote with ease. His noble head, as he sat on the platform, was the most distinguished in the whole company. The foreigners included a number with whom I was already acquainted and others whom I was now glad to meet for the first time. From France came the venerable Gosselet of Lille, A. de Lapparent, and Charles Barrois; from Germany, Baron F. von Richthofen and Prof. von Zittel; from Italy, Capellini of Bologna; from Portugal, Delgado and Choffat; from Switzerland, Renevier and Heim; from Sweden, Dr. Torell; and from America a strong muster of geologists, including G. K. Gilbert, C. D. Walcott, Professor Marsh, G. H. Williams, and others.

The experience of this meeting deepened the impression made on my mind at the previous Congress in Berlin, that not much of real scientific importance is accomplished at these gatherings, and that their chief practical value is to be recognised in the opportunities afforded for friendly intercourse among geologists brought together from all parts of the world. Attempts were made at some of the business meetings to draw up and enforce a uniformity of stratigraphical nomenclature, but they met with no general concurrence. Those of us who did not believe that any such unification could be of universal application, except only for the larger divisions of geological time, were not a little amused at the evident disappointment and vexation of some of the older terminologists, who came for the most part from small countries. Ferdinand Zirkel did not attend the Congress, having gone to the East, but I saw much of De Lapparent, who grew more charming the longer and more closely one came to know him. He became at once a delight to my children, and he was pleased to find that they could chat with him in French. He, Dr. Torell, and G. K. Gilbert were my guests at Harrow during the time of the Congress.

It was a peculiar pleasure to have the company of Gilbert, whom I regarded as the most able and accomplished geologist in the United States. I had studied his published papers with the greatest interest and profit, and personal contact now enhanced my estimation of his powers. I was about to make a tour of inspection in the north of Ireland and invited him to accompany me. As a citizen of the United States, he knew a good deal about the Irish element among his own population, and he had a strong desire to see the Irishman at home in the Emerald Isle. My tour gave him an opportunity to make acquaintance with all aspects of Irish life, from the broad and busy streets of Belfast to the squalid cabins of Donegal. He was initiated also into the dirt and discomfort of the little inns in some of the remoter and less visited places, into the pleasures of the Irish car, and the vagaries of its jarveys, as well as into the geological problems that were puzzling the brains of the Irish staff.

Gilbert also wished to visit Scotland, chiefly for the purpose of seeing the famous Parallel Roads of Glen Roy. As I had to be in Scotland after the Irish inspection was completed, and had planned to cross direct from the north of Ireland, he accompanied me. Taking the Giant's Causeway and the Antrim cliffs on the way, we sailed from Larne and made a rapid journey to Fort William, whence I conducted him to the Parallel Roads and the great moraines of Glen Spean. In the course of our rambles we had occasion to ford the River Treig, which happened to be fairly full at the time. Gilbert, with his trousers well tucked up, and his boots and stockings in his hands, was cautiously feeling his way across the rough boulder-strewn river-bed, when he dropped one of his boots, which was instantly swept down by the rapid current for some distance, till luckily caught on an islet. When we resumed our walk on the other side, he remarked that it was a pity so fine a stream as the Treig should have no fish in it. I assured him that it yielded both trout and salmon. He looked incredulous, and replied, " But there are no poles along its banks," gravely adding that in the case of the rivers of the Far West, to which he was accustomed, it was the general belief that no fish need be looked for, save in those streams whose banks bear trees that can furnish fishing-rods ! Leaving my friend in Glen Spean to continue his detailed examination of the Parallel Roads, which greatly interested him in connection with his studies of the lake-terraces in Western America, I went on into Argyllshire.

The Duke of Argyll had found certain markings in the rocks of his countryside, which he was persuaded were relics of once living organisms, and as the discovery of fossils in these ancient rocks would, if confirmed, be of great importance in the investigation of Scottish geology, he was naturally much excited over the discovery. He had long been interested in geological questions, and used to attend the meetings of the Royal Physical Society in Edinburgh, when Hugh Miller read a paper. It was at one of these meetings that I first saw him. Having been enabled to bring to notice the occurrence of Tertiary plants among the basalts of Mull, he had some claim to be regarded as a geologist. He was a firm adherent of the " convulsionist school " in geology. As far back as 1867 he had marked me out for criticism, and in a letter read at the meeting of the British Association in Dundee that year, he denounced in good set terms certain opinions which I had published. He was utterly opposed to the importance which I attributed to denudation in the gradual evolution of the topography of a land-surface, where he, on the other hand, could only see proofs of primeval convulsions and cataclysms. My views as applied to the elucidation of the history of Scottish scenery, he ultimately stigmatised as the "Gutter Theory." I did not reply to his attacks, which recalled Juvenal's line, "si rixa est, ubi tu pulsas, ego vapulo tantum." I was therefore somewhat surprised to receive from him a courteous invitation to come to Inveraray and see the great geological treasures which he had unearthed from the rocks of his ancestral domain. As I bore him no enmity, in spite of the strength of his vituperation of my opinions on denudation, and as I was as keenly anxious as he could possibly be that fossils should be found in the rocks from which his "find" had been taken, I had no hesitation in accepting his invitation.

It happened that the day of my arrival at Inveraray was that of a County Meeting, to which landed proprietors came from all corners of Argyllshire. In the evening they were entertained to dinner at the Castle by the Duke. Of course the Campbell clan mustered strong, clad in their distinctive tartan. The Princess Louise and the Marquis of Lorne were residing with the Duke at the time. On coming into the drawing-room the Princess went round the company, shaking hands with each one. This ceremony over, the Duke led her into the dining-room, and we all followed. One of the lairds, Donald Nicol, Younger of Ardmarnock, an old acquaintance of mine, was good enough to sit next me at dinner, and from him I was able to learn much about my fellow-guests.

After dinner I was happy to meet the Sheriff of the County, Mr. Forbes Irvine of Drum, who, with his charming wife, was one of our most genial friends in the old Edinburgh days. Among the array of kilted Campbells in the company, there was one who specially attracted my attention, as, in addition to the usual adornments of the Highland garb, he bore a conspicuous silver key. I learnt that he was Campbell of Dunstaffnage, an estate which he held from the head of the clan, the MacCallum Mohr, and that the key was that of the ruined castle, which on demand he was bound to present to the Duke. Of Campbell of Kilberry I had heard much from his son Angus, who was a school-mate of my son at Harrow, and often at my house there. I was glad now to make acquaintance with the father.

Next morning after breakfast the Duke took me to a small room in which he had laid out his supposed fossils. I examined the whole of the specimens with care and with the strongest desire to discover some trace of organic structure in them. But I was compelled to confess that to my eyes not one of them showed any trace of organic origin. They were all common inorganic structures found among the schists. The Duke, however, remained convinced that he had made an important discovery. In the course of his subsequent correspondence with scientific authorities in London, he found one palaeontologist who had the hardihood to declare that some of the specimens were in his opinion relics of organisms. It so happened that some years before this time I had an opportunity of testing this palaeontologist's scientific acumen. Among the coast cliffs of Berwickshire I had one day come upon a large vertical face of smooth hard shale which the country folk used as a target for rifle-shooting. The face of the rock was consequently pitted with bullet marks, and loose bits of lead could be picked up from the grass below. On looking at the pittings on the face of shale, I noticed in some of them a rudely stellate arrangement of the bruised stone, made by the bullet as it flattened out at the moment of impact. These markings might at first sight suggest an organism of some kind, and I succeeded in flaking off a few pieces of the shale which contained them. Eventually I submitted these specimens to the palaeontologist in question for his opinion. He at once, without hesitation, pronounced them to be true fossils, probably of Cambrian age. I saved him from further committing himself to a surmise of the probable genus to which they belonged by telling him frankly what they were. I could not mention this incident at Inveraray, but it came to mind when the opinion of this palaeontological authority was cited and relied on. The Duke, however, had made up his mind that the things were of organic origin, and he communicated his discovery to the Royal Society of Edinburgh—another "mare's nest" chronicled in scientific literature.

Towards the end of 1888 the ninth edition of the *Encyclopaedia Britannica*, which had been in course of preparation and publication for some fourteen years, was completed. The publishers, in celebration of the event, invited the large body of contributors to a dinner, which was appropriately given in the hall of Christ's College, Cambridge, under the auspices of the distinguished and much-esteemed editor, W. Robertson Smith, then a resident Fellow of the College. As the author of the longest article in the whole series of volumes, I was called upon for a speech.

In the year 1889 inspection duty in the field took up more time than usual, involving the examination of much ground in each of the three kingdoms. This work began in February and was continued at intervals, in one area after another, the last of these not being completed until a few days before Christmas. In England the work lay chiefly in the southern counties from Newton Abbot to Eastbourne. The most novel and interesting feature of the re-examination of this coast-line by the staff, was the discovery of a hitherto unnoticed form of dislocation among the Secondary formations of the southern counties, similar to the great thrust planes among the most ancient rocks of the north-west Highlands, but on a much smaller scale. This structure had been detected in the chalk on the coast of Dorset by A. Strahan, and at Eastbourne by Clement Reid. The lower members of the Cretaceous series have there in some places been pushed over the higher; the Gault and Greensand, for example, have been torn up and driven bodily over the chalk. At Eastbourne, where some of this remarkable dislocation is well exposed on the beach, I induced Professor Huxley to join me in examining it. He had come to live permanently at this coast-town, having found the climate of the place more favourable to his health than that of London, and he was now in the course of building himself a house, which he liked to show to his friends, conducting them skilfully up and down the scaffolding. He was greatly interested in the proofs of the stupendous upthrusting of the strata laid bare on the beach. I dined with him and Mrs. Huxley after our walk. He then looked more like his old self than he had done for some time past, and was as bright and brilliant as ever.

During some weeks in the spring of this year I made another attempt to ascertain whether the details of structure which had now been fully worked out in the north-west of Scotland could be used in the unravelling of the arrangement of the crystalline rocks of the west of Ireland. As the area to be examined I chose the wild rocky region that forms the western parts of the counties of Mayo and Galway, from Benwee Head to Galway Bay. Besides members of the Irish staff, Mr. B. N. Peach, who was thoroughly conversant with the minutest details of the mapping in Sutherland, accompanied me, in order that his experience and knowledge should be available for our purpose. We traversed the coast-line of Mayo, from Erris Head to Achill Sound, crossed Achill Island to its great western precipice, and examined the southern shores of Clew Bay. Crossing Murrisk to Killary Harbour, we spent some days on both sides of that fine inlet. From Clifden and Recess we explored the rocks of Connemara, and ended our researches at the town of Galway.

The results of this tour were of interest, though they threw less light than we had hoped on the most ancient rocks of Ireland. We found a large area of County Galway to be occupied by a gneiss so similar in all its lithological characters and topographical features to the typical Lewisian gneiss of the north-west of Scotland that had I been brought blindfolded to this region I would at once have supposed myself to be in Sutherland, or among the Outer Hebrides. We could not devote the time necessary to trace out the relations of this mass to the quartzites, limestones, and schists which so closely resemble those of the central and south-western Highlands that one naturally assumes that they are a continuation of these rocks, and have been affected by similar intense dislocation, crushing, and metamorphism. It was to me a subject of much regret that the tracts of Mayo and Galway which we examined had already been mapped by the Survey, and the maps of the ground had been published, while the key to the structure of the country was still undiscovered. Possibly the key has been destroyed by the complex plication and metamorphism which the rocks have undergone, but the search for it would be a worthy task for future geologists.

From Ireland I crossed to Scotland, and while engaged for some weeks in the west Highlands, had occasion to visit the island of Iona, where the ruins and historical associations on the one hand, and the geological interest on the other, present unusual attractions. The comfortable little inn on the islet possessed a copy of Bishop Reeve's edition of Adamnan's *Life of St. Columba*, which I read from cover to cover. Wandering over the lonely scene, and tracing the topographical features more especially connected with the Saint, one could enter so fully into the spirit of the place that it seemed as if it would hardly have been a surprise to come upon Columba himself and his white horse. The remarkable Celtic verses descriptive of the scenery of this little western isle, which have come down to us as the composition of the Saint, are so tenderly true to the life that we may well believe them to have been written by him. Their author was certainly endowed with great appreciation of the charms of Nature, and had the poetic gift of vividly depicting them as they filled his eyes and ears with delight. We can picture him sitting "on the pinnacle of a rock" and taking his fill of a scene of which he loved every detail—"the heaving waves of the wide ocean when they chant music to their Father"; "the level sparkling strand," with "the song of the wonderful birds"; and "the roar of the surrounding sea" with its "monsters, the greatest of all wonders," and "the thunder of the crowding waves upon the rocks."[1]

It was strange that this quiet spot, so redolent of memories of the early Celtic Church and monastic solitude, could on a sudden be converted into a crowded scene of gaping tourists struggling to make the most of the hour between the arrival and departure of the steamboat that brought them from Oban. On the excursion day I crossed to the western shore, with its little bays and rocky promontories, laved by the clear waters of the open Atlantic, and with no sign or sound of man.

An interesting discovery had recently been made in Scandinavia, of Palaeozoic fossils in rocks now metamorphosed into schists. Anxious to leave no evidence unexamined which might throw light upon the problems of metamorphism with which the Survey was confronted in Scotland and Ireland, I made a brief excursion to Norway in the July of this year. My friend Hans Reusch, Director of the Geological Survey of that country, by whom the fossils had been found, kindly furnished me with an itinerary to guide me to the most important localities in the Bergen region, and I was to meet him afterwards at Trondhjem. The Bergen ground was easily examined in a week. The more northern area was much more extensive, and required longer time. On reaching Trondhjem I found no trace of Reusch (who had never received my letter) and began the exploration inland by myself. By mere chance he turned up one evening at a little inn, where I happened to be lodging for the night. We made the rest of the tour in company. It was a most instructive experience in problems of metamorphism, and my observations completely filled a notebook, though I did not succeed in finding the solution of some of the difficulties in Scotland and Ireland for which I was seeking. From the Trondhjem Stift I went by train to Christiania, where two days were spent with

[1] A translation of this beautiful poem will be found in Skene's *Celtic Scotland*, vol. ii. p. 92.

Professor Brögger, who led me over the ground which he has so minutely studied. At the Newcastle meeting of the British Association in September I gave a brief outline of this work in Norway, intending to prepare a longer paper from the ample details which I had collected, but the leisure for this task never came. Eight years later, on the occasion of a brief visit to Norway, I had another opportunity of studying the Christiania rocks, again under the guidance of Professor Brögger. On this occasion we were accompanied by Dr. Nansen, and on some excursions into the interior, it was interesting to see the popularity of the Arctic explorer among his fellow-countrymen. In quiet out of the way villages through which it was known he would pass, he was received with an ovation of flags and cheers, and the children were given a holiday that they might see the great Norwegian.

At the Anniversary of the Royal Society on 30th November, 1889, I was elected Foreign Secretary, an honour for which I was unprepared, and which, had I been consulted beforehand, would not improbably have been declined, from a belief that its duties could not be properly discharged by one whose official work led him to be so much absent from London. The duties of the Secretary for Foreign Correspondence proved, however, to be much less onerous than I had supposed. They include conducting the correspondence with foreign parts relating to the business of the Society, returning thanks to foreigners for presents made to the Society, and forwarding their diplomas of election to Foreign Members. The most important part of the functions of the office consists in the share which the holder is called upon to take, with the other officers, in considering and arranging the business to be discussed by the Council. The office is held for four years. Eventually I was able so to plan my inspection work as to miss few of the stated Council meetings. These four years of experience in the business of the Royal Society proved of considerable value in the subsequent much more intimate connection with the whole work of the Society which I was destined to enjoy.

The year 1889 also brought a number of diplomas of honorary membership of foreign Academies and learned Societies. These came mostly from Germany, and included the Berlin Academy of Sciences, an election largely brought about, as I heard, by my friend Professor Lossen, and the Royal Society of Sciences of Göttingen, where I succeeded to the place left vacant by the death of my venerable Swiss friend Studer. The Royal Society of Edinburgh now awarded to me for the second time its Makdougall-Brisbane medal.

During the latter half of this year the quiet happiness of my home was again overshadowed by illness in the family. Roderick, as the outcome of a severe cold, developed symptoms which, in our medical man's opinion, were serious enough to make it advisable that he should winter on the Riviera. He had done well at Harrow School, having been moved up into the upper sixth, the highest form. At this time he stood seventh among his class-fellows, and this place was to be kept for him. He was so much younger than the other boys, that there seemed a possibility that he would eventually become Head of the School. But it had for some time been evident to us that a home-boarder, though enjoying the advantages of family life under the parental roof, never obtains quite the full benefit of mingling intimately with the other boys, of learning discipline and subordination, and of acquiring a sense of responsibility and a capacity for control. I therefore arranged that on his return to Harrow he should enter the house of Mr. F. E. Marshall. Meanwhile, on 17th October my wife took him and his sisters to Cannes, there to join her mother in the same pleasant villa as before.

The mention of Harrow reminds me that in the spring of 1889 I devoted a good deal of time and thought to the formation of a collection of specimens to illustrate the fundamental principles of Geology. These I arranged in a series of wall-cases and table-cases in the Butler Museum, which forms one of the school buildings at Harrow. To make this collection useful to the boys, I wrote a hand-book to it which, with many illustrative wood-cuts, was printed and published in June.

The family being abroad and our house at Harrow in charge of a caretaker, I lived during the winter mostly at the Athenaeum, with sleeping-quarters not far distant. As the days grew shorter and the London fogs became denser, the end of the week often brought me into the country. Especially pleasant were occasional visits to one or other of the Universities.

At the beginning of 1890 the retiring President of the Geological Society informed me that there was a general desire that I should be his successor in the Chair. Some years previously I had been sounded on this subject, but had declined the honour of the Presidentship on the ground of my frequent absence from London. But there was no Society to which I felt so strongly bound as to this. It was the first learned body to elect me into its membership at the age of three and twenty, and my associations with it were many and pleasant. I was well aware that the duties of its Presidentship were much more onerous than those of the Foreign Secretaryship of the Royal Society, and the work of the Geological Survey still continued to require my frequent absence in the country. But all my predecessors in the Survey had filled the Chair of the Geological Society, and to continue to decline the office might be interpreted as indicative of some dislike of the Society. I therefore now accepted the invitation, fully appreciating the honour, and hoping that it would be found possible adequately to discharge its duties, if the Council would permit me to delegate some of these duties to a Vice-president, when the public service required my absence from London. At the Anniversary on 21st February I was duly elected. The office is held for two years. During that time I was able so to arrange the inspection work that the aid of a deputy was seldom required. I began badly, however, for it had long been settled that I should join my family at Cannes, and in fulfilment of that arrangement I left immediately after the election, and did not return to London for some weeks.

This year in the west of Ireland I met with one of the accidents to which travellers in the national one-horse outside-cars are always liable. We were trotting down a not very steep hill when the horse suddenly tripped and fell, the car and driver being shot forward over the prostrate animal. My companion was pitched out on one side of the road, while I with my despatch box was launched forth on the other. Except for a few scratches and bruises and a plentiful coating of dust on our clothes, we were none the worse. The shafts, however, were broken and the horse considerably damaged, so that further progress with that car was impossible. We were on a lonely road, still some six or seven miles from the little town in which we meant to pass the night, and evening was rapidly settling down. In the end a cart came up, bound for a neighbouring farm. Its driver was persuaded to transport us and our baggage to our destination, where we arrived when night had set in. The light from the candle in the window of a little shop revealed to the local constable the unusual contents of the cart. He accompanied us till he saw us and our belongings transferred to the little inn, and doubtless made enquiries as to what we were and what our business might be in this remote corner of Ireland.

Sir George G. Stokes, having held the office of President of the Royal Society for the usual period of five years, would retire on next St. Andrew's Day. The most outstanding man to succeed him was another illustrious physicist, Sir William Thomson, afterwards Lord Kelvin. When the subject was mentioned to him, he was unwilling to have his name proposed. I had several interviews with him, and found that his main objection lay in his loyalty to the Royal Society of Edinburgh, of which he was then President. The Council of that Society, he said, had been so good to him that he could not resign this Presidentship in order to accept another, and that he set a high value on the honour of the post which he now held. To remove this obstacle, the only course to be followed was to ascertain whether the Council in

Edinburgh would be willing to help us in the matter. As an old member of that Council, and having friendly relations with its leading members, I was asked to undertake the mediation. The application was speedily successful. The Council agreed not to nominate Sir William at the next annual election, on the understanding that after he had served the usual time as head of the Royal Society of London, he would be willing to resume his former place in the Scottish Society. The way was thus opened for his acceptance of our nomination, and he was duly elected on the following St. Andrew's Day.

The Duke of Argyll with indomitable enthusiasm, continued this year, more vigorously than ever, his search for fossils in the rocks around Inveraray. He reported to me every fresh discovery, and sent up specimens to London that should convince the unbelievers in the south. Early in the year he wrote that he had found in the Glen Aray schists, worms so distinct as to make him, he said, "independent of all experts." When samples of these "worms" reached me I found them to be pieces of crushed amygdaloidal diabase—an ancient igneous rock, wherein no organisms were to be looked for. My opinion was independently confirmed by all the staff of petrographers and palaeontologists at the Jermyn Street Museum. In order to aid him in looking for the kind of rock in which organic remains might possibly be found, I sent him specimens of some of the fossiliferous schists which I had collected in Norway. The limestones on the east side of Loch Awe are in some places so little affected by metamorphism that it is hard to believe that fossils will not ultimately be found in them. My Norwegian specimens, however, directed the Duke's investigation into a new and equally hopeless channel, and led to a bombardment of letters and telegrams from him. He had found what he believed to be a "bed of corals." I could not make out anything like a fossil in his drawings, though one of them seemed to suggest that he might really have hit upon something organic. He urged me to come and see the place with my own eyes. It would be an important discovery in Highland geology, and a new feather in the Highland bonnet of the MacCallum Mohr, if he were successful this time. So, heartily wishing that he had lighted upon some indisputable fossils, but quite prepared for another disappointment, I went with him to the place of his great "find." But the rock in which his fossils lay proved again to be a basic igneous rock which was associated with limestone, quartzite and coarse pebbly grit. The supposed "corals" were only slicken-sided veinings of calcite, without a trace of anything organic about them. Being on the ground, I carefully searched a mass of limestone near the head of Glen Aray, which showed the little-altered character of these rocks. On its weathered surface I detected projecting minute spherical bodies that recalled the manner in which the foraminifer *Saccammina* appears on a weathered face of Carboniferous limestone. But though specimens were submitted to the petrographers and palaeontologists at headquarters, their united experience would not vouch for the organic nature of even these suspicious globular projections.

At the anniversary of the Geological Society, held in February each year, it is the duty of the President to give an address in which some account is presented of the progress of geology, or of some department of the science with which the speaker is more specially connected. As has been recorded in the foregoing chapters, ever since entering the Geological Survey, for more than five and thirty years, I had been studying the history of volcanic action in the British Isles, and had from time to time published the results of my researches. In 1867 I had given a summary of all that was then known on the subject in an address to the Geological Section of the British Association. Since that date, in the course of journeys in all parts of the three kingdoms, I had never lost an opportunity of visiting every volcanic area within reach, until, in the end, few or none of the known localities were left which I had not seen with my own eyes. I therefore chose this subject as the theme of my first discourse from the Geological Society's Chair at the Anniversary in 1891.

So ample had my observations grown that in that address only the earlier portion of the volcanic record of our islands could be given, from Archaean time to the close of the Silurian period.

Another of the necessary duties of the President at these annual reunions is to read obituaries of Fellows, Foreign Members and Correspondents who have died since the previous Anniversary. Among these notices I gave an account of my esteemed colleague, Andrew Crombie Ramsay, who died on the 9th of the preceding December. This sketch of him brought from his widow a grateful letter of thanks. When a year had passed after this time, and no movement had been initiated to have a fuller and more adequate biography of him, I proposed to her that if no one else thought of taking it up, I would be glad to write it. She at once and cordially assented. The materials which the family supplied, and what could be collected by myself, enabled me to compile, with my whole heart, the record of my friend's life which was published in 1895 as an octavo volume of some 400 pages.

The most important incident in the field-work of the Survey this year was the discovery, in the north-west of Sutherland, of a certain trilobite (*Olenellus*), characteristic of the lower parts of the Cambrian system. This addition to our knowledge showed that the quartzites and limestones of that region are of Cambrian age, and that the Torridon sandstone, which had hitherto been labelled "Cambrian" was immensely more ancient. So crucial a discovery had to be substantiated by detailed mapping of the ground and the accurate determination of the geological horizon of the fossil. In the course of my inspection during July, I had an opportunity of visiting the scene of the first "find." It lay among the mountains to the south of Little Loch Broom, eight miles from any road fit for wheeled traffic. All food and other supplies had to be transported on the backs of ponies. The same mode of conveyance served also for the portmanteau of any visitor, while he himself would usually make the journey on foot along the same rough track. At this remote spot

I found that my colleagues, Peach and Horne, had made their camp, and had been joined by three others of the staff, all eager to see in detail the strata which had yielded the trilobite. In spite of heavy rain, with fair intervals in which the midges and horse-flies were active, we made a merry party, and saw with the utmost satisfaction and pleasure the "Olenellus-zone," as developed in the north-west of Scotland. Since that time the "zone" has been followed for many miles, occupying a definite position in the group of fossiliferous strata which overlie the massive quartzites.

A new departure was now made by the Geological Survey. The great coalfield of South Wales had been surveyed long ago by W. E. Logan, whose maps were adopted and published in the early days of the Survey by De La Beche. Logan's mapping, however, on the scale of one inch to a mile, though remarkably good so far as it went, could not adequately represent in detail the structure of the ground. Moreover, since his time sixty years had passed away, during which a great development of underground workings had brought much fresh information to light. The old maps were quite out of date, and of little or no practical utility. In compliance with the public demand, the Government of the day determined that this wide area, so important in the mineral industry of the country, should be entirely re-surveyed upon maps on the scale of six inches to the mile. We lost no time in carrying out this resolution, and continued the work for some years until maps of the whole coalfield on the larger scale were available to the public.

The British Association by a happy coincidence met this year at Cardiff, under the presidentship of Sir William Huggins. I attended the meeting, and gave at the Geological Section an account of the discovery of the "Olenellus-zone" in Ross-shire, and of the latest work of the Survey among the most ancient gneisses of the north-west Highlands. At this meeting I was elected President of the Association, which would meet the following year at Edinburgh. A succession of honours came to me this year. One of the most agreeable was election as Correspondent of the Academy of Sciences in the Institute of France. The University of Cambridge conferred on me the honorary degree of Doctor in Science. The most unexpected of all came in the following letter :

Private. 28th May, 1891.

MY DEAR SIR,

I have much pleasure in being authorised to inform you that the Queen has been pleased to confer upon you the honour of Knighthood, on the occasion of Her Majesty's Birthday, in token of Her appreciation of your valuable services as Director-General of the Geological Survey of the United Kingdom.

It is a matter of great satisfaction to me to be the means of making this communication to you.

Believe me,
Yours faithfully,
SALISBURY.

In due course the royal command reached me to appear on 30th July at Osborne. The doings of this day were fully described in a letter to my mother-in-law which has been preserved, and from which a few sentences may here be quoted :

" My name was called first, and when the door of the audience chamber was opened, there stood Her Majesty—a bowed and bent old lady, but still queenly, supporting herself with a stick in her left hand. Her back being to the light, I could not see her face so well as I could have wished. On one side stood the Princess Beatrice, and on the other the equerry with his drawn sword, ready for execution ; while Lord Cranbrook, nearer the door, held the list in his hand from which to name the visitors in due succession.

" Bowing low, I advanced in front of the Queen and knelt on my right knee. She took the sword, and tapping me gently first on the left and then on the right shoulder, said in a deep hollow voice, ' Rise, Sir Archibald.' That voice, so different from what I had fancied it to be, gave me a momentary shock, but I quickly recovered, and pressed gently to my lips the aged wrinkled hand, half covered with a mitten, which the Queen held out to me.

"Though I had seen her Majesty at various public functions, I had never been so near her before, and had not realised that she was so feeble. The whole incident was over in almost less time than I have taken to write this account of it. I rose, and bowing again, retired backwards and so outwards into the corridor."

# CHAPTER VIII

## MISCELLANEOUS DUTIES OF THE DIRECTOR-GENERAL (1892–1896)

The Anniversary of the Geological Society on 19th February, 1892, when my tenure of the Presidentship expired, was remarkable for an incident of a kind as interesting as it was unusual. Three weeks before the meeting I received the following letter from Miss Yule, daughter of the late Sir Henry Yule, the eminent geographer, and an old friend of Sir Roderick Murchison, dated from her residence in France, 28th January, 1892:

"I observe that the Annual Meeting of the Geological Society will be held on the 19th February That day, as you may remember, was Sir Roderick's birthday, this Anniversary is the centenary of his birth, and having a very strong affection for Sir Roderick, I should like to lay a small 'stone on his cairn' that day, if you will kindly enable me to do so.

"It has occurred to me that the way of marking the day, most consonant with his own habits and taste, would be to make the occasion the excuse for giving a little help to some unprosperous brother of the Hammer. And I am desirous to devote to that object a sum of at least fifty pounds—more if I find I could afford it. I should wish to offer it anonymously, simply 'in grateful remembrance of Feb. 19, 1792,' or the like. Now the favour I would ask of you is:

"1st. That, if you approve of the idea, you would select a suitable recipient, by preference from our own dear country or from Siluria (you may interpret the latter designation either geographically or geologically). I do not absolutely exclude South Britain, but confess I should prefer a candidate from any other part of the kingdom, or even from France.

"2nd. That you would kindly secure that it should be presented officially through the President at the Anniversary Meeting.

"You would be quite at liberty to bring my name before the Council if required, though not before the public. I suppress my name; first, because I would wish to avoid all stain of self from this act of piety, and also that this little homage to our friend's honoured (but fading) memory should have a chance to be attributed to some one less insignificant than the present writer."

I had known Miss Yule for some years, and on many occasions had noticed her veneration for Murchison, as well as her love of Scotland. A week after the receipt of her letter she came to England and had an interview with me, at which it was arranged that her gift, now increased to £100, should be divided between two geologists who were following in Murchison's footsteps, and had been doing admirable work in unravelling the complicated geological structure of the Scottish Highlands—Benjamin N. Peach and John Horne, both on the staff of the Geological Survey of Scotland. To her money gift, Miss Yule added for each of the recipients, a framed portrait of Murchison. It was my happy privilege to present her gift to my old and trusty friend B. N. Peach, who attended the Anniversary, and at the same time to hand to him the other half of the generous donation to be conveyed to his colleague John Horne.

These two members, whose names are so often mentioned in these chapters, were the Castor and Pollux of the Geological Survey, always closely associated in their scientific work, and coupled together in the appreciation of the value of that work expressed both by their colleagues and by the outside world. Yet no two men in the whole staff were more strongly marked off from each other. Peach, with the eyes of a hawk, the enthusiasm of a youth and the perception of a veteran, was, in my opinion, the ablest field-geologist in this country. Nothing escaped his notice. He had the intuitive gift of realising swiftly, and almost always correctly, the meaning of what he saw. Even where his very rapidity of perception led him sometimes into error, his mistakes were those of a genius. In the full plenitude of his remarkable gifts, he dearly loved to argue in a disputed question of geological structure, becoming more and more excited as the argument grew warmer, his strong voice growing so loud that the replies of his opponent could hardly be heard. At the same time his good nature was never ruffled, and his ready helpfulness was always forthcoming for any one who asked it. The only serious defect under which he laboured was his repugnance to the use of the pen. He would rather walk a mile than write a letter. He lacked the literary fluency which would have enabled him to obtain the full advantage of his exceptional abilities. Had he been able and willing to be at the trouble of putting his observations and conclusions into written form, he would undoubtedly have taken a very high place among the geological writers of his time. Though below middle height he was stoutly built, and when in his prime, was of great physical strength and endurance. He is still alive and enjoys fairly good health; his thick crop of hair, now grown snow-white, makes his head a conspicuous and attractive feature in any company. Unfortunately he has been liable to occasional failures of health, especially to attacks of sciatica and rheumatic gout, brought on or aggravated by the enthusiasm with which, through rain and mist and bitter winds, across bogs and streams, often up to the knees in water, he would continue his field-work as long as daylight lasted.

John Horne is of a wholly different type. Without the originality and observational powers of Peach, he is unquestionably a highly intelligent, patient, and indefatigable investigator, slow to draw conclusions until he has exhausted all the obtainable evidence, accurate and careful in his work, and happily possessing the art of describing, in clear and succinct language, his observations and the deductions to be drawn from them. Prompt in correspondence, and methodical in all business requirements, he was naturally selected to be the Director of the Scottish branch of the Survey, and his senior colleague Peach fully recognised the propriety of the appointment. It is mainly to Horne's persistent and imperturbable efforts to extract from Peach the story of his work in the field, and to place that story fully before the world in clear language, that science owes the series of conjoint writings with which their names are associated. After Horne retired from the public service, on the pension which he had so well earned, his scientific reputation, tact, and business capacity led to his election as President of the Royal Society of Edinburgh, an office which he worthily filled for several years.[1]

To return to the Geological Society's Anniversary; while the proceedings were in progress at the Society's apartments in Burlington House, Miss Yule, as I afterwards learnt, was engaged at the Brompton Cemetery with a pail of water and a scrubbing brush, piously removing the London soot from the gravestone of Murchison and his wife. Her devotion to the memory of the old chief was still more strikingly manifested in later years. After Sir Henry Yule's death, when she had realised his estate, she resolved to make her home in Scotland, and the house which she chose as her residence was the mansion-house of Tarradale, Murchison's birthplace at the head of the Beauly Firth. She obtained a lease of this residence and lived there. Eventually she persuaded the owner, Mr. Baillie of Dochfour, to sell the property to her. In that remote spot, consecrated by her romantic attachment to the memory of the illustrious man who was born there, she spent the later years of her beneficent life, until her death on 24th August, 1916.

[1] Since these lines were written the two geologists have again been bracketed as the recipients of the approbation of their geological brethren. At the Council of the Geological Society held on 5th January, 1921, it was unanimously agreed to award the Wollaston medal in duplicate to B. N. Peach and John Horne.

But though she has passed away, she has left a permanent memorial of her regard for Murchison, and of her benevolent nature, as well as of the *perfervidum ingenium Scotorum* which throbbed so warmly in her heart. By her will she bequeathed her house of Tarradale, together with the grounds and contents and the residue of her money, to be held in trust by her executors, the Lord Lieutenant of Ross and Cromarty, and the Chairman for the time being of the County Council of Ross and Cromarty, for "the rest and refreshment of poor scholars or other students, preferentially but not exclusively for those no longer young and for other causes unable to seek such refreshment unassisted. These must be British subjects, and of Scottish, preferably Highland birth or descent. If the house were kept open only three or four months in the year, the funds hereinafter assigned should be sufficient to pay the necessary expenses of maintenance and upkeep and a resident janitor or housekeeper. I do not wish the Memorial to be eleemosynary, but to afford some help to those who will help themselves. The bequest is to be known as the 'Murchison of Tarradale Memorial'."

When Miss Yule died her funds were found to be considerably less than they were at the time when she made her will, owing partly to the fall in the value of securities, and partly to encroachments on her capital which her generous nature prompted her to make on behalf of various war objects. Some delay may therefore be necessary before the executors are in a position to carry out her benevolent scheme.

Chiefly for educational purposes it had been arranged that my wife and daughters should spend the winter of 1891-92 in Paris, while for those months I should take up bachelor life in London, with the Athenaeum Club as headquarters. After the beginning of the new year an ailment attacked me that turned eventually into jaundice, by which I was completely disabled. My old friend Dr. Lauder Brunton, after full examination, insisted that I must stop all engagements for two or three months and go abroad, so as to be quite away from the temptation of work. For the first time in thirty-seven years passed in the public service, I was compelled to apply to head-quarters for sick-leave. The two months asked for were at once granted.

I joined the family in their pleasant rooms in the Avenue Marceau, as soon as all duties at the Geological Society were cleared off. The medical order to do no work was obeyed somewhat regretfully, but it did not prevent one from enjoying pleasant rambles with one or other of the home circle in Paris, nor from meeting one's French friends, and increasing their number. As a Correspondent of the Institut de France I attended the weekly meetings of the Académie des Sciences, and met there many of the more eminent men of science of the day. Pasteur charmed me by the quiet dignity, courtesy and modesty of his manner. One day he drew an amusing contrast between the scanty accommodation accorded to science in the Palais de l'Institut with what he called the palatial quarters provided for the scientific Societies at Burlington House. I was able to assure him that we had already almost out-grown the quarters assigned to us, and would before long be as cramped for book-space as they were at the Institut.

One of my most esteemed friends in Paris was A. Daubrée, Professor of Geology at the Muséum d'Histoire Naturelle. We had long corresponded and exchanged publications. It was now possible to cultivate more intimate personal relations. He was at this time seventy-five years of age. His active career was practically ended, but he continued to take a keen interest in the advance of his favourite science. From this time onward till his death, our friendship became increasingly close, and when we met in the street he always saluted in French fashion by kissing me on each cheek. I never met a simpler, more unaffected and kindly nature than his, or one where eminent success had left inherent modesty more unimpaired. It was worth while to be invalided in Paris to enjoy his genial company.

Another friend, Jean Albert Gaudry, was Professor of Palaeontology at the Muséum. Some thirteen years younger than Daubrée, he was still in full vigour, living with his accomplished wife in a pleasant roomy mansion not far from the Institut, where he loved to entertain his friends. There was a neat spruceness about his manner and dress which made him look younger than he really was. He greatly valued his cordial relations with English-speaking men of science, and took care to cultivate these relations.

With Ferdinand Fouqué, now in his sixty-fourth year, Professor of the Natural History of Inorganic Bodies at the Collège de France, my acquaintance had grown into a friendship which was to become henceforth more intimate. He was a thoroughly domesticated man, living in the midst of his family circle in a large house in the Rue Humboldt. Under his hospitable roof my daughter Lucy and I passed many pleasant hours, in the midst of a cultivated, genial and characteristically French household. As a man of science he stood at the head of French petrographers. He had also made important contributions to the study of earthquakes. His daughter, who afterwards married M. Lacroix, inherited her father's geniality and ardour for science, and has shared in her husband's explorations at home and abroad. M. Fouqué has passed away, but his affectionate regard for me has been continued by his daughter and his son-in-law. M. Lacroix has risen rapidly in reputation and is the acknowledged successor of Fouqué in the same line of research. In 1914 he was elected Secrétaire Perpétuel de l'Académie des Sciences.

Closely associated with Professor Fouqué was his friend Auguste Michel-Lévy, Director of the Carte Géologique de France, another of the same brilliant band of investigators. He welcomed me to his office, where he showed his latest results in the study of the microscopic structure of rocks, and to his family circle at his bright home in the Rue Spontini. His tall and broad figure suggested that of a strong man, and his friends looked forward to his continuing for many years at the head of French official geologists, but he died in 1911. He was at this time wonderfully active and alert, quick in perception, forceful and emphatic in the affirmation of his opinions, but always courteous and friendly.

With these magnates should be classed Professor A. de Lapparent, to whom reference has already been made. The acquaintance with him, begun at Berlin, had grown into warm friendship, with an active exchange of letters. It was pleasant now to meet him under his own roof and in the midst of his family. He was certainly one of the most accomplished Frenchmen I ever met. To a full share of the typical graciousness and courtesy of his fellow-countrymen, he united a sincerity and depth of feeling altogether remarkable. He was genial and witty as a companion. As an impromptu speaker, he had few rivals in scientific circles for the readiness, grace, and humour with which he could reply to any unexpected call. A devout Catholic, he must often have had a difficult part to play in the midst of the irreligious politicians of the day, and I heard, but never from him, that his orthodoxy had stood in the way of his appointment to one or other of the official posts which he was particularly well qualified to hold. There could not have been a higher testimony to his ability, tact and personal charm than his being chosen in 1907 by the Academy of Sciences their Secrétaire Perpétual—an honourable position which he had only held for a few months when his career was cut short by death.[1]

At one of the sittings of the Academy a curious incident occurred. A Professor having resigned his Chair, which was a Government appointment, the Academy was called upon to furnish the Minister of State with the names of two candidates, either of whom might become his successor. I sat beside the retiring Professor, who told

[1] Since these lines of appreciation were written an excellent "Notice" of A. de Lapparent by Professor A. Lacroix (who succeeded him in the Secretaryship) was issued in December 1920 in the *Memoirs of the Académie des Sciences, Paris.*

me that his assistant would be appointed. When the voting papers were emptied out of the urn, the President, A. T. D'Abbadie, with octogenarian deliberation, unfolded and read them aloud. It was soon apparent that not the assistant, but the other candidate was the choice of the great majority of the Academicians. I turned to the Professor and asked the meaning of this voting. It had evidently surprised and vexed him, and he gave vent to his feeling by exclaiming impatiently "Stupide!" As the result of the scrutiny the other man was shown to have the great majority of votes. His name was accordingly placed *en première ligne*, and that of the assistant was put below. Normally the Minister would accept the first name on the list as that of the candidate recommended by the Academy. But in this case the assistant had powerful political friends, whom the Government would rather not displease. So the assistant duly became Professor, to the regret of men of science.

After an absence of six weeks in France I returned to duty on 7th April restored in health, and by the end of the month the family was once more established at home. The main objects for which we had resided six years at Harrow were now accomplished. Our daughters had come through the various ailments of childhood, and now needed greater facilities for education. Our son had enjoyed the advantages of home life during his earlier years at the school, and, to enable him to reap the full advantages of public-school life, he had been placed in the house of one of the assistant masters, of which he was now head, as well as prospective Head of the School. My official duties had grown more onerous, and to them were added the claims of commissions, committees, councils, boards, and other organisations which could not be put aside. For some time it had become increasingly obvious that to be resident in London would save much time, though at the loss of fresh air and of the garden which we loved so much. In October 1892 we moved family and household goods to Chester Terrace, Regent's Park.

On returning to London from Paris I found such an

accumulation of arrears and of new work as detained me at headquarters until the end of July. I had been appointed member of a Royal Commission on the water-supply of London, the object of which was to ascertain whether the basins of the Thames and Lea could for a time continue to furnish an adequate supply of water for the needs of the metropolis, or if it would soon be necessary to seek a further supply outside the watersheds of these rivers. The meetings of this Commission, of which Lord Balfour of Burleigh was Chairman, began in May and were continued at frequent intervals, except in August and September, up to the end of the year. After hearing much evidence we came to the conclusion that there was no immediate necessity to go outside the present sources of supply. The intercourse with the Chairman, which the meetings of this Commission brought about, led to the foundation of a pleasant friendship with Lord and Lady Balfour.

About the same time that the Water Commission was appointed, I was asked to serve on a Departmental Committee of the Board of Agriculture to enquire into the conditions of the Ordnance Survey, especially with reference to the expediting of the one-inch map, to the arrangements that should be made for the continuous revision of the maps published by the Survey, and to possible improvement in execution, form, information conveyed, and price. The meetings of this Committee, of which Sir John Dorington, Bart., M.P., was Chairman, were also numerous, and the work to be accomplished, like that of the Water Commission, was full of interest as well as of great public utility.

Another duty outside one's normal official life came this summer. Professor Tyndall, who had for some years represented the Royal Society on the Board of Governors of Harrow School, intimated his resignation of the office, owing to failing health, and the Council asked me to succeed him. My interest in the School was so keen that I gladly accepted the post. At the close of the prize-giving on 29th June, Mr. Welldon, the headmaster, made a kindly allusion to this appointment and to Roderick, who was now to become Head of the School. For thirty years the duties of this Governorship were a continual source of interest and pleasure to me, and only the advance of old age compelled me in 1922 to relinquish them. As representative of the Royal Society on the Governing Body, it was more especially incumbent on me to watch over the teaching of science in the School, and great has been my satisfaction to mark the advancement made in this branch of the tuition, chiefly through the strenuous efforts of the succession of able science masters with whom I collaborated.

One of the most agreeable of the occasional interludes in the whirl of this summer, was that of the celebration of the tercentenary of Trinity College, Dublin, on 5th July. My frequent visits to Ireland had made this venerable institution more and more dear to me, not only as a centre of intellectual life, but as a haven of rest and friendly intercourse in the midst of all the political turmoil of Ireland. I gladly availed myself of the invitation to attend the festival. The proceedings were arranged with admirable forethought and skill, and I found in later years the advantage of the lesson which they furnished in the methods whereby such gatherings may be most efficiently conducted. I learnt that this perfection of arrangement was mainly due to Dr. Daniel John Cunningham, the Professor of Anatomy, who afterwards succeeded Sir William Turner in the corresponding chair in Edinburgh University. One of the preliminary parts of the festival was a march in procession from the College to a service in St. Patrick's Cathedral. The route lay partly through a rather squalid thoroughfare, where the long array of bright robes and decorations attracted a motley populace. Just in front of me on the march, the President of the Royal College of Physicians of England had somehow strayed from his row in the ranks, and was marching along by himself in his gorgeous gown of black silk and gold lace, bowing to the cheering spectators on either side. As we passed a group of bareheaded women at a doorway, I heard one of them exclaim, as she pointed to the doctor's splendour, "And who says that Oirland's a poor counthry?"

The hood of the Doctor of Science of Trinity College has a facing of green silk, and some of the processionists were discussing what this tint of green should be called. John Evans, Treasurer of the Royal Society, overhearing the question propounded, immediately interposed with the remark that "it must be *sage*-green." His wit was always ready and never malevolent. A few weeks afterwards, as one of the recipients of the degree, I received a gift of the hood with a friendly note from Professor Samuel Haughton, one of the wittiest Irishmen of his day. It was one of the amusements of Dublin society to listen to his speeches at the breakfasts in the Zoological Garden. They were among the drollest discursive displays to which I have ever listened, and their effect was heightened by his strong Irish brogue. If one told him a humorous anecdote in London, he would chuckle a little and remark, "Faith, that will be a good story by the time I get back to Dublin."

The British Association for the Advancement of Science met this year in Edinburgh from 3rd to 10th August. The presidentship of this great assembly was not a post to which I had ever aspired, and nobody could have been more surprised than I was when the Council nominated me to fill it. To return to my native city at the head of this great scientific gathering, after the passing of an interval of thirty-seven years since, as a lad, I had for the first time attended one of its meetings, could not but be a source of gratification. Not without a little difficulty my wife was persuaded to accompany me, for she shrank from publicity. But once back in Edinburgh, heartily welcomed by her old friends and acquaintances, and much made of everywhere, she found to her surprise that she really enjoyed the whole meeting from beginning to end. We were guests of the Principal of the University, Sir William Muir, the eminent Indian official, Arabic scholar, biographer of Mahomet, and historian of the

Caliphate. A few notable foreigners came to Edinburgh on this occasion, including Helmholz, Arrhenius, Wiedemann, Ostwald, Goebel, the Prince of Monaco, and Baron F. von Richthofen with his charming wife.

For the subject of the Presidential Address I chose the centenary of the publication at Edinburgh of James Hutton's epoch-making "Theory of the Earth"—an event that seemed to me to be fitly celebrated by a meeting of the British Association in Edinburgh. Sketching the salient features of Hutton's teaching, I indicated how much had sprung from it in different branches of enquiry. The excursion which I led to the interesting rocks on the shore at North Berwick and Tantallon Castle revived the recollections of my Ladies' Class in the old days, and it was actually attended by several of the old pupils.

The presidentship of the British Association vividly revealed to me a feature of public life in this country which I had never before sufficiently realised. Rival press-cutting agencies bombarded me for some days with clippings from scores of newspapers and journals from John o' Groats to Land's End, containing reports, criticisms, biographical sketches, and portraits of the President. Among the papers there came a phrenological journal with an article extending over two pages of print composed of vague generalities. Their character may be inferred from the phrenologist's first sentence—"The organisation of this gentleman is most favourable for almost any useful sphere of life."

Another new experience came with the year of office at the head of the British Association. Besides attending his own Council and committees, the President is called upon, as a kind of public representative of science, to take part in many other meetings. Never before this year had so many extraneous demands been made on my time and attention. The meetings in daylight were numerous enough, but they were not unfrequently followed by dinners in the evening. To one who strongly disliked tobacco smoke, and shrank from speech-making, it was often a true infliction to sit through a long public dinner. But on such occasions one has to be guided by the excellent maxim of *noblesse oblige*, and to go patiently and cheerfully through it all.

Among the duties that devolve upon the President of the Association is that of acting as the medium of invitation between the Council and the person who is thought fittest to be the occupant of the chair after the next meeting. The Council selected the Marquis of Salisbury, who was at that time out of office, and it was hoped that he might be induced to fill a position which his well-known practical interest in science might make agreeable to him. His reply to my letter expressed doubt of his fitness on the ground of his not being a scientific man. In a second letter I informed him that it had been the practice of the Association, from the beginning of its existence, to elect from time to time as its President, not a specialist in science, but a man of affairs who could treat of the broad bearings and wide relationships of science; that it was of great service, both to the Association as a body and to the position of science in the community at large, that we should be occasionally presided over by such a man; that the time seemed eminently appropriate for reverting to this practice; and that at Oxford, where the meeting in 1894 was to be held, there could not be a more fitting President of the Association than the Chancellor of the University. The great statesman sent me a prompt answer accepting the nomination.

Towards the end of December 1892 the new Vice-President of the Committee of Council on Education, Mr. A. H. Dyke Acland, favoured me with a long conversation on the subject of the Geological Survey, of which by virtue of his office he now had charge. In the course of the interview he asked whether I was satisfied with the Annual Reports of Progress of the Geological Survey, to which I promptly replied in the negative. These Reports had all along been mere bald statements of statistics, chiefly numerical, of no special interest to anybody; but such was the wish of the authorities by whom these Reports were included with those of the other branches of the department in the annual blue-book. Mr. Acland, entirely agreeing with me, said that, as he was now head of the Department, he would be glad if I told him frankly all that I thought should be done. In reply I advised that, besides the necessary statistics, the Report should contain a sketch of the more important or interesting work done by the officers of the Survey in the course of the year. He cordially approved of my proposals, and authorised me to make a beginning with the Annual Report of the year just ending. Henceforth it was my earnest effort to make these Annual Reports a succinct journal of the progress of the Survey in each of the three kingdoms, which should be obtainable by the public in separate form, and as early as possible in each year. Besides making the work of the Survey more widely known, such a publication would have the advantage of, at least in some measure, removing or lessening an old grievance. The observations and discoveries made by the members of the staff in the course of their official duty were obviously the property of the State, at whose expense the work of observing and discovering is conducted. It was against the rules, therefore, for any officer to publish, without permission, any of his Survey work. The most appropriate channel of publicity was naturally the official Memoirs of the Survey. But these Memoirs were often unavoidably delayed until the whole of a sheet of the map, or the whole survey of a given district, was completed. Many months, or even some years might elapse before the proper official channel of publication was ready. In these circumstances it had been customary to grant permission to publish in some scientific journal, facts of interest or importance which members of the staff had discovered, together with the deductions drawn from them. It was always my desire that, as far as practicable, the Survey should have the credit of the first publication of its discoveries and observations, but I never hesitated to sanction prompt publication in an outside journal when the importance of the subject demanded it. The Annual Reports, as now enlarged and published without delay, even if they gave only a summary of the work done, would go far to secure the priority which is naturally a matter of importance to observers.

To this year belongs another enterprise introduced into the work of the Survey—the preparation of monographs descriptive of each great division of the Geological Record. From my earliest efforts among the Liassic deposits of Skye, I had found the want of such compendious works of reference.[1] The Survey in its maps, memoirs, collections, and manuscript notes had accumulated a vast amount of material which, combined with full reference to the work of previous observers, might be used in the preparation of a series of most useful volumes. The Jurassic rocks of Yorkshire, which in many respects differ from their equivalents in the rest of the country, had just been completely surveyed. Instead of issuing the results of this mapping in a series of local memoirs, as had been the practice, we gathered all the material into one volume, which should be the first of a series descriptive of the Jurassic rocks of England. This first instalment, devoted to Yorkshire, and prepared by C. Fox Strangways, was issued in May this year as a handsome octavo volume of 550 pages, well illustrated, and accompanied by a thinner volume composed of elaborate tables of fossils and a copious bibliography. This series of volumes was continued by my colleague, Horace B. Woodward, with characteristic painstaking industry and accuracy, and with all the editorial help that could be given by me. The whole series was ultimately completed in five volumes, forming a compendious account of each subdivision of the Jurassic system of formations in the length and breadth of England.

Among the private dinners, which the return to residence in London now permitted me to attend more frequently, there was one which I shall ever vividly remember. It was given at his own house by a well-known Member of Parliament whose friendship I enjoyed, and who on this occasion invited me to meet

[1] So had Hugh Miller: see his letter, *ante*, p. 58.

Mr. and Mrs. Gladstone on 19th April, 1893. I was placed near the Prime Minister, who sat at the head of the table beside the hostess. It was thus possible to watch the great politician, and hear every word he said. He was in excellent spirits, and discoursed volubly on all manner of subjects. The question of bishoprics having arisen in the course of conversation, he affirmed that he had never proposed to the Queen the name of a man who had asked for the office. The subject of the Waverley Novels, started by him, led to the mention of various preferences, but he maintained the pre-eminence of *Old Mortality*. When the dessert was served he helped himself to a Tangerine orange, and stripping off the skin, which he crushed in his hand and then held to his nose, he announced with emphasis that in his opinion it was the most delicious perfume in the world. After the wine had passed round he introduced, to the obvious surprise of his listeners, the subject of castor-oil. He enlarged on its qualities, and wound up by assuring us that it was good for man and beast. I remarked that it certainly was good for beast, as had been proved a day or two before at the Zoological Gardens. He at once enquired what that occasion was. I replied that Her Majesty had been presented with an ostrich, but not knowing how to feed it she sent it to the Zoological Gardens, where it began to pick up and swallow straw, gravel, nails, and all sorts of rubbish, and grew so ill that they administered seven ounces of castor-oil; and it recovered. "Indeed," exclaimed the Premier, "and how did they get it down?" I answered that I had not heard details of the treatment. "Ah, well," he said in a little while, "I'll tell you how I get it down." Then taking up an empty wine-glass, he proceeded with great gravity to say: "First, I put in a *couche* of water" (pausing a moment to allow us to comprehend the action), "then I pour in the castor oil" (with another pause as he glanced round to see that we followed him), "and lastly I put another *couche* of water on the top," smacking his lips with a kind of satisfaction, as he set down the wine-glass again. I then ventured to interpose by asking how he got the upper layer of water to remain above the much lighter oil. He at once saw the dilemma, and with great readiness replied " Ah, I admit, it requires to be done with great caution."

Next day, at the Royal Society, I told this anecdote to the two secretaries, Lord Rayleigh and Michael Foster, when Lord Kelvin, then President, who was writing a note at a side table, turned round to ask what was the joke at which they were laughing. I had to repeat the story to him. The great physicist, who had at this time become an active politician in opposition to Home Rule for Ireland, emphatically exclaimed, " Well, well, the man's physics are as bad as his politics."

Another interesting dinner was that of the Royal Society Club to commemorate the 150th anniversary of the Club's foundation. To give a special character to this festival it was resolved that the price of the dinner should be one shilling and sixpence, which was the cost in the Club's earliest years, and that the menu should be as nearly as possible the same as that of 24th March, 1747, the oldest " bill of fare " that has been preserved in the Club's records. The company numbered sixty-seven, and the quaint menu ran as follows: " Cod's Head, Smelts, Calve's Head, Pidgeon Pye, Bacon and Greens, Fillet of Veal, Chine of Pork, Haunch of Venison [sent by Sir Andrew Noble], Plum Pudding, Apple Custard, Butter and Cheese "—a most fleshly banquet to modern taste!

The amount of leisure available to me at this time for private literary work had become much reduced. In the course of the year, however, I completed and published a third edition of the *Text-book of Geology*; gave an address on the work and methods of the Geological Survey to the Federated Institution of Mining Engineers; contributed a paper on the Precambrian Rocks of the British Isles to Professor Chamberlin's *Journal of Geology*; and supplied an article on " Landscape and the Imagination "[1] to the *Fortnightly Review*. The preparation of the biography of my old companion and chief, Sir Andrew C. Ramsay,

[1] Reprinted in my *Landscape in History*.

commenced at the beginning of this year, filled many spare moments of my time. It was completed and published towards the end of 1894.

The pressure of extraneous work, however, now began to lessen. The year 1894 released me from a number of duties which, though in themselves for the most part pleasurable, necessarily demanded much time and thought. Thus the four years' tenure of the Foreign Secretaryship of the Royal Society had terminated, likewise the Presidentship of the Geological Society and of the British Association, as well as the Royal Commission on Water Supply and the Departmental Committee on the Ordnance Survey. After clearing off all arrears of official work at Jermyn Street I was able to devote more time to the field-work that required attention. Various questions had arisen as to the mapping of portions of the southern counties from Hampshire to the west of Cornwall, and others had occurred in Leicestershire and Derbyshire. Ireland needed careful consideration, not only in the ground that was in course of being mapped, but in areas of which the maps had been completed, but where the explanatory memoirs remained to be written. In Scotland the inspection comprised a greater extent of country than ever before, for, in addition to the work in the Highlands, it had been found necessary to re-examine some portions of the Southern Uplands for the purpose of the descriptive memoir of that region.

"Three weeks without an address, a letter or a telegram, a comfortable berth in a roomy steam-yacht, some pleasant companions, and a cruise wherever you would like to go among the Western Isles." Such was the pleasant prospect which had been held out to me for two or three years; but which, owing to official and other ties, it had not been possible to accept. The invitation came from Henry Evans, banker in Derby, who having a lease of the deer forest of the island of Jura, had there built himself a commodious house and laid out charming grounds commanding a fine sea view. My inclusion among his yachting comrades was suggested to him by my friend,

Alfred Newton, Professor of Zoology in the University of Cambridge, who had long been an intimate associate of Evans, and had accompanied him for some years in his summer cruises. At last in 1894 it was found possible to take advantage of the privilege. Henry Evans was not only a good man of business, but had literary tastes, and, above all, the instincts and enthusiasm of a true naturalist. His special affection for birds and his wide knowledge of them led him to protect them by every means in his power, so that under his fostering care Jura became a sanctuary for bird-life. He was a keen sportsman and good shot. Some years before my first visit to his island home, the contents of a friend's rifle were accidentally discharged into one of his legs, which had to be amputated. But though disabled from climbing the steep and rough quartzite mountains, he continued to be as keen a deer-stalker as ever, mounted on his stout hill-pony.

Alfred Newton, or " Alfred the Great " as Evans used playfully to style him, was an enthusiastic naturalist, especially devoted to the study of birds. His wide range of knowledge, his well-stored and accurate memory, his fund of reminiscence and anecdote, and his delightful sense of humour made him a most pleasing companion. His annual cruise with Evans was looked forward to by him with almost boyish exuberance, and he afterwards loved to recall its various incidents. Nowhere could he be seen more completely in his element than on board of his friend's commodious 250 ton steam-yacht the *Aster*. Clad in the light-grey tweed suit that did duty on these occasions, but without topcoat or waterproof, he would sit for hours on some exposed part of the vessel, smoking innumerable pipes and watching for every variety of sea-fowl that might show itself either in the air or on the water. In the course of a few days sun, wind, rain and salt spray told on his complexion, which became " ruddier than the cherry." The sharpness of his eyesight in the detection of birds on the wing, even when he had reached the age of seventy years, was always a wonder to his companions.

Our host reckoned on almost always getting good cruising weather about the longest day. The Professor and I therefore travelled north together, and appeared in Jura before the day of the summer solstice. During a succession of years these annual arrangements held good, and each cruise lasted about three weeks. The *Aster* was fit to go anywhere on the open sea, and yet of such light draught as to be able to enter many little inlets, where she could anchor for the night. We thus became acquainted with the whole north-west coast of Scotland, including the Orkney and Shetland Islands, and on two occasions extended our cruise to the furthest northerly headland of the Faroe archipelago—*Occidentis ad ultimum sinum.* I was thus able to see the geology of the islands with remarkable ease, the vessel being piloted to every cliff that I wished to observe close at hand or to examine from a small boat.[1]

The International Geological Congress met in September 1894 at Zurich. I read a paper in which attention was called to the remarkable segregation of minerals in bands in the Skye gabbros of Tertiary age, to the occasional curvature and plication of these bands, and to the possible light thrown by them on the origin of the more ancient gneisses. When the day came for the reading of my communication, I was entreated by several German members not to give it in French. They would rather have it in English or German. This was not the only sign of the German jealousy of France, even among students of science. But I had written my paper in French, which was the official language of the Congress, and I read it in that language, to the manifest contentment of the French and Swiss members. I am not sure, however, that some of my German associates quite forgave me.

One of the closing incidents of this year was a dinner

[1] Many references to the places visited in our cruises will be found in my *Scottish Reminiscences*. The geological observations among the Faroe Islands are contained in the paper on "The Tertiary Basalt-plateaux of North-Western Europe," *Quart. Journ. Geol. Soc.*, vol. lii. (1896).

given by Messrs. Macmillan & Company in honour of Norman Lockyer, and in celebration of the issue of the fiftieth volume of the weekly journal, *Nature*, of which he had been editor from the beginning. As a contributor to the first number of this most useful publication, and to many subsequent numbers, I received an invitation, and found the company to comprise the leaders in almost every branch of science. Huxley, who was in excellent form, proposed the toast of the evening.

In 1895 the inspecting duties of the Geological Survey took up altogether ten weeks in the field between the beginning of March and the end of October. In England the most notable feature of the field-work was the detailed survey of the Isle of Man by Mr. G. W. Lamplugh. This accomplished geologist, after having made some able researches in Yorkshire, had recently been appointed to a place in the Geological Survey. The Manx problems were many and difficult; but he successfully grappled with them, and in the course of a few years completed the investigation of the whole island. In Ireland the inspection involved a renewed examination of some of the complicated ground in the west of the counties of Mayo and Galway. In Scotland the areas traversed lay in the west of Ross-shire, Inverness-shire and Argyle on the one side, and among the uplands of Galloway on the other.

While engaged in the area between Fort William and the West Coast I took the opportunity to climb Ben Nevis and to spend the night of 14th June at the Observatory on the top of the mountain, which is 4406 feet above sea-level, the highest elevation in the British Isles. At the summit I found the observer then in charge, Dr. W. S. Bruce of Antarctic fame, stretched on a lately-fallen sheet of snow, carefully collecting the numerous and varied forms of insect-life which, borne upward by warm air currents from lower levels, were chilled by the cold surface on which they had come to rest.

As the weather was clear and dry, the whole vast panorama of mountains, glens, rivers, lakes and the sea was visible to its utmost limits—a scene not often so fully displayed to the chance visitor, and, when thus beheld, never to be forgotten. As evening wore on, the whole range of the Outer Hebrides stood out against the after-glow of the western sky. It was now nearly the longest day. In this northern clime, when the sky remains unclouded, there is hardly any night at midsummer. On this occasion the stars at midnight shone with a radiance seldom seen in our misty lowlands, and the air became piercingly cold. Next morning a group of icicles hung from the roof above my window, and a wonderful change had supervened upon the landscape. Only the higher hill-tops rose up into the sunlight. All the region below them was shrouded under a vast sea of white mist. It was a rare piece of good fortune to look on such a phantom representation of how the Highlands would appear as a great archipelago, during a time of serious submergence. I had long recognised that they are probably the remnant of an ancient tableland which was planed down below sea-level by the various agents of denudation, and that on this platform, when subsequently upraised into land, the glens and straths have been carved out by these same eroding agencies. The undulating surface of the wide sea of mist might be taken to represent the general surface of that tableland. But the mist vanished as the sun rose higher in the sky, and revealed once more the familiar landscape. I descended again to the lower earth after an experience at the summit which would ever remain vivid in memory.

In this year I was elected into the Literary Society, and, until I ceased to reside in London, its monthly dinners were a source of great pleasure. The Society was founded in the first decade of last century as an association of men of letters in the widest sense of the term, for the members included men of note in the Church, in science, in law, in art, in medicine, in politics, and in public life, but all with more or less of literary tastes; and this original character it still maintains. Among its first members probably the most illustrious was William Wordsworth, while Walter Scott was elected into its honorary list. The number of members was limited to forty. They dined together at an hotel once in each month from January to July and in November and December. Among those most frequently present about the time I joined were Sir Spencer Walpole (President), Sidney Colvin (Secretary), Dean Bradley, W. E. H. Lecky, Alfred C. Lyall, Theodore Martin, E. H. Pember, Canon Ainger, D. Mackenzie Wallace, Henry S. Cunningham, Austin Dobson, Hon. Arthur D. Elliot, R. Henn Collins, and W. J. Courthope. Sir Theodore Martin and I often sat next each other, and a warm friendship was soon formed between us. From time to time he told me many incidents of his early life in Scotland, his association with W. E. Aytoun, and his own literary career. With Dean Bradley also I had many pleasant talks. He had a remarkably retentive memory, had seen many notable people, and could graphically tell his recollections of them. When our party broke up I used often to escort him over the street crossings that lay between our dining place (which for some years was the Bristol Hotel) and St. James's Park. He would never let me cross the Park with him, as he said there was no longer any danger, and he could easily find his way back to the deanery in Westminster. Charming too were the long interviews with Ainger—a delightful commingling of seriousness with humour and wit. He loved a good anecdote, and when he met me his first question often was "What is the last good story from the north?" My constant journeyings over all parts of Scotland gave opportunities of picking up anecdotes and incidents which amused him. For a time one was puzzled to account for their appearance shortly afterwards in the pages of *Punch*, with the accompaniment of illustrations by Du Maurier, until I found that the Canon and the artist lived as neighbours at Hampstead.

The yachting cruise in 1895 began with a preliminary circumnavigation of the island of Jura. I had often heard of the raised beaches on the west side of the island, but had never visited them, and was quite unprepared for the number, freshness, and bareness of their successive terraces. They are certainly the most remarkable example of this geological feature in the British Isles. Four or five flat-topped banks or terraces, some of them many acres in extent, may be traced one above another up to the highest which, on landing, we measured and found to be 110 feet above high water-mark. Each of them is formed of coarse water-worn shingle of white quartzite. The stones, all well rounded, lie closely packed, with hardly any intermingled clay, sand or gravel to bind them together. The absence of any compacting material, and the sterile nature of the quartzite, have checked the growth of vegetation on these platforms. Consequently the bare white terraces look as fresh as if the waves had been rolling their materials only a few weeks ago. It is hard to believe that the oldest of these terraces may perhaps be more ancient than the infancy of mankind in Britain. In our walk along the shore I was delighted to find beautifully ripple-marked surfaces in the hard quartzite on which were many well-preserved worm-burrows, like those of the quartzite of the north-west Highlands.

On the way to St. Kilda the yacht encountered extensive banks of summer fog so dense that the prow could not be seen from the stern. It was late before we dropped anchor in the only available anchorage in this remote group of islands. On account of the uncertainty of the weather, and the necessity of leaving the place if the wind should blow hard into our little natural harbour, I next day made a rapid traverse of the main island, to observe its geological structure and take samples of its constituent rocks. The most wonderful sight at St Kilda is certainly the amazing abundance and variety of its bird life. There are probably more gannets here than in all the rest of the British Isles put together. It was a wonderful sight to watch the different companies of these birds, each consisting of hundreds of individuals, as they kept wheeling round in wide circles and plunging in rapid succession into the sea, each bird like a white shooting star, rushing down into the deep-blue water and sending up a spout of white spray. There may have been some thousands, plunging and splashing, within sight at a time. Besides these larger birds aloft, innumerable companies of smaller fowl flew at lower levels—kittiwakes, gulls, fulmars, cormorants, puffins, and many more. The air, to use Milton's fine line, seemed to be

> Brushed with the hiss of rustling wings.

While the yacht's engines were going dead slow, and we were watching this scene of bird movement, three whales appeared, two together and one behind, puffing out their bursts of spray and showing their black backs as they quietly moved through the area where the gannets were diving; one could speculate as to what would happen should one of them be harpooned by one of the strong beaks that were rushing headlong into the water all around them. Whales are probably more plentiful in our Atlantic waters than is commonly supposed. In these cruises in the *Aster* we continually met with them. This feature of our seas would seem to have attracted the notice of early navigators. Horace alludes to the "ocean teeming with monsters (beluosus oceanus) that roars around distant Britain." The number of them might lead to exaggeration of their size. Juvenal assigns a notable bulk to the British whale. It was doubtless with these old legends in his mind that Milton pictured the body of Lycidas, swept by the whelming tide beyond the stormy Hebrides, and visiting "the bottom of the *monstrous* world." It was a sad descent from ancient poetry to modern prose when, in the course of our cruises among the Hebrides, we discovered that one or two Scandinavian companies had established whaling stations at various places among the islands, and were doing their best to exterminate the "monsters." At these stations carcases of the animals, occasionally eight or nine together, could be seen in all stages of decomposition, and filling the air for miles around with the most disgusting stench.

The voyage to the Faroe Isles is often hampered in summer by sea-fog, and we came in for our share of this impediment. But in successive visits we saw the whole archipelago in detail up to the furthest northern headland, often landing at the settlements and villages and obtaining a close view of every sea cliff which presented points of special interest.[1] The Faroe community was evidently well-to-do and industrious. The great industry being fishing, one of the chief employments of the villagers consists in looking after the drying of the fish on slopes of rough causeway, and in being ready to spread tarpaulins over them should rain threaten to fall. When the drying process is completed the fish are piled up into heaps ready for export. Everybody seemed busy and cheerful. Their cottages, neat and trim, with painted fronts and white-curtained windows, present a singular contrast to the wretched cabins amidst the filth and stench of the fishing stations in the Hebrides, such as Castle Bay. One tried to find a reason for this amazing contrast. Why should the Celtic population be content to remain in squalor, dirt and laziness, while the Norse communities live in such cleanliness, comfort and activity? There may be something in difference of race, but at least one cause of the contrast probably arises from the fact that the herring industry, in the Hebrides, is in the hands of men from the east side of Scotland who come to these western islands only for the season of the fishing. They try to make as much money as they can, and to spend as little as possible on the accommodation of the natives, whom they employ in the processes of the curing. At the end of the season they return to the mainland and leave the resident community to clear away the mess—a task not congenial to the Celtic temperament. Among the Faroes, on the other hand, the cod fishing is largely carried on by the native population, which is inherently cleanly and industrious, and retains a large share in the profits of the industry.

In this same year (1895) I had occasion to cross to the Continent three times. They were brief visits, but they

[1] The geological observations made in these cruises are described in *Ancient Volcanoes of Great Britain*, vol. ii.

helped further to cement friendships in France. Professor Fouqué at his laboratory most obligingly showed me all his latest material, in the admirable researches he was conducting in the elucidation of the microscopic structure of rocks. He then mentioned that he had acquired the apparatus and results of the experiments made at the close of the eighteenth century by Sir James Hall on the fusion of rocks. These interesting materials could not have fallen into more appreciative hands, though one could have wished that they had been retained in Scotland, as a valuable relic of the founder of experimental geology. In my veneration for this original investigator I had proposed many years before to search for materials from which to prepare a biographical account of him, and I consulted Principal J. D. Forbes, who then told me that he had formed the same intention, but had not yet been able to obtain the necessary documents. Of course I dropped my project; but Forbes passed away, and the memoir remains unwritten. In justice to Hall and to Scottish science, some attempt should still be made to supply this gap in geological literature.

During this brief visit to France I spent some days in Auvergne for the purpose of again studying certain volcanic features of that region, with a view to ascertain whether some of their details would throw light upon certain problems presented by the Tertiary volcanic area of north-western Europe. The places which I specially wished to see lay rather out of the way, and led me into pleasant intercourse with the peasantry of the remoter uplands. In the course of my many sojourns in France how often has Goldsmith's graphic couplet been recalled by the manners and ways of the simple country folk!

> They please, are pleased, they give to get esteem,
> Till, seeming blest, they grow to what they seem.

The second short sojourn on the Continent came at the end of summer, when I took my daughters, Elsie and Gabrielle, from their school at Eastbourne to join the rest of the family who had been for some time in Switzerland.

Before leaving them I had time to climb with them from the Rhone valley to Finhaut, to gather gentians at the Col de Balme, and to take them up to Zermatt and the Gornergrat, that they might have a glimpse of the glacier world, and the giant sentinels that guard it, from Monte Rosa to the Matterhorn.

The third traverse of the Strait of Dover involved an absence of only six days. It was occasioned by the celebration of the centenary of the Institut de France, to which, as one of the Correspondents, I received an invitation. The fêtes, occupying from the 23rd to the 26th of October, were characteristically French and thoroughly enjoyable. According to the official programme the proceedings were to begin with a reception at the rooms of the Institut at half past two o'clock on the Wednesday. But among the various papers awaiting each foreign guest there was, at least in my packet, one requesting his presence at a religious service in the church of St. Germain des Prés at 10 o'clock. It was a Mass for the repose of the souls of all the members who had died since the foundation of the Institut. The outstanding feature of this function was the presence of Monseigneur Perraud, Bishop of Autun, who presided and gave an address of remarkable eloquence, closing with a touching reference to the recent death of Pasteur. Turning to the audience, he added, "Your presence here, gentlemen, only confirms my words. It is good to see the flower of intellect affirming so well that it is impossible to drive out the idea of religion." The learned prelate concluded the ceremony by announcing to his fellow-members of the Institut that the Sovereign Pontiff could not let the centenary pass without sending his benediction to the Five Academies, which the Bishop, with the mitre on his head, now pronounced.

It was a strange surprise to find out in the course of the day that the ecclesiastics had stolen a march on the men of literature, science and art, by issuing their invitation to this morning service. One well-known member of the Academy of Sciences was furiously angry, not only that the service should have been held, but that the note of invitation to it bore in large letters at the top of the page, and in the proper official type—INSTITUT DE FRANCE, as if it had been actually issued as part of the official programme.

The opening reception at the Palais de l'Institut brought together a somewhat confused crowd, in which, however, people could move about and greet friends and acquaintances. As soon as dear old Daubrée caught sight of me he hurried forward, kissed me in his usual fashion, and presented me to M. Ambroise Thomas, then the President of the Institut. In the evening there was a more formal reception, with music and acting, at the Ministry of Public Instruction, where M. Raymond Poincaré, the head of that Department, received us. This eminent man, then in his thirty-fifth year, and full of capacity and energy, was the most prominent and effective of all the Ministers who attended the centenary festival. Nothing could exceed his courteous attention to the foreign members. Always alert and ready, he could make an excellent impromptu speech and fire it off with great animation. It was no surprise to those who watched him at this time, that in later years he should be chosen President of the French Republic—an office which during the strain of the Great War he filled with dignity, tact and success.[1]

On the morning of Thursday an interesting ceremony, not inserted in the original programme, took place in the conservatory of the Elysée, where M. Felix Faure, President of the Republic, received the officials of the Bureau of the Institut and the Foreign Associates and Correspondents, to the number of fifty-five. The President passed round us as we stood, courteously shaking hands with each, and stopping to say a word or two, when he recognised the name and had heard of the work of the man whom he was addressing The whole ceremony lasted only half an hour.

[1] Since this sentence was written his vigour and patriotism have continued to place him in the forefront of his fellow countrymen, and he has now (1924) been Prime Minister for several years.

The central function of the Celebration took place in the afternoon of the same day, when a public assembly of the whole Institut was held in the great amphitheatre of the Sorbonne. In this noble chamber the first object to arrest the eye was the vast wall-painting by Puvis de Chavannes behind the far platform. In front of this great work of art sat the President of the Republic in the Chair, with the Ministers on the left and the ambassadors on the right, their secretaries and attachés crowded behind, all ablaze in uniforms, orders, and decorations. In the body of the amphitheatre the members of the Institut were placed immediately in front of the platform. Behind them professors, judges, lawyers, teachers and many more, in their various robes and hats—a wonderful kaleidoscope of colour. There were some ladies present and I was glad to be able to find a place for my daughter Lucy, then resident in Paris as a student of art in the studio of M. Courtois.

The chief oration was given by the illustrious statesman, philosopher and man of letters, Jules Simon, then in his eightieth year. He discoursed with great eloquence on the history of the Institut, and the influence of its five Academies on the progress of literature, philosophy, science and art. He closed his address with an impassioned peroration, in which he told us that we had here before our eyes the Congress of Peace, the Congress wherein truth is loved for its own sake, whatsoever may be the land from which it comes; the country of eternal truth and of eternal beauty is likewise the country of peace. Then, turning to the Foreign Associates and Correspondents, he declared that not only would we carry thence a remembrance of the warm sympathies with which we had been welcomed, but that both we and our French hosts would derive from this fraternal concourse a redoubled love of peace, of the sciences whereby peace is fertilised, and of the arts by which it is embellished. We would all continue at work, each in his own chosen corner of the universal workshop, for the prosperity of the house, that is to say, for the welfare of humanity. Less than twenty years after this eloquent forecast, the country in which it was spoken became the theatre of the greatest war in the whole history of man.

With characteristic tact and good taste interludes of music were introduced to relieve the attention which the addresses demanded. When Jules Simon sat down, the orchestra played a part of the *Mors et Vita* of Gounod, who had been a member of the section of music in the Académie des Beaux Arts. At the end of this most impressive composition M. Poincaré addressed the audience. Sketching with graphic power the first sitting of the Institut, which lasted four hours, he traced the part which the organisation, then set on foot, had since played in the intellectual progress of France. As a literary effort the address was remarkably fine, and it was pronounced with the grace and fire of an accomplished orator.

Of course no celebration of the kind is considered to be complete without a banquet. The evening of the chief day of the Centenary was accordingly devoted to a sumptuous repast in the Hotel Continental with much splendour and many speeches.

On Friday morning another little ceremony, not on the printed programme, took place in the strip of garden on the south side of the Louvre, for the purpose of unveiling the statue of Meissonier. There were speeches, of course, but these were short, the chief of them being made by the indefatigable Minister of Public Instruction, who, again to the fore, summed up in terse and eloquent words the characteristics of Meissonier's art. The last speaker was M. Mounet-Sully, who recited with great effect some verses written for the occasion. The whole proceedings took up just one hour.

In the afternoon of the same day a delightful and most characteristically French entertainment was given at the Théatre Français, where scenes from Corneille's *Cid* and Molière's *Femmes Savantes* and *École des Femmes* were rendered with a finished perfection nowhere else to be found. The reception in the evening at the Elysée brought the formal functions in Paris to a close. The

President again shook hands with the members of the Institut, and seemed glad to welcome them. The French instinct for organizing a series of festivities was never more happily displayed.

Saturday was made a kind of holiday after all the formalities that preceded it, and was given up to a visit to the Chateau de Chantilly. This noble mansion belonged to the Duc d'Aumale, fourth son of King Louis Philippe. He had accumulated in it many priceless treasures of art which he desired to leave to the State as a national possession. Looking around for the body or institution that would be most likely to endure amidst the political vicissitudes through which France might yet have to pass, and most able to administer the gift for the public benefit, he made choice of the Institut de France as offering the best prospect of continuance and ability. He was now residing in the Chateau, and it was on his invitation that the members of the Institut came to visit the princely abode of which in the future they were to be custodians. Still suffering from a recent attack of gout, the Duc sat in the centre of the great hall, wrapped in black velvet, with a cap on his head, looking the very type of a grand seigneur of the olden time. With easy vivacity he shook hands with the foreigners as they were successively presented, altering his language according to that of the country of the visitor, and making appropriate little speeches with consummate grace and charm. As I drew near I noticed that his fingers were knotted with chalk-stones as large as walnuts, memorials of what he must have endured from his malady. In excellent English he welcomed me to France. He was himself the impressive centre of the whole mansion and its many treasures. I vividly retain in my memory the impression of this striking scene—a kind of living relic of the *ancien régime*. The Duc was then in his seventy-fourth year. He died two years afterwards from shock caused by the tragic death of his niece, the Duchesse d'Alençon, at the conflagration of a charitable bazaar in Paris.

My last evening in *la ville lumière* was spent at the

Comte de Franqueville's interesting old mansion of Muette, almost surrounded with the houses and streets of modern and growing Paris. My acquaintance with this charming Frenchman, now begun, became more intimate as years passed. He often visited England with the Comtesse and his family, and on these occasions was a frequent inmate of the Athenaeum. In celebration of the Centenary, he had just published two sumptuous quarto volumes, in which he gave a valuable detailed account of the first century of the Institut.

While the true man of science finds that the successful prosecution of his researches is its own highest reward, he cannot but be gratified and encouraged by the recognition of his labours on the part of his scientific brethren at home and abroad. It was assuredly with no small pleasure that at the anniversary of the Geological Society in 1895, I received from my fellow-geologists the Wollaston medal —the highest award in the Society's gift, and next year from the Royal Society, one of its Royal Medals. The brotherhood of men of science is in no way more marked than by their generous recognition of the labours of their fellow-workers in other lands. For many years past honorary membership of foreign academies and societies had been conferred on me, and the number of these distinctions continued, until in the course of years I found myself enrolled in most of the more important of these institutions both in the Old and the New World. There is still another mode in which the sympathy of scientific men is shown. They take pleasure in sending to each other copies of their papers and memoirs, which might otherwise for a time escape notice. In this way geologists in all parts of the world have enriched my library, and have put me in possession of much material which has been of the greatest service in the preparation of successive editions of my *Text-book*. The academies abroad usually send their publications to those whom they elect as foreign correspondents or associates, and these volumes are numerous and often bulky. When to these serial publications, there were added the numerous separate memoirs sent by their authors, the whole formed a rapidly increasing part of one's library, until it became difficult to find house-room for it all. It was ultimately desirable to bestow a large part of it elsewhere. While filling the Chair of Geology at Edinburgh, I had begun the formation of a class library for the use of the students, and I now relieved my shelves of all the geological literature which could be spared and sent it to Edinburgh for incorporation in that library. This practice has since been continued from year to year.

Owing to a great increase of work at headquarters, the year 1896 was spent mostly there. Besides the daily demands of my own institution, a number of official questions arose, involving both letters and interviews; the public showed an increased desire for information and guidance in geological matters; and my correspondence had consequently much increased. The inspection duty in Scotland had to be restricted to a single visit. The survey of the Oban district was now in progress, and the detailed structure of Strath in Skye, and of the islands of Raasay, Scalpa, and Pabba, was being worked out upon maps on the scale of six inches to the mile. It was with much interest that I returned to see the results of more precise investigation of the ground on which my own explorations were begun at a time when no reliable maps were available. On this visit to Skye I found one of my colleagues in a small farmhouse by the sea, with a howling tempest of wind and rain outside and a jug of milk and scones inside, quite contented with his homely accommodation, looking forward to be joined there in three days by his wife and children, and except for the multiplicity of basalt-dykes, which delayed the progress of his mapping, as cheerful and merry as usual. Another member of the staff was wrestling with the Liassic strata, which he knew well in England as a regular undisturbed formation, but which were here cut up in the most complicated way by these dykes, thus adding greatly to the number and complexity of the lines which he had to trace. A third colleague had found quarters in one of the wildest corners of Skye, where, however, the ruggedness of the ground and the intricacy of the geology were in some measure compensated by the existence of a good garden, into which, the family at the "big house" being absent, he could be turned loose by the gardener, among strawberries, raspberries, and gooseberries, to compete in the struggle for existence with thrushes and blackbirds. His luxuries were further increased by unlimited facilities for ablution, seeing that he could have a plunge in the sea before breakfast, and on his return at the end of a day of mountaineering, a natural shower-bath under a waterfall.

The summer of this year brought with it the jubilee of Lord Kelvin at Glasgow University, an event which assembled a notable company of eminent physicists from all parts of the globe, to do honour to the great philosopher. For fifty years he had been Professor of Natural Philosophy in this University, and had long been looked up to by the whole world as the most distinguished physicist of his time. He was now in his seventy-second year, but still full of mental vigour, and he continued to fill the chair for three years longer.

The annual cruise in the *Aster* this year lay through Orkney and Shetland. Though I had already visited these islands more than once, it was only now, by living among them, and sailing in and out through their sounds and voes, that I came to know them well. As the yacht was always placed completely at my disposal, I was able in the course of years, and during repeated cruises, to become fairly familiar with them from end to end. We sailed close under the vast sea-front of Hoy which, from the warm yellow hue of its component sandstone, seems always as if lit up by sunshine. Sometimes we anchored for the night at the quaint haven of Stromness, or at Kirkwall, where the fine old Norse Cathedral of St. Magnus was always visited. But the Shetlands were usually our chief cruising ground. Their wonderfully abundant bird life was a perpetual delight to our host and the Professor, while their varied geological structure, their singular, sometimes almost weird, topographical features, and their astounding examples of marine denudation were a constant source of interest to me.

The year 1896 was saddened for me by the death of some of my most cherished scientific friends. Joseph Prestwich had been slowly failing, and was now confined to his picturesque home at Darent-Hulme, where, though feeble in body, he retained such vitality of mind as to give hope that his life might still be further prolonged. Queen Victoria had conferred knighthood upon him at the New Year, and his interest in the outer world was maintained by the many congratulatory letters sent by his friends, which his wife read to him. His eighty-fourth birthday on 12th March brought him such overflowing greetings from old friends at a distance, and from neighbours and villagers at his own home, as to make it, he said, " a very happy birthday." But he continued to sink, and quietly ceased to breathe on 23rd June. Prestwich was held in high respect everywhere for the value of his geological work, but it was his own simple, modest, gentle and sympathetic nature that so endeared him to all who came to know him. From the time when we climbed Vesuvius together in 1870, I had become more intimate with him, my affection increasing as the beauty of his character unfolded itself in our intercourse. He was certainly one of the most simple and lovable men I have ever known. His widow asked me to select a few books from his library, which I was glad to obtain as a personal memorial of him. A duty which I willingly undertook at her request was to furnish a sketch of his scientific work for insertion in the biography of him which she published in 1899.

Professor Daubrée likewise passed away in the same month of June. His distinction in science was combined with an almost child-like simplicity of character which gave him an irresistible charm. From the first time we met he had taken such kindly interest in me and my work as to win my devoted regard. He took my family also into the circle of his attachment. When Lucy was living in Paris as a student of art he was unceasingly kind in his attention to her. One of the attractions of

Paris to me was his presence there, and since his death it has never seemed to be quite the same place.

Another grievous loss came in the death of Alexander Henry Green, one of my early associates in the Geological Survey. We used to " chum " in the same lodgings in London, and grew to be fast friends. He left the Survey in 1874 when he was appointed Professor of Geology at Leeds, but we continued a helpful correspondence, and met occasionally. He was one of the men who, from one cause or another, never quite do themselves justice. I regarded him as one of the most accomplished geologists of his day in England, and from the long talks we used to enjoy in the old Survey years, one looked forward to his rising into eminence by the introduction of mathematical methods into the discussion of geological problems. When he obtained the Oxford Chair of Geology in 1888 his friends sincerely hoped that the comparative leisure of the position might give him the opportunity to do what had been expected of him. But his health began to suffer, and he died in August this year at the age of sixty-four.

By the death of Professor Green the Chair of Geology at Oxford again became vacant. Various applicants for the post came forward, but the University authorities were in no hurry to appoint a successor. Late in the autumn one of the Board of Electors again urged me to apply for the Professorship. I assured him that this was quite impossible. Early in December another of the elective body wrote to me still more strongly pressing the same proposal, and pointing out that I would be sure of a Fellowship, which with my Civil Service pension, would bring the small professorial stipend up to the value of my income in the Geological Survey, and he added, " some day you will feel the bucketing about from Land's End to John o' Groats a little severe, and will be glad to have a *pied à terre* in Oxford." But I remained deaf to these allurements, and advised the electors to choose Professor Sollas of Dublin, who had just returned from his coral island expedition. At the same time I urged him to send in his application as a candidate, but this for a time he hesitated to do, owing to the inadequacy of the emoluments of the Oxford professorship. Eventually he applied and was elected. The University thereby obtained one of the most brilliant geologists of the day. The financial conditions of the post have since been improved. With an augmented income, a growing attendance at his lectures, and leisure for research and for the production of fresh contributions to geological literature, he has become one of the most eminent leaders in the science to which he has devoted his life.

During this year a correspondence arose about a proposal which was made to me by Dr. D. C. Gilman, President of the Johns Hopkins University of Baltimore. In his first letter he wrote: "The widow of the late Professor George H. Williams, who was Professor of Geology in this University until his death nearly a year ago, proposes to maintain a lectureship in this place to commemorate his name. It is the desire of my associates, and of those geologists whom I have consulted, as well as of Mrs. Williams, that the opening course in this series should be given by you. We are sure that your coming to this country will give a fresh impulse to the studies which you represent. Among your hearers there will be not only our own students of geology and the allied sciences, but also, as we hope, several members of the U.S. Geological Survey, and possibly others from a distance. The lectures may be delivered at any time between the 15th of October and the 1st of May."

Between the geologists of America and myself there had long been the friendliest regard. My first visit to the United States laid the foundation of many personal attachments, and there had arisen a copious and continuous correspondence between us. I had therefore a strong desire to renew personal contact with the old friends, and to make the acquaintance of many more whom I knew only by their papers and their letters. Many difficulties had to be surmounted before I was at liberty to accept the invitation. The whole journey to America and back had to be comprised within the time of my annual holiday, as I was unwilling to apply for any extra leave, seeing that there was the prospect of another long tour abroad which could only be undertaken by special authorisation of my official chiefs. It was at last arranged that I should appear in Baltimore, and give the first lecture on 21st April, 1897.

The greatest gratification which came this year to my wife and myself was the success of our son Roderick at Cambridge, where he obtained the Winchester Reading Prize, and took his B.A. degree, with a first class in the Historical Tripos. He also took a prominent part in the Cambridge dramatic performances, and gained a good reputation both in English comedy and in Greek tragedy.

The more sedentary life of 1896 enabled me to complete a good deal of literary work. In May the Geological Society published my paper on "The Tertiary Basalt-plateaux of North-Western Europe"[1] This paper summed up the additions which I had been able to make to the subject during the seven years that had elapsed since the appearance of the quarto memoir of 1888. It included the results of the further study of the Inner Hebrides and the Faroe Islands. I contributed also an article on "Scottish Mountains" to the *Scottish Mountaineering Journal*. But the chief occupation of my leisure hours was the preparation of two works which were published in the following year. The first of these consisted of the course of lectures to be given at the Johns Hopkins University. The subject selected by me was one which had never been adequately treated in English, and yet possessed much interest to all geological students. The title "Founders of Geology" sufficiently indicated its general scope. The second work was the gathering together and amplification of all that I had published in scattered papers and memoirs on the history of volcanic action in this country. It was published at the beginning of 1897 in two large octavo volumes entitled *The Ancient Volcanoes of Great Britain*.

[1] In *Quart. Journ. Geol. Soc.*, May, 1896.

# CHAPTER IX

## THE "WANDERJAHR" OF A GEOLOGIST

(1897)

THE year 1897 was altogether an exceptional one in the course of my life. It was truly a "Wanderjahr," for it included a much longer absence abroad, and a more varied range of foreign travel than ever came in my way, either before or since. The spring found me in the United States, and the autumn in Russia, from the Baltic to the Caspian Sea.

The early months of the year were spent at head-quarters, where much official work had to be arranged, but brief visits were paid to the offices in Edinburgh and Dublin, in order that my absence abroad should lead to as little inconvenience as possible. During these months I also completed what remained to be done to the two volumes on the *Ancient Volcanoes of Great Britain*, which were published on 9th April. They were prefaced with a dedication of them to Professor Fouqué and M. Michel-Lévy, as "distinguished representatives of that French school of Geology, which by the hands of Desmarest founded the study of Ancient Volcanoes, and has since done so much to promote its progress."

Here may be intercalated mention of an incident which took place later in this same year. On 27th November a letter, together with a box, came to me from the French Embassy in London. The letter expressed the desire of the French Minister of Public Works that I should accept the accompanying piece of china in recognition of my services to French geologists. The box contained a beautiful *coupe* in the finest Sèvres ware. I was much puzzled by this wholly unexpected gift, for though I lived on the most friendly terms with my scientific brethren in France, I was not aware of any service of mine that could call for such public acknowledgment. It was not until the following spring that the matter was explained by the following letter from Professor Fouqué:

23 Rue Humboldt,
26 Février 1898.

CHER ET HONORÉ CONFRÈRE,

Il y a quelques mois, M. Michel Lévy et moi avons demandé à notre Gouvernement de bien vouloir vous accorder la décoration de la Légion d'Honneur. M. Linder, President du Service de la Carte Géologique de la France, s'est associé à nous et a transmis notre demande officielle à notre Ministère des Affaires Étrangers.

M. Hanotaux a, suivant l'usage, fait part à l'Ambassadeur de la Grande Bretagne de l'honneur qu'il avait le dessin de vous conférer, et celui-ci a répondu que, tout en sachant gré à notre Ministre de son intention, il le priait de n'y pas donner suite. Il a donné pour motif que le Gouvernement Anglais n'aimait à voir ses nationaux accepter des décorations étrangères. La dessus, M. Hanotaux, désirant néanmois vous donner en notre nom une marque publique d'estime, a eu l'idée de vous envoyer un objet venant de la manufacture de Sèvres.

Je regrette vivement pour mon compte que votre Gouvernement soit aussi exclusif, car c'eut été pour moi un grand plaisir d'établir entre nous un lien de fraternité de plus.

Veuillez agréer, avec mes regrets, mes salutations les plus cordiales.

F. FOUQUÉ.

Years passed, and when both my kind friends Fouqué and Michel Lévy were no longer living, the Cross of the Legion of Honneur was handed to me in St. James's

Palace by the President of the French Republic, when he held a reception during his visit to London in 1913.

The invitation to inaugurate the course of lectures established in memory of Professor Williams at the Johns Hopkins University gave me peculiar pleasure. I had great respect for the general body of American geologists, as eminent pioneers alike in the stratigraphical, petrographical and palaeontological branches of the science, while for some of them, whose personal acquaintance I had made, I entertained the warmest regard. The work of such explorers as Newberry, Powell, Dutton, King, Gilbert, Hayden, Iddings, Leidy, Cope, Marsh and others filled one with admiration. Among these senior men I had formed some fast friendships during my former visit to the United States. But a new generation had arisen since then, and many active geologists had come into prominence in all departments of the science. The opportunity of meeting some of these younger men, and of renewing personal intercourse with the veterans who still survived, was welcomed by me as a great privilege. The letters which reached me from Baltimore and from other parts of the States showed that great pains were being taken to gather geologists, young and old, from all parts of the country to meet me at Baltimore.

Not a little consideration was needed to choose a subject that would be likely to interest the cultivators of each of the many departments of our science. Obviously it was undesirable to select a limited portion of this wide field, even were it the part on which the lecturer himself had chiefly worked. As already mentioned, I made choice of the history of geology, as revealed in the life-work of a few of the more eminent observers out of whose labours the framework of the science has been gradually built up. The short but excellent historical summary, given by Lyell in his *Principles of Geology*, was the best compendium of the subject in English. But a much fuller statement, with the inclusion of other co-workers in the same wide field, seemed to me desirable. I therefore set to work upon the older literature of the science which comparatively few geologists have taken the time to read. The title chosen for the lectures, "The Founders of Geology," indicated their partly biographical character.

Leaving Liverpool on 10th April in the *Campania*, one of the Cunard fleet, I reached New York in good time for a reception with which my friends of the New York Academy of Sciences honoured me, and thereafter hurried on by rail to Baltimore. During the stay there I was the guest of W. B. Clarke, Professor of Geology at Johns Hopkins University, and Director of the Geological Survey of Maryland. A goodly company of geologists, some sixty in number, had been gathered together, including many old friends and acquaintances. Especially welcome was the presence of the intrepid pioneer, J. W. Powell, and of C. D. Walcott, Director of the United States Geological Survey; W. H. Niles of Boston, Samuel Calvin of Iowa, J. F. Kempe of Columbia University, Edward Orton of Ohio, J. C. White of West Virginia, L. V. Pirsson, H. S. Williams of Yale, and F. C. Adams from Montreal. A few lady students diversified the audience. At the end of each lecture the listeners came up in swift succession to be introduced to the lecturer. It was gratifying to meet so large a gathering of professional geologists, as well as of students from so many widely scattered colleges and seminaries, and to reflect that some of the young men who had come to hear me would doubtless be leaders in the science in after years.

The social functions prepared in my honour with characteristic American hospitality, filled up every day, terminating with a dinner and reception in the evening. At one of these receptions I met Mr. Bonaparte, grandson of Jerome, Napoleon's youngest brother, who married Miss Patterson, a rich heiress of Baltimore. He retained the unmistakeable family features, and much of the courteous French manner. He was, however, a thorough American, wealthy, and one of the most prominent and useful men in the city.

There is much in Baltimore to remind one that it is a

city of the South. The vast number of negroes at once arrests notice; "darkies" of every shade of nigritude from boot-blacking to the weakest café-au-lait. They drive carriages, wait at table, act as porters, messengers, and in many other capacities. The house of my host was entirely served by women of colour, and nothing could be more praiseworthy than the quiet orderly way in which they did their work. The dark children seemed to me to be specially interesting; they look so merry and mischievous, with their bright twinkling eyes and broad grin. I would sometimes stop to watch a group of them at play—"innocent blacknesses," as Charles Lamb called the boy chimney-sweepers of his day. Another indication of the southern latitude of Baltimore appears in the wealth of fruit and vegetables that comes to table. At the time of my visit strawberries were in full plenitude, and huge dishes of them appeared at breakfast, together with melons and other luscious fruits. With this attractive course the breakfast began, and then marched onward through porridge to piles of hot rolls, steak, chicken, eggs, etc. As breakfast has always been my chief meal of the day, I fully appreciated this voluptuous type of the repast.

My course of lectures ended on Tuesday, 27th April, and it had been arranged that the rest of the week should be devoted to geological excursions. After a day spent in cruising in Chesapeake Bay, and landing at one or two places to look more narrowly at the "bluffs," we embarked in a short train put at our service by the railway company, to make a tour of three days through the western part of Maryland and the eastern borders of West Virginia. In this trip we descended into a coal mine, visited places of geological interest in the valleys of the Potomac and Shenandoah, and concluded with a visit to the battlefield of Gettysburg.

On the last day spent in Baltimore I was invited to take part in a conference in memory of James Joseph Sylvester, then recently deceased, who had been Professor of Mathematics at the Johns Hopkins University from 1877 to 1883. I had a profound respect for his scientific prowess, but unfortunately my intercourse with him in London and Oxford was too slight to warrant me in making any contribution to the reminiscences which were called forth on the occasion. Moreover, our acquaintance was largely connected with some of the foibles of this great and really lovable man. He was in the belief that the mantle of Milton, as a sonneteer, had descended upon him, and he used to read to me some of the effusions on which he based this belief, and to consult me as to alterations which he proposed to make on some of them. He sincerely thought that, since the time of the author of *Paradise Lost*, no sonnets written in the English language had been so fine as his own. It almost seemed as if he were disposed to pride himself more upon his sonnets than upon his important contributions to mathematics.

Leaving Baltimore on the morning of 3rd May, with the heartiest good wishes of my friends there, I went on to Washington to fulfil several engagements. Of these one of the most interesting was to pay a visit to the office of the Geological Survey of the United States—a much more extensive establishment than ours in Jermyn Street. I found that besides the geological staff, it included a large corps of topographers, engravers, printers, as well as the officials who superintend national irrigation, water-supply, and forest-lands. With the Director of the whole organisation, Mr. Walcott, I had long been in correspondence, and I had formed a high opinion of his ability, both as an original man of science, and as a director of the work of others. What perhaps interested me most, as evidence of his personal character, was a small room, full of boxes, cabinets, books, and papers, with hardly space left in which to turn round. To this little sanctum he retired whenever a few moments could be snatched from the day's work, in order to resume his labours among the Cambrian trilobites—a department of palaeontology in which he has become the highest authority.

While in Washington, I was the guest of Mr. Arnold Hague, one of the senior members of the Geological

Survey. One afternoon at the hour for tea, a visitor called, whose entrance into the drawing-room was like a blast of the north-west wind. He appeared to be a familiar friend of the house, for he kept walking to and fro, loudly and graphically describing some recent experiences of his in New York. This proved to be Theodore Roosevelt, President of the New York Police Board, and afterwards for six years President of the United States. His pictures of police work were drawn with a vigour and volubility that quite carried one away.

Mr. Walcott took me to the White House to introduce me to the President of the Republic at one of his usual receptions. We were ushered into a small, rather stuffy room, with two windows (both closed) looking from the second story into the gardens. About a dozen callers were there, waiting their turn for an interview with Mr. Mackinlay, who stood at one end and received them as they were successively presented. He shook hands with each, heard his story, and when necessary moved somewhat nearer to the windows, so as to listen more in private. He was evidently of a kindly nature, but one could hardly say that decision of character appeared strongly marked in his physiognomy. With conspicuous courtesy of manner, he combined a quiet businesslike way of getting through the work. In the course of my brief interview with him he remarked that I would notice the difference of style between his receptions and the more formal state of my own country. I alluded to the tax on his strength which this part of his official duties must involve. He admitted that it was so, for the reception lasted three hours. I tried to picture to myself the mental strain of being at the call of everybody who has, or imagines that he has, a grievance, and of listening attentively and giving in each case the proper answer, without unwarily committing himself or raising undue hopes. The caller who preceded Mr. Walcott and me was a negro woman who wanted redress from some injury or other. Altogether the scene had a curiously primitive or patriarchial aspect. Had the President been seated, wearing a silk robe and a turban on his head, the place might have passed for the court of some caliph of medieval times. We remained for only a few minutes of interview, and then retired to make way for further callers.

The few days spent in Washington were each terminated by a dinner, where I had the pleasure of meeting some of the leaders in science and literature, and some of the men of mark in the political world. But I was not allowed to escape without giving a public address, and holding a reception and shaking hands with several hundred listeners. From Washington I paid a brief visit to the Bryn Mawr Ladies' College on the invitation of the President, Miss Thomas. About half an hour's railway journey from Philadelphia, this admirable institution stands in a picturesque, green, rolling, well-wooded country, like much of our Surrey. It was attended by some three hundred students, who were housed and taught in a group of buildings designed after colleges in Oxford and Cambridge. The girls were in the midst of a game of ball, which they seemed thoroughly to enjoy, for they made quite as much noise as a corresponding number of Harrow boys could have done. In the evening after dinner I was asked to address an assemblage of the students; so for half-an-hour I transported them to my beloved Western Isles, their scenery, customs, superstitions, traditions, and geology, dwelling, towards the close, on the pleasure and interest which some knowledge of the geology of a country gives to the intelligent visitor. With the teacher of geology at the College, Miss Florence Bascom, who was one of my auditors at Baltimore, I had next morning a long talk over the practical difficulties she found in enlisting the interest of the average student in her subject.

A halt had to be made at Philadelphia, where the American Philosophical Society wished me to discourse to them. As one of its honorary members I was glad of the brief opportunity of meeting this Society, probably the oldest scientific institution of the kind in the United States, having been founded by Benjamin Franklin in the

Colonial days. It still meets in its old rooms, where the Declaration of Independence was signed. A kind of antique aroma hangs about the place, which I have found nowhere else in America, save in the older parts of Boston and its suburbs. It was a matter of some interest to compare the type of the audiences which one addressed in the different centres of population. It seemed to me that the type of Philadelphia could be differentiated from those of Baltimore, of Washington and of New York. There was, of course, the usual commingling of professions—judges, professors, merchants, physicians, lawyers, bankers, and others. Many, perhaps even most of those who formed the gathering in Philadelphia may have been members of the learned Society in whose apartments we met. There may, therefore, have been some pervading influence from that distinguished body, which gave a special gravity of demeanour to the company. In talking with the ladies, of whom a large number were present, I was surprised to find how many of them were familiar with geological literature. One dark-eyed maiden, who from her attire might have been supposed to be chiefly concerned with the dressmaker and the ball-room, thanked me heartily for my *Text-book of Geology*, which, she said, had been of great service to her in her studies of the science. But the general impression left on my mind was that more interest in literature and science seemed to be taken by the society in which I had been moving in the United States than would be found in corresponding circles at home. I was sometimes astonished to find both men and women referring to published papers of mine which I should hardly have supposed them likely to have even heard of.

In New York I was the guest of Mr. Dodge, a wealthy merchant, full of intelligence, and well known for his wide philanthropy. He assembled at dinner examples of what was best in the cultivated society of the city. They included one or two politicians of note—Mr. Carl Schulz, formerly Minister of the Interior, and Mr. Whitelaw Reid, U.S. Ambassador to France. With the latter I had long and pleasant converse. He told me much that was interesting about his experiences in Paris, and the people he had met there. He said that he had lately been offered the Embassy to Great Britain, but had reluctantly to decline it on account of his health. He seemed to be suffering from some affection of the throat or lungs which necessitated his wintering in Egypt, and latterly in Arizona. A few years after this time he did come to London as American Ambassador, and gained the esteem and affection of all who met him.

Another member of the party was Judge Daly, a delightful octogenarian, who at the time was President of the Geographical Society. I was greatly impressed by one of the younger guests, Professor Henry Fairfield Osborne, who seemed to be one of the ablest men I had yet met in America. As a skilled palaeontologist he was devoting himself to the study of the huge fossil vertebrate animals of the Far West. The impression formed at the dinner table was amply strengthened next day, when, at his request, I met him in the great Museum in Central Park, and was conducted through his astonishing collection of the remains of extinct vertebrates, and his ingenious restorations of them. Since those days he has steadily risen in eminence, till he is now in America at the head of the branch of science to which he has devoted his life.

Leaving New York I made for New Haven, in order to spend a day or two with Professor O. C. Marsh at his "wigwam," as he named his curiously shaped residence there. He narrated to me many of his experiences in the hunt for fossil skeletons in the Western States and Territories, and I urged him to write them down. In subsequent years we met from time to time in Europe, and I repeated my wish that he would commit to writing the story of his western adventures. But so far as I know, the record of them died with him. On the 14th of May I left New York in the *Etruria*, and by the 21st I was once more back in London.

This journey to America having consumed the whole of the annual vacation allowed by the rules of the

Geological Survey, the rest of the year would normally have been entirely spent by me in official duties. But there was a further call for my presence abroad at the end of the summer. The Seventh International Geological Congress was to meet at St. Petersburg, and there were special reasons, political as well as geological, why the invitation of the Russian Government should be fully accepted. The Royal Society had appointed me, in association with Professor Hughes of Cambridge, its delegates to the meeting. My official chiefs, recognising the importance of the occasion, most liberally granted me special leave of absence to attend.

It appeared that the Emperor of Russia, Alexander III., had taken keen interest in the sending of an invitation to the Geological Congress to visit Russia, and had given instructions that, when the visit came, every effort should be made by the Russian authorities to ensure its success. He did not live to see his desire carried out, but his son, Nicholas II., who was now Emperor, took care that the paternal wishes should be obeyed. Consequently the Congress, during its sittings at St. Petersburg and throughout every one of its excursions to different parts of the vast empire, was munificently fathered by the Government. No other meeting of this International Association, either before or since, has been so lavishly supported by the State. Without halting on the journey to St. Petersburg, I arrived there on the morning of Saturday 28th August.

Before the Congress had entered on its business, representatives of eighteen countries which had sent delegates to the meeting were invited to Peterhof, the sea-side retreat of the Imperial family, where the Grand Duke Constantine, Honorary President of the Congress, received them and presented them to the Emperor and Empress. One could not but be struck with the remarkable family likeness between the Czar and our Prince of Wales, now King George. He spoke to me in good English, with a slightly foreign accent and occasional hesitation for a word. The Czarina, who stood at a short distance, captivated me, with her winsome looks, her sweet smile, and the little twinkle in the eyes that betokened a keen sense of humour. She spoke English perfectly, and asked a number of questions about the Congress and its excursions.

The business done at the daily meetings was of the usual type, and is fully narrated in the official *Comptes-rendus*. Such leisure time as it afforded was given to some of the many attractions of the Russian capital, more especially to the magnificent collection of pictures and antiquities in the Hermitage Palace. The evenings were filled with receptions and banquets, at some of which it was possible to have conversation with a few of the more prominent citizens. I was more especially interested in an interview with the Minister of Agriculture, who discoursed most intelligently and patriotically on the natural wealth of Russia, which lay undeveloped for want of capital. He pressed upon me the desirability of inducing rich Englishmen to invest some of their wealth in his country, which would be greatly benefited thereby, while at the same time a good and steady return would be forthcoming to the investors.

One of the pleasantest entertainments was a dinner to which a few of us were invited at Oranienbaum, opposite to Cronstadt, the country palace of the Dukes George and Michel of Mecklenburg. Duke George, the senior brother, had made a morganatic marriage with a lady of the court, the Countess Carlow. At dinner the hostess engaged me in much interesting conversation. She spoke English fluently, and I found her charming, not only in her frank and gracious manner, but in her widely cultivated and keen intelligence. She gave her views on education, the position and tendencies of women in Russia, the trend of modern literature, the French stage, and other topics. She had an English nurse for her three little girls, who spoke English with the nurse and Russian with their parents. After dinner, during coffee in the drawing-room, seven Russian musicians, in a sober dark uniform, discoursed national music on curious triangular instruments, resembling zithers or guitars, but with only three strings—an invention of Duke George. The players and their instruments were graduated in size. At one end of the row sat a big, burly, bearded Russ, with the largest instrument, while the other end terminated in a small boy playing on a little toy-like reproduction of the same type. Much of the music being of a dance kind, one of the ladies at last rose and began alone a Russian dance, a sort of running galop, with pauses and stamping of the feet in time. By and by Duke George joined her, and the dance was ended by their wheeling round with joined arms, and his finally depositing her upon an ottoman. The evening left a pleasant memory of a little glimpse of home-life in the higher Russian society.

The remarkable programme of excursions and entertainments organised by the Russian authorities attracted from the neighbouring countries, more especially from Germany, many young men who joined the Congress in order to take advantage of this favourable opportunity of seeing the interior of Russia. One unfortunate result of this liberality was soon made manifest. These youths continually pushed themselves forward, took possession of the best seats, and filled the carriages provided for the members of the Congress. When we steamed down the Neva and landed at the Peterhof quay, every carriage in waiting was promptly seized by these youths. But for the intervention of an elderly geologist who, recognising Dr. Torell of Stockholm and myself, still on foot, insisted on room being made for us in the last carriage, we should have been left behind. Some of the company would appear to have been thieves as well as roughs, for I was told that a cap which had belonged to Peter the Great was missing after the party had left the palace.

Of the various excursions provided by the authorities for the Congressists after the close of the meeting, I made choice of that which would best reveal the interior of Russia, and afford the most varied aspects of the physical features, the geology and the inhabitants of this vast country. It was arranged that those who selected this excursion should travel by Moscow to the Volga, descend that river to Tzaritzin, cross the Steppes by Rostov to Vladikavkaz, traverse the Caucasus to Tiflis and the Caspian Sea, returning by rail to Batum, crossing the Black Sea to the Crimea, and finally breaking up at Odessa. At every halting-place throughout this extended journey, more or less elaborate preparations had been made for our entertainment. At the towns we were feasted by the civic authorities. Even in the sparsely peopled regions through which we had to pass, infinite trouble had been taken to make things pleasant for us. A number of the English members of the Congress took part in this excursion. It was especially pleasant to have the company of Professor and Mrs. Hughes, John Murray of the *Challenger*, and a few more fellow-countrymen, and I greatly enjoyed the opportunity of coming into closer and more continuous contact with Ferdinand Zirkel; we shared the same cabin in the steamboat, the same carriage across the mountains, and the same compartment in the trains.

The sail of some eleven hundred miles down the Volga in a roomy steamboat, continued for six days and nights, was an impressive experience. The tameness of the scenery, with its persistent bluffs and cliffs on the west side and low shores on the opposite horizon, would be somewhat monotonous but for the vastness of the great body of water, the occasional picturesque towns on the banks, and the unceasing traffic of the varied craft that move up and down the stream. We halted at some of the towns, and here and there at localities where landings were made to examine geological sections. At some of these halts, fresh supplies of oil were taken on board for the engines. There is always more or less leakage in this replenishing of the fuel, and in consequence an iridescent film of oil may be noticed on the placid surface of the river. It is said that the sturgeon and sterlet now ascend the Volga for some hundred miles less than they used to do, and this change has been attributed to the effect of this extremely attenuated film in interfering with the proper oxygenation of the water. The river was low at the time of our voyage; but was still in some places three miles wide. Here and there it is so shallow that navigation requires a good deal of caution and the use of a long pole for sounding. We stuck on a sand-bank for some hours one night. The nights at this time were marvellously beautiful. It was nearly full moon, and such stars as were visible seemed to swim overhead. One could easily read large print by the moonlight.

Saratow, where we had a great reception from the civic authorities, stands at the level of the sea, so that every mile of the rest of the voyage brought us further below that level into the great hollow of the Caspian basin. At Tsaritzin the voyage came to an end. We then crossed the Steppes by railway to Rostov, passing for some three hundred miles through a bare country with few trees, not many visible towns or villages, only scattered hamlets and farms, and a scarcity of definite roads. At the stations where the train stopped there was generally a muster of peasants to look at the strangers, and be driven back by the officials. The obedience of the peasantry to their head-men was specially conspicuous at the halts made on the Volga, when we landed at places where there would sometimes seem to be no habitations in sight. Before we had been many minutes ashore a few women and children would be joined by others who came scrambling down the rocky bluffs, until quite a crowd assembled to watch us. If the crowd approached too near, their officials, with stout cudgels, and uttering loud imprecations, drove them back, and the peasants obeyed implicitly. Notice had been sent to all these halting-places that our steamboat would call, and that every facility was to be given to us.

From Vladikavkaz, which is the terminus of the railway, we crossed the chain of the Caucasus by the Dariel Pass in well-horsed carriages, in the first of which Zirkel and I were sent off. The long procession was headed by a mounted and fully-armed Cossack as guard and escort, who was replaced by another at every military post. We had thus a good opportunity of seeing the different types of men who form the troops that maintain order in this mountainous region. The drive across the mountain chain involved several nights in rough inns or caravan-series. At the first of these, when we awoke next morning, we saw the majestic snowy cone of Kasbek in front of us, gleaming in the morning sun under a deep blue sky. This graceful mountain, though not the highest peak in the chain, rises to 16,546 feet above the sea, and thus surpasses Mont Blanc in height. It was interesting to note the endless moraine-mounds here scattered over the bottom of the valley, showing how large the glaciers must once have been. I observed that one of the lava streams had overflowed a moraine, so that glaciers and volcanic activity must have been at work here contemporaneously in comparatively recent time.[1]

In ascending the course of the Terek River, one is struck by the scanty traces of human habitation. Here and there an old fort with a few houses around it tells of the strife of which the valley was once the scene; and an occasional lonely and melancholy graveyard, crowded with tombstones, is a still more impressive record of the struggle that went on before the Russians subdued the hardy mountaineers. As the journey continued on the southern side of the watershed, evidence of a still older warfare was to be seen in the round tower which rises in the midst of each hamlet or village. At Ginvani I examined one of these towers. It was strongly built, without windows, and with only one small entrance or doorway some ten or twelve feet above the ground, which was no doubt reached by a ladder that could be pulled up inside. These buildings resemble the Irish round towers, and were probably built for the same purpose of affording protection for human life or property, at the time of a raid by one or other of the many distinct Caucasian tribes. But as

[1] In the *Times* of 25th March, 1919, a telegram announced that Elbruz, the highest peak of the Caucasus, was on 16th March "discharging smoke and sending blocks of ice to the foot of the mountain in numerous torrents."

the necessity for shelter of this kind has ceased, these curious towers are now allowed to fall into decay. The long carriage drive across the Caucasus, which was enlivened by many amusing and interesting incidents, ended at Tiflis, where further municipal attention awaited us.

The railway eastward from Tiflis to Baku crosses a wide tract of desert, across which strings of camels may here and there be seen on the march, or at rest. Eventually, in the far distance, a cloud of black smoke comes into sight rising from the ground into the clear desert air, and within it a forest of tall tapering chimney-like stacks gradually appears. When the train stops the traveller finds himself in all the noise, bustle, dirt and ugliness of a manufacturing town. But what a strange commingling of old and new!—camels, mules, Moslems, Greek priests, Armenians, Georgians, Cossacks, Russian officials and European workmen, all elbowing each other in the streets. After seeing all that was essential, from the geological point of view, in the oil-workings, we enjoyed a short cruise in a small steamboat on the Caspian Sea. It was soon obvious that the inflammable gas which rises so abundantly from the ground in many places, likewise finds its way to the open air from under the floor of the Caspian. Every here and there the surface of the water is broken by the uprise of bubbles of gas. We threw some pieces of lighted tarry tow on places where the ebullition was briskest, when the gas at once burst into flame, which continued on the surface of the dancing waves, sometimes shooting up into the air, until blown out by the wind. I noticed the same oily iridescent film on the surface of the Caspian Sea as was observed on the Volga.

At Balakhany, a few miles east of Baku, the escape of gas is still more strikingly exhibited. While we were there a column of black oil, water and gas rushed, like a geysir, with great energy up the tapering wooden shaft which had been built over the orifice of discharge. The eruption, rising inside the woodwork into the open air above to a height of some seventy feet, descended again in a black rain, much of which ran down the outside of the shaft or was blown by the wind over the surrounding ground. These ejections succeeded each other at brief intervals.

A memorable feature of this trip from Baku was a visit to the ancient temple of the fire-worshippers at Sourakhany. Many generations ago a battlemented wall had been built round an irregular piece of ground which included several places whence gas issues continually. Along the inside of this wall a number of cells had been constructed with inscriptions on them. We were told that until only some eight or nine years before the time of our visit, priests from India lived here and kept the fire continually burning. Tall stone chimneys had been erected over the chief vents of gas. We threw a lighted match into the opening at the base of one of these chimneys, and the gas instantly became alight, the flame rushing upward with a roar like that of a blast furnace. Before returning to the main party, I strolled across a little space of the bare, dusty desert to a gently rising ground where a couple of swarthy half-naked workmen (the thermometer stood at 90° in the shade) were heaping up a large cairn of blocks of limestone, like other similar heaps in the immediate neighbourhood. They had selected the spots where the emanation of gas was briskest, and when they threw a lighted match into one of the cairns the whole pile was instantly enveloped in a blaze of roaring flame, which would soon turn the stones into marketable lime.

Returning to Tiflis, we went on by railway to Batum on the Black Sea, whence a commodious steamboat conveyed the party to the Crimea. The mountain scenery visible in the course of this traverse included the glorious heights of the Ararat range to the south, and the vast chain of the Caucasus to the east. We slowed down in one part of the voyage to let John Murray and the Russian geographers take soundings and temperatures, and make other observations on the interesting physics of this great inland sea. Landing at Kertch we were able to visit the famous mud volcanoes, and came in for a further display of municipal hospitality. The Governor, a naval officer, presided at the banquet. When Mr. Wardrop, the English Consul, introduced me to him, he at once claimed me as an old acquaintance, for he had been educated in my school books.

After a most interesting and varied week in the Crimea, in which geology was merged in promiscuous sight-seeing and continued official hospitality, we completed the last scene of the excursion by sailing from Sebastopol to Odessa. A memorable incident in this voyage was the oncoming of a storm. We watched a ragged fringe of cloud dipping down in long curtain-like folds that reached the surface of the sea, which was whisked up and mingled with the cloud. There was a hissing sound in the air as the hurricane advanced swiftly towards us with a distinct front, tearing the tops off the waves and beating the sea into foam. When it reached us rain fell in a torrent, and the tempest shrieked through the rigging of the vessel. Just before the tornado struck us two merlin hawks came abreast of the ship; one of them, so tired that it perched on the bowsprit, was easily caught. A woodcock was also driven past us, and one or two small birds. As already stated, the excursion was to come to an end at Odessa. Most of the members returned across Russia by the most direct route. With Professor and Mrs. Hughes and one or two more friends, I arranged to travel homeward by Constantinople and the Mediterranean.

The skill and success wherewith every detail of this long and somewhat complicated excursion was planned and executed could not but call forth unstinted praise from the excursionists. Before we separated on our several homeward journeys, we sent from Odessa a telegram of sincere thanks to the Emperor, and another to the Grand Duke Constantine, our Honorary President. A gracious reply came at once from the Grand Duke, but none had come from His Majesty when we left Odessa. At Constantinople, however, one of the first boats to come off to our steamboat carried a Russian official waving a paper in his hand, and shouting my name. The document he brought was a courteous and kindly telegram from the

Czar, at Hesse Darmstadt, expressing his pleasure in the success of the excursion, and wishing us well.

The steamboat, *Emperor Nicholas II.*, which conveyed our diminished party from Odessa to the Turkish capital, and thence to Athens, was a fine large steamship built on the Clyde, but its Russian owners were allowing its paint to wear off, its ironwork to rust, and its interior to lack the tidiness which is found in our own mercantile marine. At Constantinople, where we arrived after midday, most of that afternoon and of the next morning was given to seeing as much, as in the time was possible, of the Turkish capital. We had arrived under a grey sky, but in leaving were favoured with an afternoon and evening of uncommon beauty. As we looked back upon the city, its famous charms came out in full glory. Above the old line of sea-walls it rose in one varied mass of buildings and foliage, its domes and minarets towering along the skyline and quivering in the mellow rays of the western sun. Conspicuous in this light, the chain of battlemented walls and towers of Roman days swept across the ridge till lost in the distance, looking majestic, venerable, and utterly apart from all the din and tumult of Turk, Greek, and Armenian behind them. Beyond these massive walls the sun-lit city shone out against the background of smoke and mist from the great harbour. On the opposite side, Scutari caught the evening light on its buildings and vast hospital. The scene fully justified all that has been said and written in its praise.

It was dark when we entered the Dardanelles. Early next morning the " isles of Greece " were in sight. To one who vividly recalled the spell which in youth this region had for him, and who now saw it for the first time, the voyage to Athens was like a dream. The weather could not have been more propitious. The clear bright sky was flecked with a few white clouds. The islands, bathed in a rosy light, stood one after another out of a sea of deepest azure that sparkled in little sunlit wavelets under a pleasant cool breeze as we sped along—Imbros, Lemnos, Tenedos, Lesbos, Chios, while landward in the distance we could see "many-fountained Ida." Now and then a group of dolphins came sporting abreast of us, leaping out of the water or rushing along underneath it, so close to the vessel as to be clearly visible, then plunging into the depths, and recalling the old legend of *inter Delphinas Arion.*

After the halt of a night at Smyrna, it was under the same conditions of weather that we approached the shores of Attica. Passing "Sunium's marbled steep," with its columns of Athena's temple, we came in sight of Aegina; beyond it Salamis, and by and by, behind a low island, came the Acropolis of Athens, with Hymettus and Pentelicon in the background, while far away to the north-west rose the snow-capped summits of Helicon and Parnassus. The classic memories of one's youth came back with a thrill of pleasure. I felt with Keats that

> The very music of the names had gone
> Into my being.

Every hour of the short stay at Athens was spent in visiting the relics still left of the ancient city. We made an excursion to Marathon, and also walked out to Eleusis along the Sacred Way, reverently entered the temple of the mysteries, sat among the marble seats of the theatre of Dionysus, still bearing the incised names of their ancient occupants, and remembered that Aeschylus was born here. While we were in Athens, the modern theatre there was opened for the first time after the end of the war with the Turks, and the drama chosen for the occasion was Shakespeare's *Macbeth.* It was the worst performance of the play I had ever seen; but there was much interest in hearing it recited in modern Greek.

From Athens the return journey was made as rapidly as possible, by way of Patras, Corfu and Brindisi to Naples. Vesuvius being actually in eruption, we halted to have an opportunity of climbing the mountain and seeing its condition. We found the new outflow of ropy lava creeping slowly downward under its pile of rugged slag, and Mrs. Hughes took excellent photographs of its most striking features. When the wind blew the steam cloud aside for a few moments, we could see a hundred feet or more down into the steaming abyss of the crater. Every few minutes there came a rumble, then a roar, when clouds of steam surged up out of the depths. Once a crowd of red-hot and also black stones went whizzing high overhead, some of which fell near us. One of the black fragments that looked quite cool blistered my fingers when I lifted it from the ground.

We hurried on to Rome, and taking the St. Gothard route reached London on the evening of 30th October.

# CHAPTER X

## LAST YEARS IN THE GEOLOGICAL SURVEY (1898–1903)

DURING my long service in the Geological Survey, it was the continued opinion of the Treasury, and of the successive Government departments under which we were placed, that this branch of the public service is of a temporary nature, and should cease to exist when the whole of the United Kingdom has once been all geologically mapped. We were therefore continually urged to push forward the completion of the mapping, an order which we did our best to carry out, so far as this could be done without sacrifice of the accuracy which might be attainable. But we knew full well that where, as often occurs, the ground is obscure, the boundary lines which we have tentatively traced on the maps may be shown, by subsequent boring or other laying open of the underground, to require alteration. Sir Roderick Murchison used to maintain that there is as much need for a permanent Geological Survey in the country as for an Astronomical Observatory, in order that a watch may be kept on fresh openings of the soil which would furnish material for making the geological maps more accurate, and keeping them up to date.

The Staff of the service was well aware that the early sheets of the map of England and Wales, on the one inch scale, surveyed by De La Beche and his colleagues, excellent for the time when they were published, might with advantage be superseded by re-mapping the ground on a larger scale. We knew that more especially would the maps of the mineral fields need to be revised and kept up to date. It was obvious, however, that no such revision would be authorised by the Government of the day, unless under strong pressure from the public, especially from the communities in the mining districts. As already mentioned, such pressure had eventually come from Cornwall and from South Wales, with the result that the ruling powers consented that these districts should be re-surveyed on maps of the scale of six inches to a mile. But a condition was imposed that the revision should not entail any increase in the annual grant voted by Parliament to the Survey. Obviously this condition would require the transference of some members of the staff from the areas on which they were at work, involving consequent delay in the completion of these areas. After some shifting of this nature was made, the revision of the old mining districts of Cornwall and of the great coal-field of South Wales had been undertaken. By degrees, during my tenure of office, the re-mapping was extended to the coal districts of North Staffordshire, Leicestershire, and Derbyshire.

There can be no doubt that from the industrial point of view Murchison's verdict was the true one. When the whole of the United Kingdom has been geologically surveyed, the maps will need to be kept abreast of the progress of our knowledge of the geological structure of the country. For this purpose no large staff may be required. But, as I have insisted in a former chapter, a body of skilled observers should be maintained, whose duty would be to watch the fresh revelations of what lies below the surface in all parts of the country, such as the information yielded by the cuttings and tunnels of new railways, the development of the mineral fields, the results of boring for water or for minerals, and the other operations which furnish fresh knowledge of the rocks of the terrestrial crust. There should be a central office to which notice of these new explorations would be required to be sent, and at which all the new information obtained would be recorded and made available for correcting and improving the geological maps.

The first shadows of my retirement from the Geological Survey were now beginning to project themselves around me. By the rules of the Civil Service, a public servant can retire and claim his pension on reaching the age of sixty, but should he still be able to perform the duties of his office he may remain for five years more. After that time he cannot be retained, unless exceptional circumstances make it desirable, in the public interest, that his term should be extended. As there were no such circumstances in my case, I would retire in less than two years, on reaching the age-limit of sixty-five. The last two years of my tenure of office were therefore largely devoted to completing as far as possible all official undertakings which depended mainly on myself. As for the working of the Survey and Museum, the whole service was now in such a state of efficiency that I might retire at any time, confident that my place could at once be adequately filled by one or other of my capable colleagues on the staff.

Among the pieces of work which it was desirable to complete before my retirement, was the editing of a detailed memoir on the Silurian rocks of Scotland, which had been in progress for some years. To my great satisfaction it was finished and published during the summer of 1899—a massive octavo volume of 750 pages, with numerous figures in the text, twenty-seven plates, and an index map. It was the combined work of three of the ablest men on the staff—B. N. Peach, John Horne and J. J. H. Teall, and is one of the finest monographs which the Survey ever produced. I count it no small honour to have helped in the planning and completion of this important treatise. I may here quote a few sentences from a letter written to me at the time by Professor Barrois of Lille, who, of all the Continental geologists, was probably the one who could best appreciate the value of the work. "Je viens de lire les Chapitres III et IV, et je suis émerveillé de cette œuvre. Je vous serais reconnaissant de dire à MM. Peach et Horne en quelle grande estime je tiens leur œuvre. C'est de la stratigraphie remarquablement précise et détaillé, dont les conclusions sont très logiquement déduites. Si Dieu me prête vie je le prendrai pour modèle dans la description de la Bretagne que je voudrais écrire."

There were two other Survey memoirs which I had myself undertaken to write. One of these dealt with the geology of central and western Fife and Kinross; the other would describe the geology of eastern Fife. These memoirs ought to have appeared many years before, but the member of the staff who surveyed most of the ground could never be induced to write out his descriptions, and was now no longer in the service. Having mapped some parts of the districts in question, and having examined from time to time every portion of them, I at last resolved to write the memoirs myself. The first was completed and published in June 1900 as an octavo volume of 284 pages, with illustrations. Besides giving a description of the coalfields, it contains the first detailed account of the volcanic phenomena for which that district of Fife and Kinross is so remarkable. The memoir on eastern Fife was in progress, but it could not be completed before the date of my retirement.

By far the most onerous official literary task in prospect, however, was the production of the large memoir to be devoted to the description of the complicated type of geological structure which the Survey had worked out in great detail in Sutherland and Ross. As a preliminary to this work, much consultation was needful with the members of the staff by whom the mapping had been done, before the plan of the volume and the distribution of its sections among the six contributing authors could be arranged. The preparation of the memoir would probably take several years. Having been personally in charge of the field-work during the whole time of its progress, I was asked to undertake the superintendence and editing of the volume, though it could not be completed until some time after my retirement. It did not appear ultimately until the middle of 1907. Thus my connection with the Survey was not finally severed until six years after I had ceased to be one of the staff.

Notwithstanding the pressure of official work, which it was my wish to complete during these last two years on the Geological Survey, I found time for a few non-official undertakings. Most important of these was a piece of literary work, entirely after my own heart, which was entrusted to me in the summer of 1897. That classic of geological literature, the immortal *Theory of the Earth*, by James Hutton, was published in 1795 in two volumes, but it was known that a third volume had never seen the light. Search had been made for the missing manuscript without success. Eventually I learnt that a portion of it existed in the possession of the Geological Society of London. Fresh exertions were then made to trace other parts, but again to no purpose. The portion preserved by the Society consisted of six chapters, one of which contained the author's account of his researches in the island of Arran, and was of special interest, for it would show, probably better than any part of the two published volumes, Hutton's qualities as an original observer in the field. These six precious chapters, bound in one volume, had come into the possession of Leonard Horner, by whom they were given to the Geological Society in 1856. As it was highly desirable that they should at last be published, I appealed to the President and Council of the Society to take the matter into consideration; offering at the same time to edit the work, should its publication be undertaken. This proposal having been accepted, I was authorised to have the manuscript carefully copied for the use of editor and printer. A certain amount of research was involved among the publications of the close of the eighteenth and beginning of the nineteenth century, and a number of explanatory notes were required. The most serious lacuna in the original manuscript was the absence of all the plates, drawings or diagrams with which the author intended to illustrate his text. In order to supply as many of them as related to the account of observations made in Arran, which formed the most valuable chapter, I returned in the summer of 1898 to the island, which was then being surveyed by my colleague, the late William Gunn. With his company and help, I succeeded, by means of photographic camera and sketch-book, in obtaining views of the structures exposed at the localities described by Hutton. The preparation of the volume, which necessarily involved a good deal of time and work, was completed in the course of the year. Type and paper were chosen as similar as was obtainable to those of the published volumes, and the book was issued in February 1899 as a volume of 278 pages. In an appendix I added an index, and also an index to the two published volumes, which had been made long before for my own use. Of all the publications with which I have been connected, none ever gave me more unalloyed gratification than the preparation of this volume, more than a hundred years after the death of its illustrious author. To be permitted to render any service, how slight soever, to the memory of this notable member of the group of founders of geology was an honour and privilege of which no geologist could fail to be proud.

On 1st June, 1898, I gave at Oxford the Romanes Lecture, taking as the subject "Types of Scenery and their influence on Literature"[1]—a theme which, among the many varied landscapes through which a geologist passes in the pursuit of his science, had long occupied my thoughts. Of the letters that came to me in reference to this lecture, I specially valued the criticism of W. J. Courthope, co-editor of Pope and Professor of Poetry in Oxford University.

"9th June, 1898. I have been reading your Lecture with the greatest pleasure. Though I am not myself inclined to assign quite the same importance that you do to Landscape as an element in poetry—that is to say, though I do not think that the very greatest poets ever leave the sphere of action for that of predominant

[1] Reprinted in *Landscape in History*, p. 76.

reflection and description—yet I fully feel the charm of the poetic landscape-painting on which you dwell with so much enthusiasm. It seems to me, if I may say so, that you have most admirably discriminated between the effects produced on the imagination by different types of scenery, and I was particularly struck with the justice of your remarks on Macpherson. I think you are quite right in your judgment as to his manner of workmanship. He was a man with a real feeling for Nature, and felt the meaning of the old fragments of Celtic poetry in their relation to the landscapes about him; on the two things together he constructed 'Ossian.' I like immensely all that you say about Scott, whom I deify. Wordsworth I am not enthusiastic about, except in places; he is too analytic for me; but I fully admit the justice of his claim to the high place you assign him among the philosophic descriptive poets. I may add that your own scientific descriptions of scenery in the Lecture give me almost as much pleasure as your poetical extracts. . . .

W. J. Courthope.

I wonder whether any poet will arise to celebrate my own dear South Downs!"

Another address was given in the autumn of 1898 at the opening of the session of Mason University College, Birmingham. Choosing the subject of "Science in Education," I took the opportunity to urge upon the science students the unwisdom of undervaluing the older or literary side of culture, and to point out the advantage they would derive from cultivating it as a source of mental refreshment, such as no branch of science can give. Sir John Lubbock, writing to me in reference to this address, remarked that "the so-called Modern Sides at public schools seem to me a mistake. It is too early to differentiate; and the two sides tend to go into extremes." My own experience as one of the Governing Body of Harrow School, led me to a similar conclusion. It was therefore with much satisfaction that twenty years after this time (in 1917) I saw the abolition of the Modern Side at that great seminary. There will be at least as much science taught there as before, probably a good deal more; but there will be no splitting of the school into two sections—an arrangement which has been apt to lead the older or classical side to disparage its younger associate. Every classical boy at Harrow is now taught the rudiments of science, in order that he may have some conception of the world in which he lives, and he is encouraged and given the opportunity to pursue the subject into more detail, if he shows special aptitude and zeal for it. One of the changes brought about by the war is the extraordinary impetus given to science-teaching. At Harrow a complete reversal of the relative position of that subject and of classics has taken place, so that if the two sides still existed, the Modern Side would far outnumber the Classical. The pendulum, to my thinking, has swung too much to the opposite side, and it is to be hoped that a more reasonable balance will eventually be secured.

Time was found in these same two years for the revision of half a dozen of my own publications of which new editions were required. These revisions included that of the *Scenery of Scotland*, which was considerably recast, with the addition of new maps and illustrations, and a geological itinerary of the country; that of the fourth edition of the *Text-book of Geology*, which had been in progress during the preceding seven years; that of the *Class-book of Physical Geography*, of which a completely revised edition with new plates was now published; and that of the *Field-Geology*. Besides these revisions the Baltimore Lectures of 1897 were prepared for the press, and were published as a small volume with the title of *Founders of Geology*. In writing these lectures I drew attention more fully than any of my countrymen had done, to the merit of some of the early French observers, who had hardly received, even from their own countrymen, the recognition which was their due. The first edition of the Baltimore Lectures was soon out of print. The amount of historical material collected in their preparation, being much more than could be put into the compass of the lectures as delivered, was made use of in the preparation of a second and much enlarged edition published in 1905.

In the spring of 1898 my last interview with my old friend Henry Clifton Sorby took place. Although we frequently exchanged letters, we now seldom met, as he did not often leave his residence in Sheffield. He came to London in April, and from his hotel sent me a note in which he reported, " I have brought with me to show you some of the results I have obtained by the long action of weak acids on sundry rocks, which are very curious, and closely like the natural products due to hundreds of years of weathering. I shall be here for a week." The subject of research to which his active and versatile mind had now turned was one that had long interested me, and has been referred to in an earlier chapter. It was therefore with peculiar pleasure that I saw how closely he had been able to imitate artificially the effects of the age-long action of the acids in rain and in the soil upon limestones and other rocks. The veteran father of modern petrography, now in his seventy-second year, was still full of enthusiasm, and maintained his markedly original manner of investigation. I little imagined at that time that in the course of a few months I would be called upon to record his death, and to write an obituary notice of his lifelong work for the *Proceedings of the Royal Society*.[1]

The years with which the present chapter deals were marked by a few events of public interest In the summer of 1899 the centenary of the Royal Institution was celebrated by a banquet presided over by the Duke of Northumberland, President of the Institution, and attended by many of the leaders in pure and applied science in this country, together with a number of scientific men from the Continent and the United States. The Jubilee of Sir George Gabriel Stokes as Professor in Cambridge University was celebrated in the same summer, when a remarkable concourse of eminent men

[1] *Proc. Roy. Soc.*, B, vol. 80, 1908.

brought to Cambridge congratulations and good wishes from colleges, universities, scientific societies and academies all over the Old World and the New. I was the guest of J. W. Clarke, Registrary of Cambridge University, who in a letter sent to me shortly after the festival wrote :—" From what I hear, most of our visitors took pleasure in the Jubilee of Sir G. Stokes, and I hope that the celebration may serve my poor University before the world. We want a little of that puffing which Oxford understands so well, and can carry out so effectively."

This same year the British Association met at Dover, under the presidentship of Professor Michael Foster. It was arranged that a contingent from the corresponding Association in France, which was holding its annual meeting simultaneously in Boulogne-sur-mer, should on one day of the week cross the Channel and fraternise with us, while on another day we should return the visit. Our President, together with the Presidents of Sections, went to the Dover quay to greet the Frenchmen on their arrival. It was amusing to see how nimbly Professor Foster, in French fashion, kissed the French President on both cheeks, before the latter could be sure which of us was the British President. The Council had nominated me to preside over Section C, and my presidential address, instead of being given at the beginning of the sittings, was deferred until the following Saturday, when the French Association was to visit us. I took for its main theme " Geological Time." Entering upon a review of the arguments of Lord Kelvin and others, I restated the geological evidence in favour of a far higher antiquity for our globe than that eminent physicist was disposed to allow. The address also alluded to the advantage which would accrue to geology if numerical data were obtained by instrumental observation and experiment, in the discussion of many geological problems, and the subject was illustrated by reference to several branches of enquiry to which such actual measurements might be applied. The audience included many Frenchmen, among them the President of the Geological Society of France, who seconded the vote of thanks to the President of Section C, which Lord Lister proposed.

In 1899, by dividing my official holiday into three parts, I was enabled to pay two brief visits to the Continent and to enjoy a short yachting cruise round Ireland. My wife and daughters, having spent the winter of 1898-9 on the Italian Riviera, had gone to Florence, where I joined them for a few days in April. In the southward journey, I was delighted to find at Modane Station Lord and Lady Kelvin fellow-passengers in the same train. We travelled together to Genoa and disembarked to spend the night there. Next morning we took a drive along the heights, in order to have a wide view over city, harbour and sea. Lord Kelvin had lately given his wife a photographic camera, the most perfect then obtainable, and she tested its capabilities by devoting a number of plates to different parts of the landscape. We had also time to look into one or two of the churches and the most famous of the picture galleries. We parted at Pisa, they going on to Rome and I to Florence.

After a short but happy time with my family in Florence, I went on to Rome. In the summer of the previous year I had been elected a foreign member of the Reale Accademia dei Lincei, and I was glad to take an early opportunity of presenting myself. From school years onward a kind of romantic love of Italy had possessed me. Roman legend and history had first of all begotten that affection, which grew deeper as the beauties of Latin literature came to be more and more appreciated. The feeling was still further fostered by the two visits to Italy in 1870 and 1873, which brought me under the spell of Rome and the Campania. To be enrolled in the membership of the venerable Academy of the Lincei, which for some three centuries had held the torch of learning alight in Rome, was a privilege on which a high value could not but be set.

An excursion was planned to Tivoli in order that Lord Kelvin might see the electric installation for lighting Rome, and I was invited to be one of the party. It was amusing to mark the care which the Italian engineers took that their illustrious guest should not inadvertently suffer injury. They told off a man to keep near him always, and watch that he did not put a finger on any dangerous part of the mechanism. He was of course greatly interested in all the machinery, and as he sometimes pressed forward and pointed with his outstretched hand to some special mechanical contrivance, his guard was ready on the instant to snatch back the hand, if it was thought to be approaching too near danger.

Later in the same year, as delegate from the Royal Society, I attended the International Geological Congress which met for a week in Paris, under the presidency of Professor Gaudry, whose gracious manner and kindly tact made everything go smoothly at the business meetings, which were more than usually interesting. To my admiration of the brilliance of Professor Gaudry's palaeontological work, there was now added an increased appreciation of his great personal charm. His roomy and picturesque house in the Rue des Saints Pères, with Madame Gaudry as its attractive hostess, was the chief centre of interest during the meeting.

Having to make some communication to the Congress, I took as its subject the desirability of promoting the development of experimental geology. This theme had been briefly alluded to in the address to the Geological Section of the British Association at Dover. What was said in that address had attracted the attention of some French geologists, and had given rise to a correspondence with them. In a paper entitled, "De la co-opération Internationale dans les Investigations Géologiques," various directions were now suggested by me in which such co-operation might be advantageously pursued. The Congress appointed a Committee to consider the whole question and to report to the next Congress, which was to be held in Vienna in 1903.

Among the various geological excursions planned to take place at the close of the sittings, I chose that which was to visit the volcanic region of Central France, under the leadership of M. Michel-Lévy. Though already fairly familiar with that region, I looked forward with no little interest to the discussion of its volcanic problems with the French geologists who had studied and mapped the ground. Michel-Lévy himself had devoted his great petrological acumen to the study of the rocks, which were at this very time being mapped in detail by Professor Glangeaud of the University of Clermont Ferrand for the *Carte detaillée* under Michel-Lévy's direction. Both of these authorities were with us. There were likewise present Professor Lacroix, of the Jardin des Plantes in Paris, with his wife, daughter of Professor Fouqué, and her younger unmarried sister; also Arnold Hague, my kind host at Washington, and Mrs. Hague. We climbed to the summit of Mont Dore, saw the famous Puys well, and had some spirited discussions of volcanic problems, which sometimes became nearly as warm as the weather. In the course of these argumentations I realised more vividly than ever what a different view of volcanic phenomena may be taken by the chemist or petrographer, working at the subject only in his laboratory, from the geologist who has studied these phenomena in the field, not only at active volcanoes, but at those of former geological periods where their internal structure has been laid open by denudation. But these somewhat heated debates formed only more serious interludes in a continued course of good humour and merriment, as we drove or walked from point to point, lunching at little country inns or halting for *al fresco* picnics at suitable spots. I remember that one of these open-air meals was taken at the foot of the hill of Gergovia, whence Julius Caesar was driven back by the Gauls. As we sat by a cool spring and were thinking of the fighting that went on there in Roman times, a toast to the immortal memory of Vercingetorix, the Gaulish leader who defied Caesar, was received with mingled amusement and sympathy by the French part of the company.

In the early summer of 1899 a letter reached me from Henry Evans, in which he wrote: "Alfred the Great says he will turn up at Jura on Tuesday, 20th June, and I hope you will do the same. You know we think of steaming round Ireland. He has been sorely plagued with phlebitis, and I have urged that the less able he is to walk about, the more is a ship to be desired. I don't know in the world how we shall be able to feed ourselves, but the Basking Shark is sometimes to be had in the West, and an Octopus, of which the eye was 18 inches across, has been taken there. We hope the *Aster* will be a little less kicklesome ; we are to weight her down deeper ; she is rather skittish when very light in coal."

To this cruise round the Emerald Isle I looked forward with no little interest. In the course of some twenty years, during which it had been my good fortune to make acquaintance with most parts of the island, I had become fairly familiar with both the geology and scenery of much of its coast-line. But it is from the sea, rather than from the land, that this scenery can be best seen and appreciated. I therefore welcomed this opportunity of renewing and greatly widening my acquaintance with the varieties of maritime landscape, which are perhaps more variously displayed by the coasts of Ireland than by those of either Scotland or England. No more favourable occasion could have arisen ; for, as before, my kind host put the yacht at my entire disposal, keeping as close to the shore as I desired, and circling round any islet, or slowing down in front of any sea-cliff that I specially wished to look at more closely. And as we always anchored for the night in some land-locked inlet, there were occasions when I could land, hammer in hand, if this was desirable. The whole cruise furnished a memorable lesson on the influence of geological structure upon landscape, and the varying results of marine denudation.

The eastern coast of Ireland being more familiar to me from frequent visits to all parts of it, I arranged to forego the pleasure of viewing it from the sea with my nautical companions, and to be landed at Waterford, whence I went westward to Tramore to meet there Mr. M'Henry, in order to examine the coast section to the west of that place in connection with the revision of some of the maps. From that area we moved northwards for the inspection of some of the ground in counties Wicklow and Wexford, which had been revised, and where we were joined by Mr. Egan, another member of the staff. While we were engaged on this duty and driving in one of the usual open Irish cars, the horse took fright and the car was overturned. I was thrown out upon the road, with no visible injury; my companions escaped with some bruises. We continued our work for a time, but before long the effects of the shock so affected my head as to compel me to return home. Dr. David Ferrier, whom I consulted, insisted that I must abandon all kinds of work for a little, and strongly advised me to "go to Scotland and lie in the bottom of a boat." I followed his prescription by making for my dear Western Isles, and spending a few restful days, some of them literally in the bottom of a boat, at Canna, with the laird of that island and his kindly family. Thence I rejoined my Survey colleagues in the Oban district, and had some interesting days in going over their work with them among the volcanic rocks of the Old Red Sandstone of Lorne.

It was my intention that the year 1900, the last of my long connection with the Geological Survey, should be spent in finishing, as far as possible, all the work that depended immediately upon myself. But this intention could only be partially realised. There had been for some time dissatisfaction in the staff as to questions of salary, travelling allowances, promotion, and other matters which lay outside our own control, and which required the consideration and concurrence of more than one official department. We now requested that a Committee should be formed to enquire into our grievances. After some delay a satisfactory Committee was appointed with an experienced member of the House of Commons as Chairman, and including representatives of our own Department and of the Treasury, together with some geologists of note from outside our ranks. The terms of reference stated that the affairs of the Survey since 1881 should be enquired into; in other words, that the whole period of my tenure of the post of Director-General should be reviewed.

As the meetings of this Committee were held in the Museum, Jermyn Street, and lasted more than two months, from the middle of May to beyond the middle of July, I was necessarily detained in London during that time. Moreover, as the meetings were held in my room, where all my papers and books were kept, progress with the official literary work which I had undertaken was necessarily much impeded. But the result of the official enquiry was eminently satisfactory. The recommendations of the Committee's Report, if carried out, would put the Geological Survey on a firmer footing than ever; the justifiable discontent of the younger members of the staff would be removed, and the status of the seniors would be improved. It was to me a great satisfaction to leave the service with the prospect that these much needed reforms would be carried out.

The prolonged detention at headquarters consumed much of the time that would gladly have been spent in the field. As soon as attendance on the Committee was no longer required, I crossed to Ireland and passed two weeks with my colleagues there, partly in resuming the work in Wexford which had been interrupted by the unfortunate accident of the previous year. Not without regret did I bid good-bye to these pleasant friends, and to active participation in the problems of Irish geology. For nineteen years I had officially visited Ireland always once, and sometimes twice, in the twelvemonth. The progress of the mapping, of the various revisions, and of the examination of ground in connection with the preparation of the explanatory memoirs had taken me into almost every part of the island. The geology, the scenery, the antiquities and the people all interested me; and I can say with truth that in no part of my duties did I find more pleasure than in those which took me to Erin.

To Scotland I was able this year to pay two visits. The first of these filled a few weeks in spring, which were passed chiefly in Fife, for the purpose of completing the materials in hand for the preparation of the two volumes on the geology of that county. As already mentioned, the first volume was completed and published in June of this year; the second could not be finished before my retirement; but it was issued to the public in 1902. September and October were chiefly given to a farewell tour of inspection round the whole staff in the field, from the north of Inverness-shire into Skye, and thence through the Perthshire Highlands. Some remarkable discoveries having recently been made by my colleagues in Arran, I ended the tour by spending a week of great interest in that island, the mapping of which had been nearly completed by Mr. Gunn. The fossil collector, Mr. A. Macconochie, had come upon Rhaetic and Liassic shales containing fossils, and also large masses of hard white chalk full of Cretaceous foraminifera. It was found that these fossiliferous strata had actually been preserved within a large volcanic vent of Tertiary age. They proved that the Jurassic and Cretaceous deposits of Antrim once extended into the south-west of Scotland. It was a curious coincidence that this tour of inspection, which was practically my last work as a geologist in the field, should have ended on the island that had been the scene of my earliest geological essay.

As the preparation of the Annual Report of Progress, which would require some weeks for its completion, could not be begun until after the end of the year, when the various officers in the field had prepared and transmitted their returns of work, the date of my retirement was postponed from the 28th December, 1900 (my sixty-fifth birthday), to 1st March, 1901. Accordingly, on the latter date, retiring from the service in which I had spent forty-four years and four months, I made way for my successor, Mr. J. J. Harris Teall. Of all the members of the staff he was, in my opinion, the most eligible for the appointment. His scientific eminence, his business capacity, his friendly relations and popularity with all his colleagues marked him out for promotion. I had therefore no hesitation in pressing his claims on the Department, and my recommendation was accepted.

The regret with which I closed official relations with my colleagues was mitigated by the spontaneous expressions of their friendly sympathy and kindliness. I am tempted to quote here the letter that came from B. N. Peach, to whom frequent reference has been made in the foregoing chapters. He had first come before me as one of the students at the Royal School of Mines, when I carried on the geological class during the disablement of Professor Ramsay. He there showed such promise that in later years I welcomed him into the staff of the Survey in Scotland, where, as told in these chapters, he rose to high distinction. His loyal service and his personal charm had made him one of my most cherished friends.

Edinburgh, 28th February, 1901.

My dear Sir Archibald,

I cannot allow the day to pass without writing to express my regret that our official connection ceases to-morrow. I feel how much I am indebted to you, not only for nearly all I know of geology, but for the constant support I have felt in having you as a friend. I trust that this latter sense will long remain with me, and that as long as you are alive I shall still look up to you, and if I survive you, I shall still have the sense of my obligation to you.

I, too, feel that it will not be long ere I have to quit the service, which I shall do with regret; for I feel that the life has been the most congenial one I could possibly have lived, and you have done a lot to make me feel so. When I have to leave, it will be a source of great satisfaction to know that the service has been much improved since I joined, which has been brought about during your administration. The staff, too, has been immensely strengthened, thanks to your judicious choice of good men. It is stronger now than it has ever been. This must be a great source of satisfaction to you, and must help to mitigate the regret that you must feel at giving up the helm; also to know that you have delivered it over to strong hands. The prospects of the staff, too, have been immensely bettered of late.

There is one point of view, however, that must be more cheering to you, and that is that you have got rid of the great load of official worry, and that you will have time now to use your unrivalled pen in more congenial work than editing Summaries of Progress and such dryas-dust work. Long may you live to enjoy your well-earned leisure, which I know you will employ for the instruction and delight of others.

I am, Yours sincerely,

B. N. Peach.

Among the letters received from abroad at this time I was specially touched by a most friendly address from the geologists of Finland, with the signatures of J. J. Sederholm and Wilhelm Ramsay at the top of the list. After expressing thanks and good wishes, they went on to hope that "you will continue to give us, as some of your great countrymen have done, the example of old age superior to youth, not only in the maturity of judgment but also in energy and perseverance."

On 1st May, 1901, a complimentary dinner, with Lord Avebury in the chair, was given to me, of which an account appeared in *Nature* on the 9th of the month. From this report a few sentences may be quoted:

"The different public departments with which the Geological Survey is most closely connected were well represented, including the Treasury, Admiralty, Board of Education, Local Government Board, Board of Agriculture, Ordnance Survey, Scottish Education Office, Stationery Office and British Museum. There were likewise present the professors of geology in London, Oxford, Cambridge, Edinburgh, Dublin and Birmingham, together with numerous other Fellows of the various learned societies. Letters, telegrams and addresses of felicitation were received from all parts of Europe and America. The following telegram from Christiania was read by the chairman: 'Also from Norway's mountains an echo of the cheers for the master of English geology—Brögger, Helland, Nansen, Reusch, Vogt.' "

One of the features of this gathering was the presentation of a beautifully illuminated address bearing the signatures of the staff of the Survey in the three kingdoms. My reply to the toast of my health closed with the following sentences :

" During my tenure of office as Director-General I have been ever supported by the loyal and unstinted devotion of the staff. It has been an honour and a pleasure to be placed at the head of such a body of men—so enthusiastic in their whole-hearted consecration to science, and so unwearied and loyal in their efforts for the interests of the service. I feel sure that in no branch of the public service could the *esprit de corps* be higher than it has been among us. You can well understand that it is impossible without regret to sever one's connection with comrades such as these. At the end of my official career I can truthfully claim to have striven to the utmost of my power for the welfare of the staff, and for the scientific renown of the service. I have sought to secure the best men whom it was possible to obtain, and I feel confident that the Geological Survey, as regards the zeal, capacity and attainments of its members, may challenge comparison with any scientific institution in any country of the world. I rejoice to think that the service is being now put on a firmer footing than it has ever held before, that the prospects of pay and promotion have been lately broadened and brightened, and that, under the guidance of my distinguished friend and successor, the Survey may look forward to a future even more illustrious and more useful than its past."

Abrupt severance from official duties which have formed the daily routine of the greater part of a man's lifetime is apt to leave him at what is popularly known as " a loose end." At such an interval, it is a momentous advantage to be able to take up another congenial pursuit which has long been familiar. Every public servant, knowing that he will be retired at sixty or sixty-five years of age, ought during his official life to cultivate some study, vocation, or "hobby," which will profitably and pleasantly fill his time after retirement comes. I have known of one such functionary who became the head of the department to which he had devoted his whole soul, almost night and day, who lived entirely for his office, to the neglect of other interests or occupations, and who, when at last he retired, became one of the most miserable of men, having nothing to do, and unable to make satisfactory use of the leisure that had now come to him. A man at sixty-five may still be in the prime of life, and even for eight or ten years longer may be almost as active as ever, both in body and mind. He will find his salvation in some suitable equivalent to replace the calling which he has had to drop.

It was fortunate for me to have more than one employment which had been constantly, though intermittently, pursued all through official life, and into which I could at once throw myself with pleasure. Geology was, of course, as open to me, and as delightful as ever it had been, though the opportunities for pursuing it were now much curtailed. But it had been cultivated so incessantly since boyhood that there was a kind of relaxation in turning to other interests. More especially delightful was it to betake oneself to literature, and to find, in that vast domain, work and pleasure, in fuller amplitude than the limited leisure of the Geological Survey had permitted. Then there was the varied work of the active scientific life of London, and the claims of the different Societies, which few men, in the full current of official duty, can attend to as they could wish. And further, foreign travel offers an inexhaustible field, when the retired official's income suffices for its cultivation.

In my case, however, the official fetters were not yet wholly unloosed, and until at least some of them were removed, it would be difficult to rearrange permanently my daily life. The completion of the memoir on Eastern Fife required that some portions of the district should be carefully re-examined. It was likewise necessary for me to meet the members of the Survey staff who had been selected for the preparation of the Memoir on the North-west Highlands. A series of conferences with them at the Edinburgh office of the Survey would be needed before the general plan of the volume could be settled, and the limits of each man's share of the description could be defined. It was obvious that it would be desirable for me to remain in Scotland for some weeks, until this important memoir was fairly launched.

Before these arrangements could be finally adjusted, an agreeable interlude gave me a few days in Paris. The Royal Society had nominated me one of a group of delegates to attend the first meeting of the International Association of Academies which was to begin in the French capital on 16th April. It was an important gathering, for it had to organise this infant institution, to define its scope, and to draw up its rules and regulations. It had been created in the hope that if the academies and learned societies of the civilised world could be brought to work together for the advance of science, and especially for influencing the various national governments towards the support of scientific undertakings, their united voice would have more weight than the counsel of the most eminent society or academy acting singly. The proceedings throughout the conference were quite harmonious, and, largely owing to the tact and energy of the President, M. Darboux, Secrétaire perpétuel de l'Académie des Sciences, they resulted in the construction of an organisation which promised to be workable, and likely to fulfil the desired aims.

One of the most prominent and picturesque figures at the meetings of this Conference was that of Professor Mommsen. On my way along the quay to the opening meeting, I saw an elderly foreigner, short in stature, clad in a long dark coat that reached nearly to his heels, and wearing a low soft felt hat from underneath which his white locks streamed in the wind. He stopped at frequent intervals to gaze around him, and entered the Palais de l'Institut immediately after me. He turned out to be no less a personage than the illustrious German historian of Rome, then in his eighty-third year. He seemed to be aware of the picturesqueness of his appearance, for when a grave discussion was in progress at a meeting of the Association, he would rise from his seat in a conspicuous part of the room, and stand or walk about, looking around the audience as if he would speak, and then resuming his seat in silence. He came up to me one evening, and in good English complained that Germans were required to speak so many languages, and that they were so little understood by foreigners when they used their own tongue.

The ninth jubilee or 450th anniversary of the University of Glasgow took place this summer, and having to be in Scotland, I had the pleasure of attending it. The festivities, which extended over four days in June, were excellently organised, and brought together a large concourse of delegates from Universities and learned Societies in many lands. Beginning with an impressive service in the fine Cathedral of St. Mungo, they included the usual presentation of congratulatory addresses, the conferring of honorary degrees, and profuse public and private hospitality. I found myself among the guests who were honoured with the ancient degree of LL.D. The opening of a new Botanical Institute in connection with the Professorship of Botany attracted a distinguished company, among whom were no fewer than three past-Presidents of the Royal Society—Sir Joseph Hooker, Lord Kelvin, and Lord Lister. In such assemblies Principal Story was always one of the most conspicuous personages. His tall form, clad in his purple robe, his noble head under an antique black velvet cap, and, above all, his dignified bearing were eminently appropriate to the president of a seat of learning. His reserved manner was combined with great courtesy and kindliness.

After these doings in the west I made for the east coast, and established myself at St. Andrews as the most convenient centre from which to make the final examinations of the ground that required to be visited, while at the same time I could correct the proof sheets of my last Summary of Progress of the Survey, and be ready to attend any conference at the Edinburgh office. No nook of Scotland is so well fitted for the literary recluse as this ancient city by the sea. Its bracing climate, its antique aspect, its crowded traditions and crumbling ruins, going back to the days of the Culdees, its varied and picturesque coast-line, its unrivalled golf-links, its admirable University library, and its agreeable society combine to give it a singular charm. There can be no wonder that men like Dean Stanley should love to escape for a while from the whirl of the busy world to the restful quiet of this quaint and venerable seat of learning. It was in these favourable conditions that I completed my last Survey Memoir, which appeared later in the year in the form of an octavo volume of 440 pages, illustrated with upwards of seventy cuts in the text, twelve plates and a geological map of Eastern Fife. The most novel feature of the book was the account of the Volcanic Necks, embodying observations made in the course of many years, and presenting for the first time a detailed description of the structure of these orifices of eruption and their relations to the surrounding strata.

While at St. Andrews, I was more than ever impressed with the almost unrivalled facilities offered by this centre and its surrounding district for demonstrating to students many of the fundamental principles of geology. Nevertheless, though in possession of such unusual educational advantages, the University had never made any provision for utilising them. Geology was not included as a distinct subject in its curriculum of study. I had often discussed this deficiency with some of the professors, and on this occasion, before leaving the place, I wrote a formal letter to Dr. Donaldson, the Principal, strongly urging that some steps should be taken to turn to account the unique facilities of the place for the practical teaching of geology. This letter met with a sympathetic reception, and two years afterwards a Lectureship of Geology was added to the teaching staff of the University. Dr. Jehu in September 1903 received the first appointment.

In the course of this year my election as Foreign Member of the National Academy of Sciences, Washington, was intimated to me in a letter from Alexander Agassiz, who was about to retire from the office of Foreign Secretary and become President of the Academy. Any fresh link that helped to attach me more closely to the great Republic of the West and its men of science was always welcome, and this further connection was made still more agreeable, coming as it did through the hands of the most eminent naturalist of the United States, who had for many years honoured me with his friendship. Another compliment which came from the Academy of Natural Sciences of Philadelphia gave me special pleasure —the gold medal of the Hayden Memorial Geological Award—for it connected me with my friend, F. V. Hayden, in whose memory the Award was founded. It always seemed to me that he never in his lifetime received from his fellow-countrymen the credit due to him for his pioneer work in the geology of the West, and from time to time I had brought his services to the notice of English readers.

The year 1902 was in large part devoted to the preparation of new editions of my educational books. In particular, the publishers intimated early in the year that the third edition of the *Text-book of Geology* was not far from being entirely sold out. Ever since the publication of that edition I had been quietly at work on the preparation of its successor. Having now regained full command of my time, and with access to all the scientific libraries of London, I had brought the revision of the book so far forward as to be able to supply the printers with "copy." The additional matter of text and illustrations for the fourth edition amounted to 300 pages. For the sake of convenience the treatise was now divided into two volumes comprising in all 1472 pages, the pagination being continuous throughout. The revision and likewise the printing of the work went on during this year and the following spring, and the volumes were published in the summer of 1903.

On the 22nd of August, 1902, the centenary of the birth of Hugh Miller was celebrated at Cromarty, his native town. A tall, gaunt pillar with a statue of him on its top had already been erected there, as a tribute to his memory, and it was round this monument that a large company gathered on that day to inaugurate a Village Institute, as another and more useful memorial of him. Being probably the only surviving geologist who had known him well, I was asked to attend and to give an address. It was an opportunity, of which I gladly availed myself, to place before the public Miller's claims on the grateful remembrance of his fellow-countrymen, and to give expression to my own indebtedness for his friendship in my youth (*ante*, pp. 24, 34, 40, 58). After luncheon and a visit to the house in which he was born, now partly converted into a museum containing many relics of his life and work, the company re-assembled in one of the churches of the town, from the pulpit of which I discoursed for an hour on the hero of the day. This Address, which had been prepared with some care, was printed as a pamphlet for circulation among those likely to be interested in its subject.[1]

Having now no official ties that bound us to London, and with all my educational books revised and in the press, my wife and daughters agreed with me that we should spend the winter of 1902-3 in Rome. They preceded me to Italy by some weeks and, on joining them in December, I found them comfortably established in Rome, on the uppermost floor of a hotel above the Piazza di Spagna, whence a noble panorama of the city could be seen stretching from the northern slopes of the Janiculan ridge to the western limbs of the Alban Hills on the south. Some of our rooms opened on the roof, which was gay with pots of flowering shrubs. Though in the very heart of the "fumum et opes strepitumque Romae," we were so high above them all, that only a distant hum reached our ears. The months spent in the Eternal City were one continued

[1] It is included among other papers in the volume on *Landscape in History*.

pleasure. For myself, the correction of the proof-sheets and the preparation of the indexes to the two volumes of the *Text-book of Geology* for a time filled up the mornings, while the afternoons were devoted to sight-seeing with the family. Thereafter, aided by the friendly assistance of Commendatore Boni, Professor Lanciani, and the officials of the British and American Schools at Rome, I studied with some care the relics of pagan Rome. Many a happy hour was spent in the Forum, on the Palatine, and in the Museums. In a succession of days we perambulated the city, keeping outside the walls and as close to them as the streets and houses would permit. As the result of these wanderings and studies, I no longer questioned the conclusion which Gibbon so well expressed: "after Rome has kindled and satisfied the enthusiasm of the classic pilgrim, his curiosity for all meaner objects insensibly subsides."

The Campagna too was a source of infinite delight. Its own mysterious charm captivated me. To all its human interest there was added, in my case, the varied and endless fascination of its geological history. That wonderful volcanic plain has never had the story of its birth and growth worked out in detail. The Italian geologists have, I think, not yet grasped how much additional information they would gain in illustration of that story, if they would carefully map its structure on a large scale. In the course of my wanderings across it I made out roughly what seemed to me to be the principal epochs in its growth. Afterwards my observations were embodied in an article to which allusion will be made on a subsequent page. But undoubtedly much remains to be discovered by more detailed investigation.

As a foreign member of the Accademia dei Lincei, I used to attend the meetings of this ancient fraternity in its spacious apartments in the Corsini Palace on the right bank of the Tiber. The Italian men of science were most courteous and friendly. With Count Ugo Balsani, member of the Section of History and Historical Geography, I formed an acquaintance which had become a warm friendship as years passed. He had married an Irish lady, owner of some landed property in Ireland, which after her death he visited every year, usually spending a short time in London, where he was elected a member of the Athenaeum Club. He spoke and wrote English more like a native Briton than a foreigner. In my attendances at the Corsini Palace one or other of the Academicians was always particular that I should sign my name as one of those present. I did not at the time know the meaning of this formality ; but it was explained later when on my return to England, I received from the Secretary a postal order for a few pounds, being so many Italian lire for each meeting at which I had been present.

The Italian geologists were also pleasant and helpful. At the office of the Geological Survey I received every assistance in consulting maps and the literature of the geology of Rome and its surroundings. Capellini came sometimes to the capital, but I presumed that it was chiefly his duties as Senatore del Regno that brought him. He certainly never appeared once at the Lincei meetings while I was there. He came to see me, however, at my hotel, and was then full of his cheery vivacity. The Italian geologist, who of all others impressed me most, was Professor Fortis of the University of Rome, with whom I had some interesting conversations. He presented me with a copy of his work on the geology of Rome, an excellent contribution from which I learnt much. By invitation of M. Boni we first met in the court of the Vestal Virgins in the Forum, where the tooth of a hippopotamus had just been exhumed from the volcanic tuff only a few inches below the pavement which these Virgins trod.

The apartments of the British School at Rome, under the genial care of Mr. Rushforth, were open to me, and at the American School I had some pleasant intercourse with its Director, Mr. Norton. These two seminaries seemed to me most admirably fitted to stimulate the love of art and of historical research among the young men and women who come to continue their studies in the inspiring atmosphere of Rome. The same stimulating influence fanned into flame the artistic enthusiasm of Lucy, my eldest, and Gabrielle, my youngest daughter. We were fortunate to secure for them the assistance of that great water-colourist, Carlandi, who gave them in the field lessons on the art of which he is so consummate a master. They accompanied him to many picturesque subjects inside and outside the city, sketching under his eye and current criticism, and watching from time to time the manipulation and progress of his brush as he simultaneously painted the same scene.

The long stay in Rome, the familiarity which that stay had given me with the crowded relics of the pagan past, and the inspiration of the *genius loci* rekindled my old enthusiasm for the Latin classics, which, though never wholly displaced, had necessarily been relegated to the background in the midst of a life dedicated to the active pursuit of science. But from this time onward, classical reading and study took up an increasing part of my attention with ever-growing delight.

Before leaving Italy, I took the opportunity to pay another visit to Naples. Besides again spending hours amid the treasures in the famous Museum, I saw the latest excavations at Pompeii. With great satisfaction I had the good fortune of going, somewhat more minutely than ever before, over the volcanic area of the Campi Phlegraei, guided by the talented young Italian geologist, G. de Lorenzo, who had made this ground his prolonged study. When we had finished our traverses I pressed him to write a paper on this famous district, which I would communicate to the Geological Society of London. He agreed to the request, and his paper, translated into English, appeared in 1904, in the sixtieth volume of the Society's *Quarterly Journal.*

We returned to England in the spring of 1903. It was pleasant to reappear at the monthly dinners of the Literary Society, which were remarkably well attended and agreeable this summer. Thus on 6th July the company numbered twenty, including Sir Spencer Walpole, the Archbishop of Canterbury, Lord Goschen, W. E.

Lecky, Sidney Colvin, Andrew Lang, Sir Herbert Maxwell, John Murray, E. H. Pember, Sir Theodore Martin, George Darwin, Professor Jebb, Hubert Parry, Briton Riviere, and Donald Mackenzie Wallace. Martin was more attractive than ever.

The ninth International Geological Congress met this year in Vienna. Being Chairman of two of its Committees, I attended it. Instead of making straight for the Austrian capital, I travelled by way of the Bernese Oberland in order to see my son who, not having been very well, had been recommended to go to Mürren for a short time. While there with him, it was my good fortune to witness, on the evening of 14th August, the most magnificent display of summer lightning I have ever seen. Watching it for several hours with intense interest, I took notes of its features, and next day embodied them in a letter to *Nature*, which appeared in that journal on the 20th of the same month. The chief facts noticeable were the total absence of any audible thunder during the whole evening, and the nearness and brightness of many of the flashes. They were unquestionably local manifestations, and not reflections from distant lightning. Sometimes they appeared behind the opposite range of the Jungfrau, which then stood out dark against them, but the most vivid filled the valley in front of that range, which for a moment was then brilliantly illuminated from base to crest.

The Congress at Vienna proved to be of the usual type. Dr. Tietze ably presided, and my two reports—one, that of the Commission on Lines of Raised Beach in the Northern Hemisphere; the other, that of the Commission for International Co-operation in Geological Research—were duly communicated. Such co-operation as I had suggested evidently required, for its effective realisation, much more active sympathy and support than I had been able to enlist. The chief pleasure to me in the Congress was to meet again the Austrian and Swiss geologists with whom I had always been on the most friendly terms. Specially agreeable was it to renew intercourse with my early friend, Edouard Suess. Though now past seventy years of age, he was still bright in mind, and fairly active in body. I breakfasted with him, and we had an interesting talk about old times and old friends. At the closing banquet I was asked to speak, and took the opportunity to refer to my first visit to Vienna, and the geologists then foremost, almost all of whom had passed away except the veteran Suess. Towards the end of the speeches Suess rose, evidently much touched by the various references to himself. With some emotion he expressed his thanks for the kind things that had been said of him and of his work, concluding with an eloquent sketch of geological aims and progress. Another of the enjoyments of the Vienna meeting was to find Zirkel there, as well and genial as ever. The Swiss geologists were well represented by Baltzer, Heim, and Lugeon, the French by Barrois and Haug.

# CHAPTER XI

## SECRETARY OF THE ROYAL SOCIETY
(1903–1908)

In the afternoon of 5th November, 1903, as I was writing in the Athenaeum Club, the following telegram from the Royal Society was put into my hands: " Are you willing to be nominated for the office of General Secretary? Huggins." There came also a note from the Assistant Secretary informing me that the Council of the Society, then in session, was " waiting for an answer by the bearer." It was a startling request. The subject had however been brought to my attention in the autumn of the previous year by Professor Michael Foster, Senior General Secretary of the Society, who then told me that he proposed to resign his office next year, and he wished to know whether I would be willing to succeed him. I replied that having lately retired from official life, I rather shrank from the idea of entering so soon upon it again in another form, but that, as I was about to spend the winter in Rome, I would fully consider the matter there and let him know my decision when I returned, or sooner, if he desired. When I came back from Italy in the spring I met the Professor several times, but I left it to him to renew his proposal. During the stay in Rome I had fully considered it. There could be no doubt that the cessation of my position in the Geological Survey would leave a large measure of free time at my disposal. The four years of experience in the affairs of the Royal Society, gained when Foreign Secretary, had shown me the inner working of that important body in which I would be glad to share, and for which I would now have ample time. I had therefore decided to accept the Secretaryship, if asked by the Council. But as during our occasional interviews, Professor Foster never alluded to the subject, I could see that he had evidently found another candidate. I therefore dismissed the subject from my thoughts. The President's telegram was thus a complete surprise to me; but, having already made up my mind, I had no hesitation in at once replying affirmatively, "by the bearer."

Great pressure was immediately brought to bear on me to retire in favour of an eminent physiologist, supported by Professor Foster and a number of other biologists. But I had accepted the nomination of the Council, and was in their hands. At the anniversary on the following St. Andrew's Day I was duly elected Secretary by a majority of some three to one.

During the first two centuries of the Royal Society's existence, no limit was fixed to the length of time during which its officers might be continued in their respective positions. Each of them might be re-elected at the Anniversary, year after year, for an indefinite period. Thus, the first President held office for fourteen years, Samuel Pepys for two years, and other successors for only a single year. The longest tenure of the office was that of Sir Joseph Banks, who occupied the chair for forty-two years, from 1778 till his death in 1820. About thirty years ago, however, the Council made a regulation that the officers should not hold their respective offices beyond certain limits of time. The tenure of the Presidentship was fixed at five years, the General Secretaryship at ten years, and the Foreign Secretaryship at four years.

For the next ten years the business of the Royal Society formed my chief occupation. This change of employment, though on the whole most agreeable, could not but bring with it some regrets. Chief of these was the cessation of the open-air life, and constant active participation in geological field-work. Yet it was doubtless well that this deprivation should come before failure of bodily strength would compel retirement. And in truth, the regret was not a little mitigated by the sense of unwonted freedom, when all the former ties and trammels of officialdom were ruptured, and I could devote my increased leisure to pursuits of my own choice, especially to studies which, though dear to me, I had never found time enough to cultivate. Resident in London, and now honoured with an important post in the very heart of the scientific life of the country, I could see another wide field of work opening before me, and I entered it full of hope and ardour. Looking back into the past from the advanced old age of to-day, I incline to believe that the ten years on which I now entered were not the least enjoyable section of my life.

Having already had the experience of the Foreign Secretaryship, I was soon able to master the details of the duties of the new post. There are two General Secretaries, one who looks after the physical sciences, while the other (who was now myself) takes the biological side. As in the case of other Societies which hold meetings for the reading and discussion of papers communicated by members, a large part of the work of the Secretaries is concerned with arrangements for the meetings, transmission of the papers to referees, and the printing, illustration and publication of the approved papers. As the Royal Society, representing the whole domain of "natural knowledge," welcomes contributions from every branch of science, this department of the secretarial duties is particularly important and onerous. For the proper consideration of the papers it has been necessary to create eight Sectional Committees, each consisting of Fellows who are recognised authorities in one branch or division of science. With these Committees the General Secretaries closely co-operate in regard to the reading and publication of papers. Other Committees deal with the domestic affairs of the Society, such as Finance, the Library, and the administration of a Fund for affording relief to necessitous men of science or their families.

But over and above what may be regarded as the administration of its own proper affairs, the Royal Society has always been applied to by the various Departments of State for advice in matters wherein science is concerned. It has played, and continues increasingly to play, the part of scientific adviser to the Government of the day. In my time the War Office, the Admiralty, the Home Office, the Colonial Office, and other Departments referred to the Society for advice, and in some cases the Society was induced to undertake, for the public good, researches of a difficult and protracted kind, such as the investigation of the cause and treatment of sleeping sickness, Malta fever, glass-workers' cataract, and other diseases. In later years, especially during the War, the Society, I believe, has been of still greater utility.

Besides applications from Departments of State, which arise from time to time, the Society has undertaken for the Government other duties of a more permanent nature. Thus, it has been entrusted with the distribution of an annual Parliamentary grant for the encouragement and support of research, and in order to secure due attention to the claims of the whole range of science, it has instituted seven Boards, each representing one great division of natural science. These Boards consider the applications for financial assistance from this grant. Again, the National Physical Laboratory was devised and built up by Fellows of the Royal Society until it became too vast an institution to be managed and financed by a scientific Society, and was then put under a Department of the State. Greenwich Observatory has from the first been placed under the wing of the Society, and its Board of Visitors has the President as its Chairman.

The premier position of the Society in this country, as representative of the whole domain of science, has been recognised in the efforts which have been made towards international action in furtherance of scientific work. In the foundation and working of the International Association of Academies the Royal Society took an active part, from the start of that Association in 1901 until the meetings ceased on the outbreak of the Great War. Throughout its history the Society has been in correspondence with foreign Academies and Universities, and this connection is now closer than ever. Personal contact with the men of science in different lands, always cultivated from the very beginning, has now grown more intimate, not only by welcoming them to this country but by sending delegations to meetings abroad. As one or other, sometimes both of the General Secretaries may be sent on these missions, the duties of the Secretaryship are not only numerous, but of a singularly varied and interesting kind.

The officers whom I now joined in the innermost executive circle of the Royal Society were all old friends. Sir William Huggins proved to be a most kindly and helpful President, with whom it was always a pleasure to be associated. Necessarily I saw much of him at Burlington House, and also at his quaint home on Tulse Hill. He and Lady Huggins were in many ways a singularly picturesque pair. Their strong mutual attachment, which showed itself in many little courtesies and attentions, gave a tone to all their relations with the world outside. The old-fashioned formality of manner, so markedly, but so differently, shown in each of them, was an outward manifestation of their love of order, neatness, and precision. They shared each other's life with a fulness as rare as it was beautiful to see, for this comradeship extended into the domain of science, where in the abstruse and delicate investigations in which her husband was a pioneer, Lady Huggins took her place at his side, not only with untiring devotion, but with the helpful sympathy of one who was competent to share in his work. Moreover, she loved literature and had some skill in art, using brush, pen, and pencil; as her friends used to realise in the pretty etchings and water-colour drawings which she sent to them at Christmas.

The Treasurer, Alfred Bray Kempe, barrister and mathematician, much impressed me as an excellent man of business, with a sound judgment, great tact, and endless good nature. I soon became much attached to him; eventually also to his wife and children. My colleague,

Joseph Larmor, who now became Senior Secretary, is a Fellow of St. John's College and Lucasian Professor of Mathematics at Cambridge, and for some years was one of the two Members of Parliament for the University. Though probably the foremost mathematician in the country, he is by nature so modest and retiring that the general world does not know half his worth. From the very first I was instinctively drawn to him; we soon became close friends, and cemented our friendship not only by the most cordial co-operation in our secretarial work, but subsequently by travel together in Italy. The Foreign Secretary, Francis Darwin, a younger son of the illustrious naturalist, was an accomplished botanist whose gentle kindly ways greatly endeared him to his colleagues, while his sound judgment was always a support to them.

These five officers, President, Treasurer, two General Secretaries, and Foreign Secretary, form the executive centre of the Council. They prepare the business to be considered at each meeting of Council, and carry out the decisions then taken. There could not have been a group of colleagues with whom collaboration would have been more agreeable. There was also a capable subordinate permanent staff.

As I now lived in London, it was possible for me to spend a part of each day at the Society's rooms, and I soon found the work so varied and so attractive as to become a source of continued pleasurable employment. I held the office of General Secretary for five years. The first two of these coincided with the last two years of the Presidentship of Sir William Huggins, who retired from office on St. Andrew's Day, 30th November, 1905. I found on joining the Council that Sir William, though in his eighty-third year, still retained not a little of his vigour of body, and seemingly all his alertness of mind. He was scrupulous in his attendance at the Council. Even in his last year of service he missed only one meeting, and would not have failed then, had not his coachman reported London to be lying under so dense a fog that a journey to

Burlington House was not to be thought of. His keen interest in the affairs of the Society led him to write many letters to me, giving his opinion on questions that arose. It was interesting to compare and contrast his handwriting with that of his wife, as indicative of character. His letters, always courteous and kindly, were expressed in the formal and precise style in which he used to speak, and were written in a rather small caligraphy which always seemed to be just the kind of penmanship which his conversation would lead one to expect. Lady Huggins wrote a larger, more irregular, yet on the whole, rather a more legible hand, with bold capitals and florid twistings, as if the pen could hardly be kept quiet enough to express the thoughts, feelings, and enthusiasms that were eager to take shape in words. The contrast between the two styles of handwriting was sometimes strikingly displayed when she added a postscript to her husband's letter. I have preserved a note of the President's dated 8th September, 1905, dealing with the Society's business, and penned in his usual somewhat cramped hand and formal style. To this note his wife added a much larger postscript, in penmanship more florid than usual, for she had to tell that she had recently been staying at Abbotsford, where she had been privileged to occupy Sir Walter's own private rooms. I look back with peculiar pleasure on the afternoons spent with them on Tulse Hill, a pleasure sometimes shared with my wife and my daughter Lucy.

Lord Rayleigh became President on St. Andrew's Day, 1905. He had already served as one of the General Secretaries for eleven years, from 1885 to 1896, and was thus intimately acquainted with the work of the Royal Society. His quiet dignity of manner, his kindly courtesy, his sound judgment, his wide range of acquirement, and his distinguished place among the physicists and mathematicians of the day admirably qualified him for the office. He was complete master of the business of the Council, and conducted it with admirable tact. He loved a country life, and in his charming rural home in Essex he combined the pleasures of farming with those of physical research, which he carried on there in a modest little laboratory with the simplest apparatus. His death in 1919 deprived science of one of its most illustrious students, and one of its stoutest champions.

Among the occasional engagements of the Royal Society, one of much interest occurred soon after the beginning of my Secretaryship. London had been selected as the place at which the second meeting of the International Association of Academies was to be held in May 1904. For months previously a good deal of correspondence was carried on with foreign Societies, especially with the Academy of Sciences in Paris, where the first meeting had been held, as narrated in the foregoing chapter. As the day of the assembly drew near an unwonted buzz filled the quadrangle of Burlington House; voices in French, German, and Italian resounded in the lobbies and rooms of the Society. The multitudinous arrangements that required to be carefully planned, so as to ensure the smooth working of the conference, necessarily threw much work on the officers and subordinate staff. Of this work, perhaps the most exacting and vexatious part was connected with the hospitalities to be provided for the visitors. For weeks beforehand an active interchange of letters took place with the delegates deputed by the various foreign academies to represent them at the conference. An important piece of information had to be obtained from each of them:—whether the visitor was coming alone or would be accompanied by wife or daughter. Answers were received to most of the enquiries, though not always with the promptitude desirable. A few never replied. When they came as bachelors, their silence was of no consequence; but when they unexpectedly arrived with wives or daughters for whom no provision had been made, some effort, and not a little tact, were needed in order to satisfy everybody. One delegate taxed our patience by bringing both his wife and his daughter.

The functions provided for the entertainment of our

visitors included a garden party at Windsor Castle, where a reception of the members of the International Association was held by King Edward and Queen Alexandra. The foreigners one and all were delighted with Windsor, and expressed their surprise to find Queen Alexandra still so young and so good-looking.

The Prince of Wales (now King George V.) having expressed a wish to make the personal acquaintance of some leading men of science, a private dinner at the Athenaeum Club was arranged for 11th May, 1904, at which past-Presidents of the British Association and a few others were present. The Prince, who seemed thoroughly to enjoy himself, chatted with small groups of us after dinner. With his private Secretary, Sir Arthur Bigge (now Lord Stamfordham), who sat next me at the table, I had much conversation. He impressed me strongly with his ability and charm. I could not but feel that the Prince was fortunate in having so capable and pleasant a right-hand man at his side. And the same able Secretary remains with him still, amid the manifold labours and engagements of the Throne.

In the spring of 1905 the Secretaries of the Royal Society took part in an interesting ceremonial—the celebration of the hundredth birthday of the distinguished Manuel Garcia, inventor of the laryngoscope, which has proved of such high value in the investigation of the voice and of diseases of the throat. The venerable musician had been long settled in London as a teacher of music. He was able to attend the celebration of his centenary and to receive numerous addresses, not only from medical institutions, but from learned societies at home and abroad. One of these addresses was from the Royal Society. Garcia lived for more than a year longer.

As already mentioned, it not infrequently happens that the officers of the Royal Society are appointed as delegates to represent the Society at Congresses or other meetings abroad. An instance of this kind occurred in May 1907, when the Society was invited to send delegates to Upsala and Stockholm for the celebration of the two-hundredth anniversary of the birth of Linnaeus. The Foreign Secretary, Francis Darwin, and myself were chosen delegates. We travelled to Sweden together, and at the end of a pleasant journey reached Upsala on the evening of 22nd May.

The Swedes are proud, as they have every right to be, of their great Linnaeus. At this anniversary of his birth, after the lapse of two hundred years, their respect for him was probably greater than ever, and they were evidently pleased that so large a company, from universities and learned societies in all parts of the world, had come to his native land to do honour to his memory. Their regard for him showed itself in many forms. I was particularly struck with the use they made of the pretty little northern flower, *Linnaea borealis*, which perpetuates his name. In the illustrated *menu*-pamphlet prepared for the great dinner given by the University on the second day, conventional wreaths of this plant depicted in natural colours adorned the pages. I was the guest of Professor Quensel, and on my dressing-table I found a delicate spray of *Linnaea*, which I have kept with other souvenirs of the celebration.

It was interesting to see how conspicuous a part the students, male and female, took in the festival. They were early astir, and with their banners and sashes added much to the life and gaiety of the streets. The quiet old town was further enlivened by the variously shaped and coloured academic robes of the delegates from academies and universities in all corners of Europe. The old King was unable, from failing health, to appear on the scene (he died only a few months later); but his place was worthily filled by the Crown Prince, now King Gustaf V. Prince Eugen likewise took a gracious part in the festival.

A long procession, headed by the grand banner of the students, and their various "nations," each with its distinctive banner, followed by the University authorities and the delegates from abroad in their various costumes, marched slowly to the picturesque thirteenth century Cathedral for the ceremony of conferring honorary degrees on the guests. At this function, which lasted a considerable time, it was interesting to find that some customs have been preserved at Upsala which have long disappeared from our University procedure. Foremost among these antique usages came the putting of tall black hats, not unlike those of the traditional costume of women in Wales, on the heads of those delegates on whom the doctorate of theology was conferred. The graduands in philosophy were each crowned with a wreath of laurel leaves, and on his left hand was placed a gold ring having his name engraved inside. We were a somewhat motley company as, after the close of the proceedings, we dispersed among the general crowd. The doctors of divinity were especially conspicuous, while the doctors of philosophy, with their unusual headgear and carrying their hats in their hands, were hardly less prominent.

While to the Upsala University, at which Linnaeus was a student, and afterwards, for many years, its most renowned professor, the chief part in the celebration of his anniversary was naturally assigned, the Royal Swedish Academy of Sciences could not be denied some share in the national tribute to its most celebrated member. The delegates accordingly repaired on the 25th to Stockholm, where fresh congratulations were offered amid further festivity. To the English delegates, a pleasing incident in the proceedings was the award of the Academy's Linnaean medal to our distinguished fellow-countryman, Sir Joseph Hooker, and the appreciative language in which the President of the Academy made the presentation. Sir Joseph at his great age could not travel to Sweden; the medal was therefore placed in the hands of the British Ambassador for transmission to England.

At Stockholm the Swedish Royal Family took a still more active part in the festivities. I was more especially impressed by the charm of Prince Eugen, who to his other gifts adds a high artistic talent which has given him a foremost place among the Scandinavian painters. Renewing a conversation which we had begun at Upsala on the fine arts in Sweden, he invited Mr. Darwin and me to his picturesque home, a little out of Stockholm, where we saw in his studio a number of his own works. And that we might have a fuller appreciation of northern art, he drove us to the house of a wealthy citizen of Stockholm who possesses the best collection of the modern Swedish school. In the absence of the owner, the Prince piloted us from room to room, and kindly answered our many questions about the subjects and the artists. It was to me a memorable revelation of the scope and characteristic "technique" of the Swedish school of painters. When we returned later in the day to our hotel each of us found a copy of the large folio volume of *Svenska Landskap*, consisting of reproductions of the Prince's paintings of Swedish scenery. Inside my copy was the inscription of my name "in remembrance of the Linné days, 1907. Eugen." The Duke of Skania (now Crown Prince) was likewise indefatigable in his attentions to the delegates. At the reception in the Palace, Napoleonic times were forcibly recalled, when one was introduced to personages still known by the name of Bernadotte. To the kind attention of Sir Rennel Rodd, the British Ambassador, my colleague and I were indebted for a pleasant little dinner in the beautiful grounds of the Djurgård outside Stockholm. A fourth member of this party was Prince Roland Buonaparte, whom I had long known in Paris, and whose presence maintained the Napoleonic associations of the place.

In the autumn of this same year I received the following letter from the Prime Minister, and was much at a loss to guess what had prompted his action :

10 Downing Street, Whitehall, S.W.,
5th August, '07.

*Secret.*

Dear Sir Archibald Geikie,

It will give me great pleasure if you will allow me to submit your name to the King for the honour of K.C.B. If you accepted there would be an occasion for conferring the honour at a meeting of the Privy Council on 12th inst.

This is of course quite private until His Majesty's pleasure is taken.

Yours very truly,
H. Campbell Bannerman.

On Monday afternoon, 12th August, I duly appeared at Buckingham Palace. There was a crowd of ministers and other Privy Councillors in the ante-chamber, all talking in groups, interrupted by occasional messages from an inner room where the King sat. His Majesty was obviously impatient. I was told he had another appointment after the Council, and wished things to be pushed on. I heard him shout my name from his apartment, and I was immediately hurried into his presence. He was sitting in an easy chair. The insignia of the Knight Commander of the Bath were handed to him, which he fixed on my breast as I knelt on one knee upon a stool in front of him. He murmured a few words, of which I only caught "long and distinguished service." The ceremony was over in two minutes. Before I had reached the door in retiring, His Majesty was calling lustily for the next business.

Among the many telegrams, letters and cards which came to me with congratulations, I was more especially touched by those sent by one or two of my former students in the geology class at the Edinburgh University. From the letter of one of these faithful friends, who has risen into eminence in the Civil Service, I may quote two sentences :—"I write mainly as one of your old students who, in common with very many of his fellows, owes to your instruction, your kindly guidance and your inspiration, a capacity for enjoyment in the field which is a permanent spring of happiness. May you enjoy for many years this new visible token of appreciation, as well as your now long-standing reward—the knowledge of good work well done."

In the organisation of national exploring expeditions,

from at least the time of the voyages of James Cook in the eighteenth century, the Royal Society has taken an active and useful part. The Arctic and Antarctic explorations of the last hundred years, and the great voyage of the *Challenger*, have usually been suggested by Fellows, and have always been effectively assisted by the Society as a body in concert with the Admiralty. In recent years, together with the Trustees of the British Museum, the Society has also superintended the publication of the scientific observations made in these explorations. It happened that during my Secretaryship, the National Antarctic Expedition (1901-1904) completed its work and brought home a large amount of material, not a little of which was placed in the hands of the Royal Society to be co-ordinated, edited, and prepared for the printer. A full share of this editorial work fell to me. I found it replete with interest, and it had the added attraction that it placed me on a close and friendly footing, not only with Captain Scott but with several of his brave companions. More especially intimate were my relations with Dr. E. A. Wilson, the junior surgeon and admirable artist of the Expedition. I am proud to have assisted him in the preparation of the *Album of Photographs and Sketches*, forming a massive quarto volume containing one hundred and sixty-five plates and accompanying text, together with a *Portfolio of Panoramic Views*. These two volumes were published by the Royal Society in 1908. They furnish a vivid presentation of the landscapes and other features of the Antarctic regions.

Among various offices outside the Royal Society which were filled by me during five years of the Secretaryship, one of the most pleasant was membership of the Science Scholarships Committee of the Royal Commission for the Exhibition of 1851. The signal financial success of that first and greatest of all the industrial Exhibitions which have been held in this country, left in the hands of the Royal Commission a large sum of money, part of which was invested in the purchase of land lying to the south of Hyde Park. Over this land, streets and residences have since been erected, with the consequent growth of a large income to the Commission. A portion of this revenue is devoted to scholarships and bursaries which are open to students who desire to prosecute science as a calling, and who are recommended by universities in this country and in the Dominions. These appointments have been the means of enabling many young men to start on successful careers, and in some cases to rise to eminence as original investigators and discoverers. The Chairman of the Committee, when I joined it, was my old and valued friend, Sir Henry Roscoe, and no one could have filled the post with more tact and energy. I had been a member of the Committee for a few years when on 3rd June, 1908, I received from Sir Arthur Bigge an official letter informing me that the Prince of Wales (now King George) desired to recommend me for election at the next meeting of the Royal Commission as a permanent member of the Commission to fill the vacancy caused by the death of Lord Kelvin. I was thus admitted into the administration of this important and beneficent trust, which has been able to advance the cause of education, not only by its annual subvention to the Scholarships Committee, but by large donations to such institutions as the Imperial College of Science and Technology and the British School at Rome.

When the Imperial College of Science and Technology was incorporated by Royal Charter in 1907, I was named one of the Governing Body, and took much interest in the organisation of this important seminary. The first Rector having unfortunately to retire owing to failing health, I was nominated one of a small Committee to seek for a suitable successor to the appointment. During my close connection with the Tropical Diseases Committee of the Royal Society I had been greatly struck by the remarkable success with which the Royal Army Medical College, in Grosvenor Road, had been planned and fitted out by Sir Alfred Keogh. Having heard that, if approached on the subject, he might not be unwilling to undertake the somewhat similar task for which we were seeking a capable head, I strongly urged that the Committee should consider his qualifications. As the result of their enquiry they proposed his name to the Governing Body as the next Rector. The proposal was eventually accepted and Sir Alfred Keogh was appointed. The choice proved a signal success. The new Rector speedily won the entire confidence of the staff of professors and teachers by his tact and sympathy, and his evidently keen desire to further the advancement of the College as an important educational institution. After the outbreak of the Great War, his ability was so well recognised at the War Office, that his services were borrowed by that Department. He was made Director-General of the whole Medical Service of the Army—a laborious post which he filled with success for several years, until the Governing Body of the College asked to have him back again to look after the interests of their institution.

In the midst of these activities the greatest man of science in our time passed away; Lord Kelvin died on 17th December, 1907. There was a general desire that he should be buried in Westminster Abbey, and the Royal Society was able to assist in the arrangements for his funeral there on the 23rd of that month. It was a day of unusual darkness. The ordinary dim religious light of the venerable building was deepened into a solemn gloom, amidst which a large and notable company assembled to pay the last tribute of respect to the illustrious philosopher. The pall-bearers, twelve in number, included representatives of some of the chief Societies and Institutions with which he had been most closely associated. Amid the mournful tolling of the funeral bell, the procession moved slowly through the darkness to the grave prepared near that of Sir Isaac Newton. And there, in the hallowed sanctuary where lie the remains of so many generations of our greatest and best, we laid to rest all that was mortal of one who was not more reverenced for his towering genius and his manifold achievements in science, than he was beloved for the rare modesty, simplicity, and goodness of his character. As we walked back from the Abbey to the Royal Society's rooms, my colleague, Professor Larmor, made a remark to me which brought vividly to the mind the greatness of Kelvin, as judged by one of the most eminent of the limited number of judges competent to form an estimate. "Conceive," he said, "a perfectly level line drawn from the summit of Newton's genius across all the intervening generations; probably the only man who has reached it in these two centuries has been Kelvin."

The British Association for the Advancement of Science met in Dublin in the year 1908, under the Presidentship of Francis Darwin. It was in many ways an enjoyable, as well as useful, meeting. To me the week was made additionally agreeable by my being lodged at the Deanery of St. Patrick's under the hospitable roof of Dr. Bernard, who since then has been Bishop of Ossory and Archbishop of Dublin, and who has now returned to his Alma Mater as Provost of Trinity College, Dublin. Amid all the charms of cultivated society, never more delightful than in the Emerald Isle, I often found my thoughts during this visit back in reverie amid the scenes of Swift's life within these same walls. A good memory may be quickened, and the appreciation of an author may be stimulated by actual residence in the home where he lived and wrote. The chequered career of the great humorist seemed at times to rise up before me as I sat where he as "Dean, Drapier, Bickerstaff or Gulliver" for half a lifetime wielded the pen of the greatest man of letters in his day.

It was a great gratification to all his many friends to see the modest, retiring Francis Darwin in the Chair of the British Association. He gave an opening address full of suggestiveness. At its close I was asked to second the vote of thanks to him. In doing so I pointed out how much help Charles Darwin had it in his power to obtain from his group of gifted sons. If there was any question of physics on which he needed light, he could apply to George, who stood in the forefront of the physicists of his time. If he desired assistance in any botanical investigation, Francis was at his side in a moment. If he needed any piece of apparatus to be devised and constructed for his biological enquiries, Horace was an accomplished mechanician ready at hand. Or if ever he had a question of finance to consider, his banker son William was abundantly qualified to advise him.

Although the duties of the Secretaryship of the Royal Society took up a large part of my time, they did not prevent an active participation in the work of some of the other Societies and Institutions with which I was connected. Foremost among these bodies was the Geological Society of London, to which, as already mentioned, I was bound by long years of friendly association. In the winter of 1903-4 the President, Professor Lapworth of Birmingham University, was disabled by illness from discharging the duties of his office. As one of the Vice-Presidents I was asked by the Council to take his place in preparing the annual address from the Chair at the Society's anniversary in February. For the subject of this address I chose an examination of the evidence for the emergence and submergence of land in the British Islands during late geological time. In opposition to the published opinion of Professor Suess, I maintained the old belief that the changes of level have been mainly due to the rising and sinking of land, and not, as he contended, to variations in the level of the sea.

This dissent from one of the most prominent tenets of the distinguished Austrian geologist, naturally expressed in courteous and respectful language, never for a moment disturbed our sincere friendship. So far from allowing any differences of opinion on a scientific question to interrupt our cordial relations, I may mention that while I was correcting the proof of my address I was at the same time in correspondence with Suess to obtain from him personal details as to his career, for the preparation of an article on him and his life-work which the editor had asked me to write for *Nature*. The Copley medal had just been awarded to him by the Royal Society, and it was proposed that a notice and portrait of him should be included in the *Nature* series of "Scientific Worthies." In reply to my application my friend wrote: "It is very kind indeed that *Nature* intends to bring my picture; but the honour is extremely heightened by the circumstance that you, my dear Master, will write the accompanying notice. According to your wish I have wrote down some lines on the non-scientific part of my life. I might have added that already in 1848, during our revolution I, then 17 years old, was caught up by the events and had many an occasion to see the different facettes of human character and life. Perhaps the scenes of that part of my life have formed a part of my education. But I have thought better not to enter so far into personal reminiscences."[1]

The little biographical notice of the Copley medallist duly appeared in *Nature* on 4th May, 1905, accompanied with an excellent portrait of him. In his letter after the paper reached him he wrote:

"How am I to thank you! I always heard that a child must have an open wish and a man must have a scope of ambition. But what further ambition can move a geologist after having read so indulgent a sketch of his course of life? So this utmost limit is reached even before my task is finished,[2] and I can only, with a heart full of gratitude, say that I feel more indebted than ever to you, my dear Master and friend."

Before many months had passed Suess took the opportunity of expressing his friendship in a characteristic but quite unexpected way. On the following 28th of December I received from him a congratulatory letter in German and in his own distinct and well-known caligraphy, to

[1] My friend, from the first year of our acquaintance, always wrote to me in English. As he was born in London, though he left England while still a child, he rather prided himself on the place of his birth.

[2] His great work the *Antlitz der Erde* was not yet completed; the concluding volume, with the copious Index to the whole three volumes, was published in 1909. He himself died on 26th April, 1914, happily before the uprise of the European War by which his country was ruptured and laid prostrate.

which he had obtained the signatures of an illustrious group of Austrian geologists. The following is a literal translation of the document:

Vienna, 24th December, 1905.

HIGHLY HONOURED MASTER,

On the occasion of your seventieth birthday, we the undersigned, send herewith, in simple but most heartfelt words, an expression of our admiration of your great scientific achievements, and at the same time our thanks for all the instruction which we have drawn from your works.

May many years of physical vigour and joyous love of work be still granted to you. This we wish with sincere respect.

E. Suess, Victor Uhlig, G. Tschermak, F. J. Becke, C. Diener, M. Vacek, E. Tietze, F. Teller, Albrecht Penck, E. von Mojsisovics [and sixteen other less well-known names].

From this digression to Vienna my narrative returns to the Geological Society. Towards the end of November 1905 the President, Mr. J. E. Marr (who has since succeeded to the Woodwardian Chair of Geology at Cambridge), wrote to me that the Society would celebrate its Centenary during the Presidentship of his successor; that he had consulted former occupants of the chair and the Vice-Presidents and other officers then on the Council; that the prevalent opinion was in favour of electing a former President, and that there was a general wish that I should be asked to fill the Chair for a second time. As the work of the Secretaryship of the Royal Society filled up so much of my time, I hesitated to accept another office, which, as I knew from experience, demanded much serious attention and would require still more for the celebration of the Centenary. I was loyally attached to the Geological Society, and could not but be much touched by this mark of their appreciation and goodwill. After full consideration I agreed to the suggestion that the Council should propose my name to the Society as President for the next biennial period, 1906-8, and it was arranged that the Vice-Presidents should as far as possible relieve me of some of the duties of the Chair.

The celebration of the first Centenary of the Geological Society duly took place in London during September 1907. It attracted a large and representative assembly of geological delegates, not only from universities, colleges, learned societies and institutes in Great Britain and in our dominions and dependencies, but from almost every country in Europe, from Egypt, the United States of America, the Argentine Confederation, Mexico and Japan. The arrangements were admirably planned and were carried out without a hitch. The subject chosen by me for the Centennial Address was " Geology at the beginning of the Nineteenth Century "—a broad sketch of the state of the science in Europe at the time when the Society was founded.

At the dinner to the delegates the company numbered nearly three hundred. Seldom had so distinguished a company of geologists from all parts of the world met together. Especially pleasant was the opportunity to welcome some of the foreign leaders in the science whose names and writings were familiar to every geological student in this country. From France we were proud to see our personal friends, Gosselet, A. de Lapparent, Barrois, Termier, Boule, Vélain and others. From Germany came the ever-welcome Zirkel, with his col league at Leipzig, Hermann Credner, also Steinmann, Rothpletz, Walther, and Bergeat. Russia was well represented by Tschernyschew, Lœwinson-Lessing, Sederholm, and Pavlow. At the head of the delegates from Austria-Hungary was Dr. Tietze, the Director of the Imperial Geological Survey. Denmark sent Steenstrup and Thoroddsen. From Scandinavia there came upwards of a dozen of representatives, who included Brögger and Reusch from Norway, and Nathorst and Sjögren from Sweden. The United States furnished upwards of a dozen delegates, among whom I was glad to welcome my old comrades W. B. Clark, Arnold Hague, and J. J. Stevenson. The proceedings ended with two simultaneous visits to Oxford and Cambridge, where honorary degrees were conferred on some of the more eminent foreigners, and then the assembly broke up into parties who made excursions to various parts of England and Scotland.

Such occasions as this Centenary make their little contribution to the comity and kindliness of nations. It had always been my earnest desire to do all in my power to promote this end, and more especially to draw closer the ties of amity between Britain and France. I inherited the Scotsman's hereditary good feeling for France; this feeling was greatly intensified by my marriage, which opened for me the way into the warm heart of French family life, and it was further strengthened by close contact with French men of science and of affairs, whose genius I admired and whose personal friendliness I greatly valued. In the year preceding the Centenary of the Geological Society I had been invited by the "Alliance Française" (an association of the English branch of which I was Chairman) to give an address in the Sorbonne on 26th February, 1906, and I chose for my subject one that connected together the progress of science in France and in Britain, as illustrated especially by the contemporaneous work of two men, Lamarck in France and Playfair in Britain, at the beginning of the nineteenth century. It was in 1802 that the great French naturalist published at Paris his *Hydrogéologie* and that the Scottish philosopher gave to the world at Edinburgh his *Illustrations of the Huttonian Theory of the Earth*.

Later in the same year (1906) there came another opportunity to aid in cementing the friendship of Britain and France. I was one of the original members of the Franco-Scottish Society—a praiseworthy association formed for the purpose of reviving the old active cordiality between Scotsmen and Frenchmen. Meetings had been held both in France and in Scotland, and the success of the Society had been steady from the beginning. This year a meeting took place in Aberdeen which was well attended by natives of both countries. What with meetings for the reading of papers, receptions and banquets by the municipal and other authorities, invitations to Fyvie Castle and Haddo House, and an excursion to Dunnotar Castle on the coast, a week was pleasantly spent in the social intercourse which it is one of the main objects of the Society to promote. Having been asked to make some communication to the meeting I chose for its subject "A French impression of Scotland and the Scots in the year 1784," basing my discourse on the two volumes of the *Voyage en Angleterre en Écosse et aux Iles Hébrides* of B. Faujas-Saint-Fond, published in Paris in 1797. The report of the lecture given in the newspapers brought me a letter from the Glasgow bookseller, Mr. Hugh Hopkins, who said he had long contemplated the publication of a new edition of the English translation of Faujas' volumes, which appeared in 1799, and asked whether I would be willing to edit such a reprint. As I had long been familiar with the book, had referred to it in various writings, and regretted that it was not better known, I willingly undertook the proposed editorship. The translation required a good deal of correction. I prefixed to the new edition a biographical notice of the author, together with a portrait of him, kindly procured for me by Professor Lacroix, from the archives of the Musée d'Histoire Naturelle at Paris, and I added notes chiefly in reference to the scientific details of the work. The two volumes, well printed, and with the original illustrations reproduced by photography, were published at Glasgow in the spring of 1907 by Mr. Hopkins.

The meeting of the Franco-Scottish Society was not the only magnet which drew a large concourse of people to Aberdeen at this time. Immediately after that meeting ended, a much more numerous company assembled in the "granite city" in order to celebrate the fourth centenary of the most northerly University in the United Kingdom. The festivities extended over four days, and nothing was left undone which the Chancellor, Lord Strathcona, the

Senatus of the University, or the municipality of the venerable town could do to make them pleasurable and interesting. Aberdeen, always grey and often swept with a keen east wind, put on its gayest attire for the occasion. It had no hall large enough to allow the vast company to dine in comfort. Lord Strathcona, not to be baulked of his design, had a huge temporary hall specially constructed for the nonce. King Edward came from Balmoral to grace the celebration by inaugurating the new buildings at Marischal College. A number of honorary degrees were conferred, and I received once more that of LL.D. This degree had now been bestowed on me by each of the four Scottish Universities—an exceptional distinction of which I am naturally proud, as a signal contradiction to the rule that a prophet has no honour in his own country.

A few months before this northern meeting (December 1905) another wholly unexpected mark of appreciation had come to me from my own fellow-countrymen. The Royal Scottish Geographical Society awarded to me its highest honour—the Livingston gold medal, which was placed in my hands at a public meeting in Edinburgh where, in answer to the request of the Council I gave an address on "The History of the Geography of Scotland." This paper was printed in the Society's *Magazine* for March 1906, copiously illustrated with a series of excellent photographs of characteristic types of Scottish scenery and rock forms, kindly supplied by the Director of the Geological Survey in Scotland, and also with a series of maps indicative of the successive geographical changes in the evolution of the present physical features of Scotland.

During my tenure of the Secretaryship of the Royal Society, continuous residence in London permitted a freer enjoyment of the socialities of metropolitan life than had before been practicable. Especially pleasant were the monthly dinners of the Literary Society, where my acquaintance with Sir Theodore Martin had grown into a warm friendship. He needed to escape as often as possible from the increasing noise of the traffic past his residence in Onslow Square, but his letters from the country kept me in touch with him. He had fulminated in the *Times* against the motor-omnibuses that were making his London home almost intolerable to him. In one of his letters sent to me at this time he gives a doleful picture of an evening at the Literary Society:

"I am very sorry I cannot be at the Coquelin conference on Friday. I only came here (Folkestone) yesterday, wanting sorely something to pick me up, and shall stay on till Monday, when I must return to town. I will go to the next Literary dinner, mainly in the hope of seeing you, for I suffered so much from the smoking at the last dinner that I had resolved never to go to another. There was a large company. Everybody, except Sidney Colvin, smoked. I bore it for nearly two hours, as nobody seemed inclined to move. Didn't I pay for it? I nearly fainted on my way home, and next day I was utterly prostrate. Is not this paying rather too dearly for seeing one's friends?"

With Sir Theodore's dislike of tobacco at and after dinner I heartily sympathised. I have often wondered if it ever occurs to prandial smokers that they may be inflicting discomfort and even pain on some who do not smoke, especially on those who are unfortunate enough to have weak eyes. The habit of smoking at private dinner parties was not common in my youth. I have watched with regret its rapid advance, and with not a little bodily suffering. I continued, however, to attend the dinners of the Literary Society as long as Sir Theodore Martin lived. But after an attack of glaucoma in the left eye in 1906, which necessitated the operation of iridectomy and greatly impaired the vision of that eye, I was still more distressed by the effects of tobacco-smoke. Eventually I reluctantly ceased to attend regularly at the Literary Society, and in the end, in order not to keep out a candidate (who I hoped might be smoke-proof), I resigned my membership.

Sir Theodore Martin lived until 18th August, 1909,

when he gently passed away in sleep at the age of eighty-three. His death was a grievous loss to me. He was ten years older than myself, yet we had many common interests and associations. We were both Scots, both educated at the High School and University of Edinburgh, and both endowed with a share of the *perfervidum ingenium* of our countrymen. We had the same literary tastes, the same love of the Roman poets, and the same social instincts that brought us into the Literary Society and, despite the tobacco, kept us there. As the years passed we became more and more closely allied, and I realised with increasing joy his largeness of heart, his keen sense of humour, and his unfailing sympathy and helpfulness. With deep regret I followed his remains to the Brompton Cemetery where we laid him beside his admirable wife, Helen Faucit.

Another member of the Literary Society by whose side I often sat was Sir Richard Henn Collins, Master of the Rolls. He had a quiet, modest manner that was most winning, and the remarkable clearness and breadth of his outlook on life made his conversation full of interest. At this time the Classical Association, which was being formed under the inspiring influence of Samuel H. Butcher and a few kindred spirits, including the Master of the Rolls himself, often formed the subject of our talk. Sir Richard was chosen in 1904 as the first President of the Association, and as long as he lived he continued to further its progress. He induced me to become a member, and afterwards I was elected into the Council, the meetings of which, held in Mr. Butcher's house in Tavistock Square, I was able for some years to attend. It was a novel and pleasant experience to come into direct contact with some of the foremost classical teachers of the day, and to hear educational matters discussed from their point of view. Mr. Butcher's unfailing tact, sound judgment, and great personal charm made those meetings memorable to me.

A few special dinners of these years dwell in my memory. One of them, the first of what was intended to be a periodical series, took place in 1905 at Magdalene College, Cambridge, in honour of the memory of Samuel Pepys, on his birthday, 23rd February. I had often been a guest at this College when spending a week-end with Alfred Newton in his picturesque quarters there, and it had been my good fortune to be permitted to pass many hours in the Pepysian Library, still piously preserved in the original bookcases which the owner constructed to hold it. It was pleasant to find that the memory of the great Diarist's connection with the Royal Society was held in remembrance at his old college.

Pleasant, too, were the parties for luncheon, tea, or dinner, which the late Sir Benjamin Stone, M.P., used to gather round him in the dining-rooms or on the terrace at the House of Commons. The company, usually predominantly political, was often varied by the commingling of representatives of science, literature and art, and there was a freedom and absence of ceremony about them, due in no small measure to the geniality of the worthy host. I have preserved a menu-card of 26th July, 1905, on which the company, numbering twenty-five, wrote their names in pencil on the back, as the cards were passed round the table. The names include Joseph Chamberlain and nine other Members of Parliament. There were also present the Dean of Westminster (J. Armitage Robinson), Sir Sidney Lee, Sir E. Maunde Thompson, Signor Marconi, Sir Norman Lockyer, Hall Caine, G. F. Warner, Sir Mancherjee Bhownagree and others.

Lord Avebury's pleasant week-end parties at his country home still continued, with their rambles through the woods and friendly talk by the way. He also still kept up the time-honoured custom of breakfasts at his town house, which at this time was in St. James's Square. He would not infrequently catch some celebrity who might be visiting or passing through London, and ask his friends to meet him.

During the five secretarial years at the Royal Society I found time for some literary work. Besides the addresses, lectures and editings already alluded to, I wrote the article "Geology" for the new edition of the *Encyclopaedia Britannica* (1910), and an article on the "Origin of Landscape" in the *Edinburgh Review* for October 1906. At the request of an American friend I put together my observations on the Roman Campagna (*ante*, p. 330) in an essay which appeared in June 1904 in the now extinct *International Quarterly*. It has been reprinted as the concluding paper in *Landscape and History*. My Italian friends, Professor Lanciani and Commendatore Boni, have been good enough to express their appreciation of this essay, and I learnt at the American School in Rome that it has there been made one of the texts for students of the history of Rome and its surroundings.

During my life in Scotland and frequent rambles in all parts of the country in connection with the work of the Geological Survey, besides studying the rocks I kept an open eye and ear in noting the habits and sayings of the people, especially in rural life. It was my good fortune to meet with many incidents illustrative of native humour and reminiscent of older generations. On returning to the south I used to amuse friends and acquaintances with these experiences of Scottish manners, customs and speech, and was many a time urged to write them down. At last in the course of a few weeks in 1903 I wrote the volume which, with the title of *Scottish Reminiscences*, was published in March 1904. It met with so favourable a reception that the first edition was sold off in a month. It was reprinted in April, and again in 1905, 1906 and 1908.

One of my last duties as Secretary of the Royal Society was to contribute four obituary notices of deceased Fellows to the *Proceedings* of the Society. It was perhaps an appropriate though saddening duty, seeing that all the four were dear personal friends over whose loss I grieved. They were Canon Henry Baker Tristram, whose contributions to ornithology, especially to that of Palestine and that of the Great Sahara, gave him an eminent place among modern naturalists; Alfred Newton, ornament of the Cambridge Natural Science School, and my beloved companion in many a yachting cruise; Sir John Evans, antiquary, geologist and man of business, from my youth upward, a leal friend, and in later years a helpful colleague; and Henry Clifton Sorby, the founder of modern Petrography, who had been my friend and correspondent for more than half a century.

# CHAPTER XII

## PRESIDENT OF THE ROYAL SOCIETY

## (1908–1913)

EARLY in the summer of 1908, Lord Rayleigh, who in 1905 had succeeded Sir William Huggins as President of the Royal Society, intimated to the Council that he wished to retire from the Presidentship at the anniversary in the following November—a step which the Council and Fellows greatly regretted but could not succeed in suspending. At the Society's evening reception on 13th May of that year, the Treasurer informed me that there was a general desire in the Council and among the Fellows who had been consulted that I should be asked to be the next President. I was quite unprepared for this announcement, for although, in the contest for the Secretaryship in 1903, the succession to the Chair had been rather tactlessly dangled before me by several supporters of the other candidate, I had never allowed myself to attach any consequence to these suggestions. I can honestly say that I had never considered the scientific work which I had accomplished to be of the quality expected in the holder of the most important and dignified post in the whole range of scientific positions in this country. But when the Council unanimously nominated me for election to the office next St. Andrew's Day, I could only accept the honour with grateful thanks. I was in due course elected on the 30th November, 1908, and I entered on its duties with the resolve to devote my energies more strenuously than ever to the welfare of the Royal Society.

From the foundation of the Royal Society by King Charles II. in 1662, the office of President has been surrounded with a special dignity and prestige. The first person who held it was named by the King in the original Charter, and in that formal document the words of the oath are given which was to be administered to him by the Lord Chancellor. The original statutes of 1663 provided the President with ample powers for the conduct of the business of the Society. He was to take place of every Fellow, preside at all meetings, and when in the Chair was to remain covered " while speaking unto or hearing particular Fellows, notwithstanding their being uncovered." It was further ordained that " when any Fellow speaketh he shall address his speech to the President and be uncovered, and the rest shall be silent." No Fellow was to leave a meeting without giving notice to the President. In the same year the King presented a massive and handsome silver-gilt mace, " of the same fashion and bigness as those carried before His Majesty, to be borne before our President on meeting days."

Some of these ceremonious formalities have been dropped in the course of time, but the President of the Royal Society is still surrounded with more ceremony than the head of any other of our learned Societies. He still takes precedence of every other Fellow, not only " at the ordinary place of meeting, but in all other places where any number of the Fellows meet as a Society, Council or Committee." The original mace, as the venerable emblem of authority, is still placed in front of him before the business of the Society or the Council is begun. The conduct of the affairs of the Society is still officially affirmed to be in the hands of " the President and Council."

From the beginning of its history the Royal Society has retained the close relation to the ruling Sovereign which was established by the Royal founder. The bold and legible signature of that monarch appears on the first page of the Charter-book with the word " Founder " appended by himself. The successive sovereigns who have followed him have inscribed their names as "Patron" in the same venerable volume. The President and Council, as occasion has arisen, have availed themselves of their privilege of personal access to the Throne, and on State occasions the President receives an invitation to attend. During my Presidentship several such incidents occurred, as will be told in later pages.

One of the stated duties of the President of the Royal Society is to read an address to the Fellows at the Anniversary meeting on St. Andrew's Day. The choice of subject is left entirely to him; but it always includes mention of the names of the Fellows and foreign members who have died during the past year, with generally some slight biographical notice of the deceased. He also presents to the recipients the various medals which the Council has awarded to them, accompanying each with a brief statement of the grounds on which the prize has been assigned. The obituary notices often involve some trouble. It is not always easy to obtain the essential data required even in the baldest notice, and in the case of deceased foreign members the difficulties and delays may be much increased. In the case of the decease of the most illustrious Fellows of the Society, for whom burial in Westminster Abbey is called for by the public voice, the President and officers have often to take an important part in the arrangements. Thus when Lord Lister died, and at his request was buried beside his wife in the West Hampstead Cemetery, the desire was widespread that some memorial of him should be placed in the Abbey. The Committee of which the President was Chairman, after a consultation with the Dean, had a medallion portrait of the great physician prepared, a similar memorial for Sir Joseph Hooker, who had predeceased Lord Lister by only two months, and another for Alfred Russel Wallace were also made and erected on the same wall in the Abbey. At a distance of a few yards from this sacred spot the beautiful window to the memory of Lord Kelvin casts its varied light upon the floor. Thus in our national Valhalla modern science has been accorded its place.

The Copley Medal, the highest honour in the gift of the Royal Society was adjudged in 1910 to Francis Galton. It happened that finding the climate of Haslemere to suit his health better than that of London, he had spent the winter months for several years close to my country home, so that I saw much of him. He wrote there his volume of Reminiscences of his life. When the manuscript was in progress he used to recount with glee how remarkably the recollection of one incident, as he wrote it down, would bring to remembrance others which he had till then forgotten. I was able to send him the first private information of the adjudication of the Copley Medal to him. When the awards were made public he wrote: "I have just read in the newspaper the announcement, for which your kind letter prepared me, of the award of the Copley Medal. No one can value such an award more highly than I do, for I am loyal to the Royal Society to the backbone, and care more for their good opinion in scientific matters than for that of the whole world besides. Pray express that feeling in suitable terms for me as occasion may arise. You know how impossible it is for me to be present at the Anniversary meeting, and will, I am sure, understand my regret. The Copley Medal is indeed a prize." In his absence the medal was received for him by his relative Sir George Darwin.

I saw Galton at Haslemere after the Anniversary, and found him with his mental faculties unimpaired and in the highest spirits. About the same time that he had been designated as recipient of the Copley Medal, he had received a letter from the Prime Minister intimating the King's desire to confer upon him the honour of Knighthood. But his brilliant race was now run. Six weeks later he quietly passed away at the great age of almost ninety years. I had known him for exactly half a century. We first met in London at a dinner of the Royal Geographical Club. At that time he did me the good service of asking me to contribute a paper to his volume of *Vacation Tourists in 1861* (*ante*, p. 91), and thenceforward I was favoured with many proofs of his genial friendship. The importance of his scientific work, so varied and so original, in diverse directions, was hardly, in my opinion, adequately appreciated in his lifetime. He early displayed remarkable gifts as a pioneer traveller and scientific explorer in tropical Africa, and wrote the best book we have on the Art of Travel. His sympathy with geographical discovery remained keen to the end. But he will probably be best remembered by his various contributions to the problem of heredity, and as the brilliant but patient founder of the doctrine of Eugenics as a system for the betterment of humanity.

At the next Anniversary of the Royal Society (1911) it was my pleasant duty to present the Copley Medal to Galton's cousin, Sir George Darwin. This prize was awarded to him in recognition of the importance of his long series of researches on Tidal Theory, including its bearing on the physical constitution of the earth, and on problems of evolution in the planetary system. Little did any of us foresee that in less than two years this brilliant son of a brilliant father would be cut off by a fell disease, leaving in the ranks of British science a gap that will not soon be filled again. At the same time that the Copley Medal was adjudicated, one of the Royal Medals was assigned to George Chrystal, Professor of Natural Philosophy in the University of Edinburgh, in appreciation of his contributions to mathematical and physical science. At the moment when the President and Council made this award it was not known that the illustrious mathematician was then lying on his deathbed. He died before the announcement of our intention could reach him. The Council felt that the award should not be cancelled, but that the King, as the Royal donor of the Medal, should be appealed to for permission to transmit it to the family of the deceased philosopher. To this request His Majesty at once gave his consent, together with a request that an assurance of his sincere sympathy should be conveyed to the family. The last time when it fell to me to present the Copley Medal (1913) I had again to place it in the hands of an old friend, Sir Edwin Ray

Lankester, to whom, as the most eminent zoologist in the country, it had been adjudged.

It has long been one of the usages of the Royal Society to hold two evening receptions in the course of the early summer, one for the male sex alone, the other to which ladies are invited. These gatherings are always held at the Society's apartments in Burlington House. A marked feature of them is the display of some of the most recent inventions and discoveries in science and its applications. The meeting-rooms, with their portrait-covered walls and the great library with its ranges of books, never look so attractive as on these occasions, when they are brilliantly lit up with electric light, and filled with a cheerful crowd in evening dress and in uniform. On the ladies' evenings the President, who receives the guests, is usually accompanied by wife or daughter who assists in the reception. For me it was unhappily impossible that my invalid wife should play her part, but my eldest daughter, with not a little trepidation, undertook the duty.

One of the most arduous duties which the President and Council of the Royal Society have each year to perform is that of selecting from the number of candidates for admission into the Society those whom they consider to have the highest claims. The number to be so selected is limited to fifteen annually. Even if the list of candidates were short it would often be a delicate task to decide between rival claimants. But as the waiting list now exceeds one hundred, the task has become anxiously difficult. Since the Society embraces the whole range of "natural knowledge," it is obviously desirable that each great department of science should be adequately represented in the membership. Hence those departments which have suffered the heaviest losses have special claims for consideration. It has been found to be a convenient practice to tabulate the names of the candidates according to the branches of science to which they have individually devoted themselves, and each member of Council is supplied with a copy of this tabulation before the first meeting for the preliminary consideration of the selection. The representatives of each department of science have opportunities of conferring together and considering which of the candidates in that science they should recommend to the Council. It usually happens that the greater portion of the fifteen is secured in this way without much difference of opinion. But one or two vacancies generally remain which may give rise to prolonged discussion. It has often been found convenient to find a place there for a man who, though hardly perhaps to be counted as a man of science, has distinguished himself by his zeal for the furtherance of scientific progress or by his eminence in some of the practical applications of science, or whose election would obviously be for the benefit of the Society. For at least two centuries the Fellowship of the Royal Society was not restricted to men of science, but was liberally conferred on eminent churchmen, lawyers, statesmen, soldiers, sailors, artists, and men of letters, and unfortunately, even on individuals who had no valid claim of any kind, but whose ambition it was to

> Shine in the dignity of F.R.S.

The tendency since the middle of last century has been to restrict the election more and more to men who have attained eminence in the prosecution of original work in some branch of science. There can be no doubt that this tendency has been greatly to the advantage of the Society. But it may perhaps be carried too far. If it should entirely exclude " the odd man," as he was called, in the selection of the fifteen, it would deprive the Society of a class of men who from the very beginning have been to it a source of strength.

In my opinion it is not desirable that the Royal Society should become entirely a professorial body. That it must be recruited mainly from the ranks of the teachers of science in our universities, colleges and schools is obvious, for these ranks include the great majority of the active men of original research in the country. But my own experience convinces me that the Society needs the presence of a small leaven of men of affairs, more worldly-wise than most men immersed in the engrossing prosecution of original research. The advice and assistance of such men may be of much advantage in the broad questions of policy which the Royal Society, from its leading and representative position in science, is called upon to consider. I sincerely hope that the practice will continue to be observed, as provided in the Statutes, whereby " persons who either have rendered conspicuous service to the cause of science, or are such that their election would be of signal benefit to the Society," may from time to time be added to the membership.

During my Presidentship I did my best to promote the occasional election of such outsiders. There was one occasion when a distinguished statesman's name was mentioned at a meeting of the Council as one of those who " in the interests of the advancement of natural knowledge " might well be proposed for the Fellowship. The late Duke of Northumberland, who was himself a conspicuous example of the kind of man contemplated in the Statutes, was then in the Council and was present when the statesman's name was brought forward. A view had been expressed by a member of Council that a man in the position of the person in question had already so many honorary distinctions that it was hardly any compliment to offer him the F.R.S. The same evening the Duke promptly sent a note to me in which he wrote : " I did not like to say anything, but I did not at all agree with a remark which was made at the meeting of the Council to-day when the statesman's name was mentioned, to the effect that membership of the Royal Society was a matter of comparative indifference to those who already enjoyed a plethora of other honours. It seems to me that while the recognition of learned Societies must be peculiarly gratifying to scientific men, and titles and honours to politicians, as being in each case the testimony of those best qualified to judge of their work ; on the other hand, favours granted by the State to men of science, and by scientific bodies to politicians and men of affairs, are a sign that the world at large, and not merely their own coterie, appreciates the services which they have rendered. And therefore, 'F.R.S.' being, as it is, the blue ribbon of science, must have a special importance in the eyes even of the most distinguished recipient." The candidate was duly elected.

It is probably not generally known how many are the public institutions wherein the Royal Society is represented upon the governing body. They number some thirty in all. Some of them are educational. Thus they include by Act of Parliament the ten chief public schools in England. From 1892, for example, I continued for thirty years to be the delegate of the Society in the governing body of Harrow School (p. 251). Even among the Universities, the Society has a voice in the election of the Lowndean Professor of Astronomy and Geometry at Cambridge, and sends a member to the University Courts of Liverpool and Bristol. In the state-aided institutions, such as the Greenwich Observatory and the British Museum, the Society has been assigned a place in the administration. As already mentioned, the President is Chairman of the Board of Visitors of the Royal Observatory, Greenwich, and in that capacity presides at the annual visitation of the Observatory by the Board. This visitation is always an interesting event. The Astronomer Royal then gives an account of the varied work of the Observatory during the past year, and an opportunity is afforded to meet the higher members of the staff at the luncheon which makes an agreeable interlude in the business. The examination of instruments and of the photographic work of the past year is sometimes a most fascinating part of the employment of the day.

The President is *ex officio* a Trustee of the British Museum. When in that capacity I first attended a meeting of the Trustees there had been for some years an intermittent but persistent agitation on the part of some prominent naturalists in this country against the system of administration of the Museum. A complaint was made that the Natural History branch in Cromwell Road was unduly fettered by the control of the older establishment in Bloomsbury. As the Trustees hold their meetings alternately at each of these centres, I made it my duty to enquire into the matter in dispute. The alleged grievances had been so constantly and so forcibly reiterated that I was prepared to find them resting on substantial grounds. The officials at both of the Museums were old acquaintances from whom I had no difficulty in eliciting the actual facts. The result was to convince me that the complaints arose from misapprehension or ignorance, and that there was absolutely no justification for them.

When a fresh outbreak of dissatisfaction arose in the public press, the Trustees, who were aware that I had been looking into the subject, commissioned the Speaker of the House of Commons, who is one of the three Principal Trustees, to write a letter to me stating precisely the points on which they wished my opinion in regard to the relations between the two Museums. This letter and my reply were written expressly for the purpose of being sent to the public press, and they appeared in the *Times* of 13th December, 1909. I stated frankly and fully the result to which my investigation had led me, and concluded by "asserting that I know of no establishment, either in this or any other country, wherein more favourable conditions have been provided for harmonious and effective co-operation in scientific work than have been devised by the Board of Trustees of the British Museum for the administration of the important departments committed to their care." There was a reply in the newspaper from one of the malcontents, of which I took no notice. The perfect harmony which for many years past has subsisted between the two Museums is the best disproof of the alleged grievances.

The representative of the Royal Society among the Trustees of the British Museum is one of a Board numbering some fifty members and meeting once every three months. At these quarterly meetings a brief resumé is given of the affairs of the Museum since the previous meeting, but as a rule little or no fresh business is then brought forward. The real governing body is the Standing Committee, consisting of twenty Trustees, who meet every fortnight during most of the year. Unless, therefore, the President of the Royal Society should be elected into this Committee he has practically little or no voice in the general management of the institution, and, finding nothing required of him but to listen while the Director reads a report of work already done, he is apt to give up attendance at the meetings; and, as he ceases to be a Trustee on the expiry of the five years of his Presidentship, he cannot be greatly blamed if, in the midst of his many duties and engagements, he allows his ornamental Trusteeship to fall aside. Fortunately for me I was saved from this pardonable neglect. The Director of the Museum intimated to me that on 10th December, 1910, I had been elected to fill the vacancy on the Standing Committee caused by the retirement of the Prince of Wales on succeeding to the Throne. He accompanied this intimation with a kindly expression of his own satisfaction that I would " not only strengthen the representation of natural science but would also be in sympathy with the literary and archaeological side of the Museum." Both of the departments had for me the strongest attraction. And my experience of the work of the Standing Committee much enhanced that interest, and made me realise how great would be my loss when at the end of my tenure of the Chair of the Royal Society I should cease to be a Trustee of the Museum. This rupture was happily for me prevented. On 2nd July, 1913, that is five months before the end of the Presidentship, official intimation came that the Electing Trustees of the Museum had elected me a Trustee to fill the vacancy caused by the death of Lord Avebury. Thus, as a permanent Trustee and as a member of the Standing Committee, I continued to attend the meetings and to take the keenest pleasure in the varied interests of the two Museums and of their officers, until the advance of old age has made frequent journeys to London hardly possible.

One of the most interesting and important of the institutions of which in my time the President of the

Royal Society was Chairman was the National Physical Laboratory—a great establishment due to the initiation of a few Fellows of the Royal Society, and most of all to the advice and strong support of Lord Rayleigh. The annual visitation of the Laboratory at Bushy was one of the pleasantest functions which the President had to attend. It was interesting each year to note the persistent growth of the place, fresh departments of scientific work being continuously embraced within the scope of an able staff, under the superintendance and inspiration of the Director, Dr. (now Sir) Richard T. Glazebrook. Perhaps the most novel and impressive addition to the appliances of the laboratory, made while I presided, was the Naval Tank, so indispensable for experiments on the models of ships, which was generously presented for the service of the nation by Mr. (now Sir) Alfred Yarrow, Bart., the well-known marine engineer and shipbuilder. The number of new departments of investigation has been largely augmented in recent. years, more particularly during the course of the Great War, where the services of the Laboratory have been invaluable. From modest beginnings, this institution has risen rapidly until it has now established its position as the supreme scientific court of appeal and advice, on all questions involving the physical properties of matter, the strength and quality of materials, gauges and standards. It has, however, become too vast an establishment to be adequately supervised and audited by the Royal Society. Accordingly from 1st April, 1918, it has been transferred to the new Department of Scientific and Industrial Research.

Reference has already been made to the share taken by the Royal Society in the organisation of the National Antarctic Expedition of 1901-1904 (p. 348). Similar co-operation was freely given by the Society to the preparations for Captain Scott's second expedition, and with these preparations I was closely associated during the Presidentship. I had come to be on terms of friendly intimacy with some of the officers who were to conduct this fresh raid into the mysteries of Antarctica. It was impossible not to realise that when they sailed we were perhaps parting with these friends for the last time. With Dr. Wilson in particular, as already mentioned, I had formed close ties of sympathy. Several letters from him, written during the outward voyage and the preliminary months on land, were full of eager enthusiasm and hope. His last letter to me was a long one written at M'Murdo Sound on 31st October, 1911, the day before the start was made for the last and fatal journey. It gave an interesting summary of what the various observers had accomplished up till then, and concluded with the remark: "but we have still much to do, and every one is anxious for another year to do it." He was buoyant over the prospect of the Barrier journey and what was to be done afterwards. Brave souls! He and his companions reached their goal, but perished in the snow on their way back. They were found eight months afterwards in the attitude of sleep, Scott with his arm lying across Wilson. The preparation and publication of the scientific results of the researches in which these intrepid explorers sacrificed their lives have been in progress for some years, though greatly hindered by the War; but the series of notable volumes, now in course of publication by the Trustees of the British Museum, will form a fitting memorial of one great department of the scientific work which they accomplished.

Sometimes duties of a more personal kind are imposed on the President of the Royal Society. Thus he is from time to time appealed to by the Prime Minister for information as to the work and standing of deceased men of science who have left widows and families on whose behalf the head of the Government is solicited for pensions from the Civil List. In all cases great care is taken to obtain accurate information for the guidance of the Minister. Now and then the President may himself be the petitioner for such state assistance. Thus, when Sir William Huggins died, leaving his widow with somewhat narrowed means, the Royal Society sent a well-signed memorial to Mr. Asquith, urging that a pension should be given to Lady Huggins in appreciation of her husband's great services to science, and of her share in them. This request, granted with commendable speed, greatly cheered the widow, who, however, did not live long to enjoy the Royal bounty.

Among the visible links that connect the Royal Society with its illustrious past, none are more important, or at least more generally appreciated, than the portraits in oil of former distinguished Fellows which adorn the walls of Burlington House, and form a noble series of Presidents, Secretaries and other notable men, including a few foreign men of mark. Many of these pictures are of value as works of art, for they include canvasses by Kneller, Hogarth, Reynolds, Lawrence, G. F. Watts, John Collier, H. von Herkomer, George Reid, W. Q. Orchardson and other artists of note. This important collection had remained for many years uncleaned, not a few of the portraits being unglazed; so that they had become dull and begrimed. With the ready assent of the Council, and with the valuable expert aid of Sir Walter Armstrong, we had every picture carefully examined and cleaned by Mr. A. H. Buttery, and all the more important of them covered with glass. It was astonishing to find how many new features were revealed in some of them after the crust of London dust had been removed. They will now probably remain clear and bright for a decade or two until some future President and Council may find it necessary to repeat the treatment.

When the Germans during the War began to drop bombs on London, the walls of the Royal Society were promptly stripped of their portraits, which were taken down to the underworld of the building, where they remained until after the Armistice. The enemy's air-raids never reached them, and they have been re-hung, without having suffered perceptible injury, either during the transference, or from having spent four years in the damp darkness of a London basement.

The history of the Royal Society and the record of its relations to the public life of the times through which it has lived have always had for me a great fascination. One welcomed the occurrence of any incident in current life that recalled the early days of the Society, when its home lay in the city of London, when its rooms were in Gresham College, and when it had close and friendly relations with the Corporation and the City Companies. Several such incidents occurred during my period of office. One of the most interesting of them was the laying of the foundation-stone of the new Gresham College on 24th July, 1912, to which ceremony the Mercers' Company, in remembrance of old times, did not omit to invite the President, who at the luncheon alluded to the hospitality with which the Royal Society, in its earliest infancy, was accommodated in the original college of Sir Thomas Gresham, where for years it held its meetings, conducted its experiments, and accumulated its growing collection of "rarities."

Again, in the spring of 1911, the President of the Royal Society was invited as representative of science to take part in the celebration of the Tercentenary of the authorised version of the English Bible. It was not intended to concentrate special attention on that particular translation, but rather to deal with the broader and deeper subject of the Bible in our national and individual life. Having warm sympathy with the object in view I gladly consented. Apart from the deep religious value of the book, its importance as a monument of "English undefyled," and as one of the great literary triumphs of our language, deserved every recognition that could be given. A representative deputation waited on the King, who took a kindly interest in the celebration. The Stationers' Company entertained the members of the deputation in their picturesque hall, with all its relics and associations.

Reference may be made to one further link connecting the Royal Society with the past. The house in St. Martin's Street where Sir Isaac Newton lived from 1710 to 1725 had long been marked by one of the little porcelain tablets which are now piously affixed to residences wherein notable inmates have dwelt. I used often to turn out of Trafalgar Square in order to look at the old building, which was still inhabited. Its association with the most famous man who ever filled the Chair of the Royal Society was not its only attraction. After Newton's time it was for some years the home of Fanny Burney. To the small-paned wooden turret on the garret, which she believed to have been the great philosopher's observatory, she used to betake herself for quiet. "It is my favourite sitting-place," she said, "where I can retire to read or write any of my fancies or vagaries."

But the day came which saw the house tenantless and dilapidated. Along with other sympathisers, I tried to have it rescued from destruction, and preserved as a valuable memorial of the past. Eventually I enlisted the London County Council in the effort to save it. But with every desire to rescue it, they found this impossible, for after careful examination the premises were found to be in such a bad and even dangerous condition, that any attempt to preserve them would have been futile. I learnt that some parts of the woodwork of the interior, particularly the staircase, were carefully removed with the view of being erected elsewhere in memory of Newton. Mr. Rupert Potter, brother-in-law of Sir Henry Roscoe, took some excellent photographs of the building before it was pulled down.

The most outstanding incident in the history of the Royal Society during my Presidentship was the celebration of its two hundred and fiftieth anniversary in the summer of 1912. Early in the previous year the Council had been reminded that this date was approaching; and after full consideration they resolved that the occasion was one which deserved to be observed in some worthy way. Having regard to the recognised position of the Society among the Academies of the world, and to the friendly relations which we had always maintained with these institutions, and with men of science in every country, we determined to invite delegates from the universities and learned societies of the Old and New World to celebrate with us our 250th birthday. Illuminated cards of invitation, expressed in Latin, were sent out in January, and the festival was arranged to extend from Monday 15th to Thursday 18th July.

Besides actively sharing in the general preparations, I specially undertook to compile and edit a new and enlarged edition of the Society's *Record*, which was to be ready before the commencement of the festival. Completed in good time, and forming a quarto volume of nearly 500 pages, it contained an account of the foundation and early history of the Society, together with the original text of all its Charters, Statutes and Trusts, and lists of its Benefactors, Presidents, and other officers from the beginning down to the date of publication, also an account of the Library, Portraits and Busts. Full details were given of the existing Committees and Boards, also, as perhaps the most valuable part of the book, a Chronological Register of the Fellows was included, showing the dates of their election from the first year of the Society's existence to 1912. In the preparation of this volume important contributions were made by Sir Alfred B. Kempe (Treasurer) and the clerical staff.

I was particularly desirous also that our Charter-book should be reproduced in facsimile in good time for the Anniversary. As already stated, this venerable volume contains on its first page the original signature of Charles II., our Founder, followed in later pages by those of his royal successors on the Throne, together with those of the Fellows and many of the foreign members during the space of two centuries and a half. The vellum leaves of this valuable historical register were admirably photographed at the Society's rooms in Burlington House, under the care of the Oxford University Press. The resulting large folio of 150 pages contains a collection of autographs probably without an equal, as representative of the science and culture of Europe during the period which it embraces. The volume also includes a facsimile of the two pages of the original Resolution of 5th December, 1660, to found a Society " to consult and debate concerning the promoting of Experimentall Learning," together with the 115 signatures attached thereto. These signatures show that the investigators, by whose energy and enthusiasm the movement to start the Society was set on foot, were a mere handful of men who had gathered around them a large company of sympathisers from the social circles of the time. The literary element was conspicuously present; it included John Dryden and the most noted poets and dramatists of the day. This widely representative character of the Fellowship of the Royal Society has been preserved down to my own time. I find on the page which contains my signature those of Alfred Tennyson, Arthur Penrhyn Stanley, Lord Dufferin, David Livingstone and Frederick William Farrar.

Before the opening of our festival a copy of each of the two volumes was presented by the President and Treasurer to King George at Buckingham Palace, who as he turned over the pages expressed the pleasure with which he received them, and at the same time his good wishes for the success of our celebration.[1]

The proceedings began by an informal reception on the evening of Monday, 15th July, being the anniversary of the day on which the Charter of Incorporation of the Royal Society passed the Great Seal in 1662, therefore strictly the Society's birthday. Next morning the proceedings were inaugurated by a short commemorative service in the ancient Abbey of Westminster, when the Dean (Bishop Ryle) read a short but impressive address in which " in the name of the whole world of contemporary Christian thought," he gave " expression to the gratitude which, as a rule, the clergy have little opportunity to render, for the amazing enrichment of human thought which has resulted from the patient researches of Natural Science, during the past two hundred and fifty, and in particular during the past eighty, years." " We are convinced," he said, " that the discoveries of Science discharge a truly prophetic office in making known to

[1] An account of the proceedings at this Festival was printed for the Royal Society, and published in 1913 in a thin quarto volume with the title, *The Celebration of the Two Hundred and Fiftieth Anniversary of the Royal Society of London, July 15-19, 1912.*

mankind the facts of the Universe, in which we believe that we may read the record of the Will of the Supreme Mind."

In the afternoon of the same day, the formal reception of the visitors took place in the Society's Great Library, the delegates being placed in alphabetical order according to the country represented. The President gave an address welcoming the representatives of the Universities and Academies from all parts of the civilised world. Thereafter the delegates presented their missives, which were too numerous to be read at the time, so that the representatives of each country were asked to name one of their number to convey their felicitations in a few words. Every country in Europe was represented, and especially numerous and eminent were the delegates from the United States of America. The universities, scientific societies and institutions of the British Isles sent a large and distinguished band of representatives—a welcome tribute to the place of the Royal Society in the history of science in this country. Our Dominions and Dependencies were also well represented. The eighteen German universities which sent delegates, instead of each presenting a separate address, with great sense and good taste combined to offer a large bronze tablet with a Latin inscription recording the congratulations of the German universities "to the Royal Society, the illustrious cultivator and protectress of the Sciences, on the happy completion of 250 years."

On the evening of this first day a commemorative dinner was given. In making the arrangements for this occasion I wished to recall some of the early associations of the Society. In the first place, from the civic authorities the privilege was obtained of having the dinner in the spacious and venerable Guildhall of the City of London. The Lord Mayor and other members of the Corporation were invited, as well as representatives of the City Companies with which, in its Gresham College days, the Society had been most closely in touch. Further, as a reminiscence of the prominent place filled by the dignitaries of the Church in the early Fellowship of the Society, the presence of the Archbishops of Canterbury and York, Cardinal Bourne, the Dean of Westminster, and the Moderator of the General Assembly of the Church of Scotland was secured. The Government of the day was represented by the Prime Minister and several of his colleagues. The Lord Chief Justice and many other legal luminaries brought to remembrance the many eminent lawyers who were among the early supporters of the Society. The whole company numbered nearly 500. The Prime Minister (Mr. Asquith) proposed the toast of *The Royal Society*, and the President responded. To the toast of *The Universities at home and abroad*, proposed by Lord Morley, replies were made by Professor Picard for France, Professor Waldeyer for Germany and Professor Winkler for the Netherlands. The Archbishop of Canterbury made a happy speech in proposing *The Learned Societies in the Old World and the New*, which elicited replies from Professor Volterra representing the ancient Accademia dei Lincei of Rome, Prince Galitzin from the Imperial Academy of Sciences of St. Petersburg, and Dr. R. S. Woodward, President of the Carnegie Institution, Washington.

On the foreign guests this banquet evidently made a deep impression. The splendour of the antique Hall, with all its historical associations in the centre of the life of London, the English habits and customs, so distinct in many ways from their own, such as the white-capped and white-clothed attendants who carved the joints at side-tables, left on their minds a picture which they assured us they would never forget. Fortunately the proceedings, every detail of which had been thought out with the greatest care, passed off without, so far as I could see or hear, a single hitch.

The morning of Wednesday 17th was devoted to sight-seeing, especially among the Museums and Art Galleries. The Archbishop of Canterbury was good enough to conduct a party over the historical parts of Lambeth Palace. In the afternoon the Duke and

Duchess of Northumberland received the delegates and Fellows with their ladies at a garden party at Syon House, their residence on the Thames where, in brilliant weather, the guests saw something of the summer beauty of an English park. In the evening a conversazione was held in the rooms of the Royal Society, in the midst of its large assemblage of portraits of successive generations of Presidents and other eminent Fellows, and the various historical relics preserved at Burlington House. Besides these records of the past, some novel pieces of scientific apparatus were shown in practical working.

On the afternoon of Thursday 18th the King and Queen gave a garden party at Windsor Castle. After passing through St. George's Chapel and the State apartments, the company assembled on the Terrace, where the President, Treasurer and Secretaries were presented to their Majesties by the Lord Chamberlain, and then the delegates grouped in countries were presented in succession by the President to the King and Queen. This ceremony over, the company dispersed among the large number of other guests invited for the occasion. The evening was given to a series of farewell dinners, at which the foreign delegates and their ladies could see a little of the domestic life of English homes.

Though the series of functions arranged by the Royal Society was now completed, the festival was prolonged next day by the Universities of Oxford and Cambridge, each of which invited a large company of Fellows and delegates. Besides luncheons and garden parties, a graduation ceremonial was arranged at each place, and a few of the more eminent foreigners received honorary degrees.

The many letters which, after the celebration was over, came from the guests both at home and abroad, amply confirmed our own conviction that the festival had been highly successful. The German delegates were notably enthusiastic, and expressed their hearty wish that the festival might inaugurate a new epoch of splendid activity for the Royal Society. It happened, a few months later that the Royal Society had a further proof of German goodwill. On the following St. Andrew's Day, presiding at the Society's anniversary dinner, I placed on my right hand Prince Lichnowsky, the newly appointed Ambassador of Germany at the Court of St. James. It was the first public dinner at which he had been present since his arrival in London. In the course of conversation with him I found that he would not be unwilling to address the audience, though we had not meant to call upon him for a speech, and as it seemed to me that it might be of advantage to the political relations between the two countries if, at the outset of his career in England, he were given an opportunity to speak to the English public, I interpolated in the programme the toast of his health, which was warmly received by the company. He spoke excellent idiomatic English, and told me that he had been accustomed to speak the language and to read English literature since his boyhood. In the course of his reply he said:

"Hume was the predecessor of Kant and Schopenhauer, and I do not believe that in any country in the world are Shakespeare and Byron more fully appreciated or deeply understood than in Germany. I am confident that this close intellectual connection will, in the future as in the past, be a powerful aid to the efforts of all those who work for the establishment of good understanding and harmony between our two kindred peoples. . . . Most luckily I can avail myself of this opportunity of stating that never between England and Germany have there been more intimate and more sincere relations than at present. Both countries are working side by side in the same cause of maintaining European peace."

It is impossible to doubt that these words expressed the Prince's true sentiments and honest belief. But events have since shown that, unknown to him, the higher powers of his country were all the while busily preparing for the war which only twenty months later they launched upon the world.

So far from there being among us at that time any

widespread conviction that Germany was bent on war, various active agencies were at work to promote cordial relations with that country. Thus the organisation of the "Albert Committee" was specially designed to promote friendly feelings towards our German cousins. Lord Avebury was its President, and in 1910 I became a member. At the same time there were far-seeing prophets who, pointing to the gigantic growth of the German army and navy, warned us to prepare and be on our guard.

The International Association of Academies met at Rome in May 1910. Having been nominated one of the delegates of the Royal Society to this meeting I induced Sir David Prain, another of the delegates, and Lady Prain to travel with me direct to Naples in order to have a few days there before the sittings of the Association began in Rome. We made some pleasant excursions, and one morning I attended a meeting of the Physical and Mathematical Section of the Royal Society of Naples, whereof I had recently been elected one of the eight foreign members to fill the place made vacant by the death of Albert Gaudry. The proceedings interested me in being so different from those of our Societies at home. The question of the publication of papers which had been read at a previous meeting was discussed, and the vote was then taken by ballot. The chairman having voted, he pushed forward to me, seated at his right hand, two saucers, one containing white balls, the other black, and the ballot box was placed in front of me that I might exercise my right to vote. In vain did I protest that I knew nothing about the papers, and could not legitimately vote either for or against them. But at last, not to hinder the march of the ballot box round the table, I boldly lifted a white ball and dropped it into the box. On my right hand, the only lady member of the Society had taken her seat. She was bright, intelligent and gracious in manner, a Russian, Marussia Bakunin by name, who had married Professor Oglialoro-Todaro of the Neapolitan University, and was elected into the Society in recognition of the excellence of her contributions to chemistry. While the voting was going on I described to her our manner of balloting which conceals on which side the vote is given. I had never before seen an actual black ball at a ballot, though familiar with the practice of " black-balling." I was presented with both a white and a black ball as a souvenir of voting in Italy.

At the end of the meeting the President of the Royal Society of Naples and Rector of the University, Pasquale del Pezzo, Duca di Cajanello, took me all over the new buildings of the University, then in course of reconstruction. Some of the old, cramped and dilapidated houses were still standing, side by side with premises far more commodious, if not quite so picturesque. When the whole of the new buildings are completed the University, installed with all modern conveniences for teaching, will doubtless take a notable leap forward among the educational establishments of Italy. Another memento of this visit to Naples reached me at the end of June, in the shape of a formal letter from the Secretary of the Academy of the Physical and Mathematical Sciences, enclosing a postal order for fifteen lire due to me from this Academy for my attendance at the meeting on 7th May.

Before my travelling companions and I left Naples for Rome a telegram arrived on the morning of 8th May announcing the death of King Edward. The news came as a great shock to the English community, who were unaware of the seriousness of his illness. The event necessarily cast a gloom over the rest of our stay abroad. It was touching to see, as we journeyed northward to Rome, how promptly and widely our national loss evoked Italian sympathy, as shown everywhere by flags flying at half-mast.

In Rome many entertainments had been arranged for the members of the International Association of Academies; but these the British delegates were naturally prevented from attending. We went to only two functions—a reception by the King and Queen of Italy and another by Queen Margherita. Their

Majesties were full of sympathy in the national loss and of admiration for the ability and sagacity of the late King. The King of Italy was much interested to find that the Director of Kew Gardens was one of our number, and in the course of conversation he showed a good deal of botanical knowledge. There were some Italian plants of which specimens were wanting at Kew, and these he promised to procure and send. Queen Margherita spoke with much feeling of the Prince of Wales, and expressed her confidence that he would be an excellent and devoted sovereign.

The business of the International Association took up most of each morning during our stay in Rome, and some useful work was done. But again, not the least valuable aspect of the meeting was the opportunity which it gave for friendly personal intercourse among the delegates. There was always a special pleasure in meeting those with whom we had come to form a warm attachment but whom we so seldom saw. At the close of the meeting Professor Larmor and I started homeward the same evening, and reached London two days before the King's funeral. I found a missive from the Earl Marshal, summoning the President of the Royal Society " to assist at the interment of His late Most Sacred Majesty of blessed memory, in the Royal Chapel of St. George at Windsor." This chapel is always an impressive place, but it probably never looks so overpowering as on the occasion of a royal interment. Although the company was large, the silence and stillness of the scene were well maintained. The feeling of sorrow for the great national loss seemed to be expressed on every face. On the return journey the train by which I travelled was so crowded that, having lingered behind, I had difficulty in finding a seat, till I heard my name called out from one of the carriages as I passed its door. There was still room in it for one more. I there found myself in the midst of a group of notable politicians—Lord and Lady Crewe, Sir Edward Grey, Mr. Haldane, Colonel Seely, Mr. Robson and Mr. Rufus Isaacs.

A few weeks after the accession of King George V. to the throne, it was again the duty of the Royal Society to approach the Sovereign. The President and one of the Vice-Presidents attended a Court, and there presented an address conveying the Society's sorrow for the death of King Edward, and its wishes for the prosperity of the new monarch and his Queen. His Majesty, in a graceful reply, said: "Your words of appreciation of the late King are very welcome to me. He always regarded with the deepest interest those scientific discoveries already made which have been of such supreme importance in the advancement of civilisation. I also have watched with close attention the work of your Society; and it is my sincere hope that its prosperity will continue, and that a Fellowship of the Royal Society will always be esteemed one of the highest honours which can be earned by devotion to the cause of science. I desire to thank you most cordially for your congratulations on my accession to the Throne, and to assure you of my sympathy and support in your beneficent efforts for the promotion of natural knowledge. I gladly accede to your request that I should inscribe my name as Patron in your Charter-book."

Embracing the whole domain of "natural knowledge" within its purview, the Royal Society remained for more than a hundred years the one great central organisation for the prosecution of scientific research in Britain. But as interest in the study of Nature increased throughout the community, the need was felt for the creation of other centres wherein those who carried on original investigation could meet for mutual discussion and assistance. Especially was this need manifest at a distance from London, for in those days travelling facilities were still undeveloped, and a journey to the seat of the Royal Society was often a serious undertaking, involving considerable expense of time and money. One of the earliest local societies to come into existence was the Royal Society of Edinburgh, which from 1782 has been the chief scientific community in Scotland. A few years later the Royal Irish Academy was established in Dublin. The continued growth of the different branches of natural science naturally led to the desire to form societies, in each of which one of these branches might be specially cultivated. When these societies began to spring up in London some of the more influential Fellows of the Royal Society regarded them as unnecessary, and encroaching on the domain of the premier Society. One of the earliest of these "upstarts" was the Geological Society, established in 1807. When it proceeded to obtain a local habitation in London, and to act as an independent body, the magnates of the Royal Society were offended, and endeavoured to have it made an assistant Society, subordinate to their venerable institution. The geologists, however, refused to adopt any servitude of the kind. They were careful to show their local character by calling their body the "Geological Society of London," leaving the geologists of the rest of the country to form as many similar unions as they found desirable. The success of their resistance opened the way to the gradual creation of the numerous existing Societies, each devoted to the cultivation of one department of science. The opposition of the conservative element in the Royal Society has long since died out. From the Presidential Chair the admission has been made that the geologists, as well as the scientific world, have good reason to rejoice over the wise and far-seeing policy which left the Geological Society free to grow and to develop its powers, untrammelled by any other body.[1] The Royal Society has retained all her prestige, while her scientific standing has been amply maintained by the members of the younger societies whom generation after generation she has elected into her Fellowship.

So cordial have the relations long been between the premier Society and the younger institutions that the fight for independence by the geologists has been generally forgotten. But it gave the prestige of priority to their Society, which is now universally acknowledged to belong not to London only, but to the whole British

[1] See Sir William Huggins' volume, *The Royal Society, or Science in the State and in the Schools*, p. 47.

Empire. For more than half a century I have been bound to this Society by the closest personal relations. It has twice elected me its President; in 1908, the fiftieth year of my membership, it made me its Foreign Secretary, and ever since that date it has annually renewed this appointment—an honour which I most gratefully appreciate.

Between the Universities and the Royal Society there have always been the friendliest relations. It has even been maintained by some Oxford men that the Society was born in Wadham College. When the tercentenary of that College was celebrated in 1910, the worthy Warden, Mr. P. A. Wright Henderson, in his invitation to the President of the Royal Society to attend the festival, asked him to come to "the house and the room where the Society was, in one sense, founded." At the dinner in the hall on 23rd June, in reply to the toast of "Letters and Science," proposed by Mr. Frederic Harrison, I took the opportunity to clear away the misapprehension about the connection of the College with the Royal Society, showing that the Society was really started in London. Owing, however, to the political troubles of the time some of its prominent members, for the sake of peace and quiet, settled for a time in Oxford, where they were welcomed by one of their fellow-members, the famous Dr. John Wilkins, who was Warden of Wadham College, and who did much to keep the fire of scientific research still burning at Oxford.

The Colleges both at Oxford and Cambridge continue to maintain their ample hospitality, and I recall many pleasant social gatherings in their halls and common-rooms. One of the last of these, while I remained in office at the Royal Society, took place at Oxford in connection with the seventh centenary of Roger Bacon. In 1913 a movement was set on foot to commemorate this anniversary by erecting in Oxford, where he studied, a statue to the memory of this early champion of experimental science; by issuing a memorial volume of essays dealing with various aspects of his work; and, if possible, by arranging for the editing and publication of those of his writings which have never been published. To carry out these objects a large committee of sympathisers was formed, and also a smaller executive committee, with the President of the Royal Society as Chairman. The commemoration was successfully held on 10th June, 1914, and was attended by a number of distinguished men from the great libraries of Europe. One of the most eminent of these visitors, the delegate from the Vatican library, was Signore Ratti, who, after holding some high ecclesiastical appointments was in 1922 elected Pope of Rome as Pius XI.[1] A statue of Roger Bacon, by Mr. Hope Pinker, was unveiled and presented to the University. The interesting memorial volume was ready in time for presentation to the subscribers. The effort towards the collection and publication of Roger Bacon's writings, however, met with only partial success, and the outbreak of war two months later put an abrupt postponement to this as well as to so many other praiseworthy undertakings.

During the same interval of years Cambridge University too had its centennial celebrations. Gonville and Caius College commemorated its four-hundredth birthday, and we had then the pleasure of listening to the long but deeply interesting account given by the learned Dr. Venn of Dr. John Caius, the accomplished Elizabethan physician to whom the College was indebted for its refoundation. The University also celebrated the hundredth anniversary of Charles Darwin. When the Vice-Chancellor invited me to give the Rede Lecture he informed me that it would form part of the proceedings at that anniversary. Obviously the lecture should in some way or other be connected with the illustrious naturalist. As Darwin began his scientific career in the field of geology and was thence gradually drawn into the wider domain which he afterwards made his own, I chose

[1] The Committee ceased to meet after 1914, but by correspondence among the members in March 1922 it was unanimously agreed to send a Latin letter of felicitation to their former associate, the new Pope. His Holiness honoured us by sending a genial reply, also in Latin, dated from the Vatican, 10th October, 1922.

to dwell upon the extent and bearings of the geological work of his earlier years on his evolutional studies.[1]

It was during the five years of my Presidentship that the Royal Society saw the creation of the British Academy—the most important addition to our literary societies which has been made in recent years. This new institution arose in the following way. At the meetings of the International Association of Academies the Royal Society, representing only Science, had only one vote, while foreign Academies, which had not only a science side but also a literary side, had two votes. It was seen that it would be of eminent advantage that the humanistic studies of Britain should also be represented in the International Association. Accordingly a few leaders in the higher departments of literature consulted the Royal Society as to the steps which should be taken to secure this representation. There was no existing incorporation which could be selected as the proper body to be a member of the International Association. Some Fellows of the Royal Society suggested that if a new institution were created for the purpose it might be linked with the Royal Society, somewhat in the manner in which the five Academies in Paris are comprised in one great body as the Institut de France. But the difficulties in the way of such an affiliation were found to be insuperable. The British Academy was accordingly founded as a wholly independent body. But it had no funds beyond the subscriptions of its members, nor any premises in which to establish itself and hold its meetings. Under these serious drawbacks one of the leading men in the new-born Academy came to me to enquire whether, as at least a temporary expedient, his Academy might be allowed to meet in the Royal Society's apartments in Burlington House. Having keen sympathy with the creation of this new institution for the promotion of the higher literary studies, I brought this request before my Council, which willingly gave its consent, and the meetings of the British

[1] "Charles Darwin as Geologist," the Rede Lecture, 24th June, 1909. Cambridge University Press.

Academy have ever since been held in the rooms of the Royal Society.

Not long after the infant British Academy had found a welcome home in the rooms of the Royal Society, the following letter, dated 21st November, 1909, reached me from Mr. S. H. Butcher, one of the founders of the Academy:

"This time it is the Classical Association, not the British Academy for which I write. The Council has always felt indebted to you for the interest you have shown in our work, and now our hope is that you may consent to be nominated President for 1910. The sole duty of the President is to deliver an address during his year of office at the general annual meeting. That meeting will be held not in 1910 but in January 1911 in Liverpool.

"We are specially anxious to make it plain that our Association is in no sense antagonistic to scientific teaching and study. The more consultation and exchange of views there is between classical scholars and men of science, the better, we believe, it will be for both. There is no one who could have so mediating an influence as yourself. It is our earnest hope and wish that you may accede to our request, and do what we think would be a public service. . . . We only ask for one public utterance —no Council meetings to be attended. Your acceptance will win our warm recognition and gratitude."

When I read the first paragraph of this letter it seemed that the writer must have addressed it to me by mistake for some classical scholar. It was true that a winter sojourn in Rome, together with the possession of more leisure time than the life in the Geological Survey afforded, had rekindled my early fondness for Latin literature. I had joined the Classical Association and attended with much interest the meetings of its Council, though merely as a sympathetic onlooker. I had gone back to my favourite Latin authors with much satisfaction; but I had no pretensions to classical scholarship, and no one could feel himself more completely unqualified to preside over a company of classical scholars. On full consideration, however, it appeared to me that if, as Mr. Butcher pressed earnestly, my coming forward as an avowed supporter of classical training would help the cause he had so much at heart, and with which I cordially sympathised, my duty would be to take my courage in both hands and agree to his request.

Interest in Latin literature and archaeology was in the air at this time. It took formal concrete shape on 3rd June, 1910, when a public meeting was held in London at which a new Society was created for the promotion of Roman studies, and to cover a field similar to that occupied by the Hellenic Society in Greek archaeology, art and history. This fresh institution, launched successfully, has continued to increase in membership and usefulness. As one of the original Vice-Presidents and a member of the Council I have been much interested in taking part in its proceedings, and I continue to look forward to a future for it not less prosperous than that of its Hellenic sister.

In considering the theme that might most usefully be chosen as the subject of the address which was to be made to the Classical Association, I sought one that might in some degree connect classical studies with the scientific pursuits which had been my own chief occupation in life. As a student and lover of Nature, I had found enjoyment in trying to trace among the authors of antiquity the mental attitude with which they regarded the landscapes amidst which they lived, and in thus seeking to ascertain how far they expressed feelings that were akin to or at variance with those which prevail in our own day. Eventually I selected as the subject of my discourse the evidence from Latin literature of the extent to which the Romans have expressed an appreciation of Nature, and the forms in which this appreciation found expression in written words during the later decades of the Republic and the first century of the Empire—the golden age of Latin literature. Nothing more than a mere outline of this ample field could be comprised within a single Address. But the tracing of this outline showed how abundant was the material for further treatment, and before the preparation of the address was finished I had resolved to continue the prosecution of the subject, and during the following year, as the result of this pleasant employment, the volume on *The Love of Nature among the Romans* took shape in my hands.

The Address to the Classical Association, and the book which grew out of it, are intimately connected with one of the saddest times in my life. Lord Collins and Mr. Butcher, to whom I looked up as my classical sponsors, and for whom I had an affectionate regard, both died before the date of the meeting at Liverpool. But a far heavier blow fell upon me in the tragic death of my only son. This bereavement came only four weeks before I had to appear at Liverpool. I determined, however, not to fail in fulfilling the duty which I had undertaken, and nothing could exceed the sympathy and kindness of my classical friends in freeing me from most of the duties of President. In like manner I felt that the wisest course for me would be to continue to perform all the service due to the Royal Society and the Geological Society. As a further distraction from my grief, as soon as it was possible, I sat down to write the volume about the Roman appreciation of Nature, planned the year before, hoping that steady work of this kind would keep me from brooding on the premature extinction of a dear young life, brilliant and full of promise. [See Appendix.]

It was not until the spring of the year 1913 that I was able to revisit Italy. After full consideration of the subject, I had resolved to visit and carefully study the landscapes amidst which a few of the Roman poets were born, or where they lived in their maturer years, with the view of discovering how far any influence from their surroundings could be traced in their poetry. It was my intention to return year after year, and to begin with the poetry of Horace in one region and of Catullus in another. I induced my colleague, Sir Joseph Larmor, to accompany me in the first year of this enquiry, when, as we had no congress or other official duties to attend to, we could arrange our movements from day to day. After a short stay in Rome,[1] during which I made a careful study of Horace's Sabine valley, we went on to Naples, chiefly for the purpose of making a traverse of the mountainous region between the Campania and the central mass of the Apennine chain. I had long wished to explore this region of the Abruzzi, with the view of gaining a general impression of its scenery and, if possible, of the relations between the forms of the surface and the underlying geological structures of the Apennines. I believed that the line of railway to Sulmona, one of the great masterpieces of Italian engineering skill, would furnish me with not a little of the information desired. After creeping up the valley of the River Volturno, this line winds among the tortuous depressions of the mountains, crossing many valleys and ravines on lofty viaducts, and piercing the ridges between the valleys by tunnels and cuttings that lay bare the rocks, until it eventually reaches a height of some 2000 feet. We halted for a night at Sulmona, a little town some 1300 feet above the sea, and famous as the birthplace of Ovid. As this poet's verses had fascinated me in my school days, it was interesting to note the physical features which he accurately enumerated in his allusions to his early home. In summer the valley is green and fertile, with many a clear rivule coursing down the declivities; but, when the hills are thickly covered with snow, the place must still deserve its ancient renown for the coldness of its winter climate. It was the abundance of the running water that seems to have been the feature of his birthplace that dwelt most vividly in the poet's memory. He alludes to his birth among the "watery Peligni," and when he revisited his paternal fields it was the salubrity due to their many streamlets

[1] It was an added pleasure to this visit to Rome to meet there my second daughter Elsie, who with one of her school friends was travelling through Italy. She had made a special study of Italian history, and came to the country with a fund of historical knowledge which greatly increased her interest in what she saw. I could act as her guide among the pagan antiquities, but in the tangled complex of Italian civil and ecclesiastical annals she was my teacher. I joined her again at Naples and on the homeward journey at Perugia.

that chiefly claimed his praise. He wrote that as Mantua was proud of Virgil and Verona of Catullus, so he would be called the glory of the Pelignian race. The little town has preserved the memory of its musical poet by naming after him one of its streets, the Corso Ovidio, and by a mediaeval statue of him.

The profusion of running water in the Sulmona valley awakens other thoughts in the modern visitor, who sees in it an immense volume of mechanical force running to waste. The natives have made a beginning indeed in the utilisation of this source of energy, by using it to light Sulmona by electricity. But the time may come when this upland valley, with its abundant source of force utilised, will lose its quiet beauty and be turned into a busy hive of factories and industry, or, as has happened in North Italy, may be converted into a lake from which the stored up energy may be obtained that will work some of the railways of Southern Italy.

From Sulmona we struck westward by Aquila to Assergi, in full view of the whole range of the Gran Sasso d'Italia. This imposing mountain rises in vast grey walls from the upland plain to a height of 9585 feet above the sea, plentifully streaked with snow, and forming a worthy crown to the whole Apennine chain. It was too early in the season to attempt to climb to the summit, even if there had been time for the ascent; but to have come close enough to scan every striking feature on the front of the great peak that overlooks the whole of Central and Southern Italy was well worth the time and trouble of the excursion.

My companion and I went on to Venice, where we unearthed my old friend Horatio F. Brown who, though a leal Scot, and laird of one of the most beautiful estates near Edinburgh, had for years buried himself among the Venetian archives. Having only recently returned from the Lago di Garda, which was a favourite retreat of his and whither we were now bound, he could not at that time go back with us, but he promised to come with me another time, which fortunately he was able to do in the following year. My main object in this detour was to visit Sirmione, the home that Catullus loved, at the southern end of the Lago di Garda. We halted on the way at Verona, the birthplace of this tenderest of Roman poets. With the thought of him in one's mind, and with some of his inimitable lines on one's lips, the hours spent in this old Roman provincial town brought him vividly into memory. The swift Adige still rushes under the arches of the antique bridge, and the eye rests on the same wide, fair and mountain-bound landscape which he loved and to which he could so fondly return,

> Multas per gentes et multa per aequora vectus.

The few hours which we spent on Sirmione enabled us to take a general survey of the topography of the island, and to walk round its coast on the flat platform which in the course of centuries has been eroded by the waves of the lake out of the limestone rocks. We sat for a while among the ruins of the villa perched upon the top of the northern cliff where perhaps stood the poet's dwelling-place. The morning was remarkably fine, and the noble array of mountains that form the background of the landscape could not have been seen more impressively. Every peak and crest gleamed clear and sharp against the blue sky and the distant white clouds which rose in splendid towers of cumulus above the plains. On our way back to Desenzano, in an open boat with oars, we were overtaken midway by a severe thunderstorm with heavy continuous rain against which we could only partially protect ourselves. But the grandeur of the scene fascinated us. The vividness of the lightning, the loudness of the thunder, the thousand reverberations of each peal from the mountains, the inky blackness of the lake lashed by the " big rain," in contrast to its smooth caerulean blue in the morning, combined to form a scene which will never be forgotten.

During my tenure of the Presidentship, at the dinners of the Royal Society Club I used to see much of Sir Andrew Noble. He had purchased a large estate at the upper end of Loch Fyne and built the mansion of Ardkinglas, which he liked to fill with his friends. In the year 1910 he included me as one of his guests, and from that time onward, as almost one of the family, I was welcomed back every year. My daughter Lucy was soon included in the invitation, and these annual visits, with the ever closer-drawn friendship of Lady and Miss Noble, became a cherished landmark in our year. Sir Andrew repaired and enlarged the picturesque old Castle of Dunderave on the opposite side of the loch, which, after his death, became the summer residence of his widow and daughter, who continue there the hereditary hospitality. Amidst scenery so impressive on every side, and with the charms of music and pleasant talk, the days at Dunderave were memorable. But, as Ovid musically sang,

Tempora labuntur, tacitisque senescimus annis.

The five years of the Presidentship of the Royal Society had slipped away, and on St. Andrew's Day, 30th November, 1913, I resigned the Chair to Sir William Crookes, who was elected to succeed me. In my final Address, after thanking the Fellows, the officers and the members of the permanent staff for the uniform helpfulness and cordiality with which they had assisted me in the discharge of the multiform duties of President, I added, " these last five years have been to me a singularly full and happy time, the memory of which will ever be a delightful retrospect."

A few weeks later an unusual and most friendly compliment was paid to me by the American Ambassador, the illustrious and much regretted Walter H. Page. I first met him soon after his arrival in London at a small dinner given in his honour by Major Leonard Darwin at the Athenaeum Club. As we then sat next each other we had much pleasant converse, and an intimacy was begun which soon became a warm friendship. He had learnt that my birthday was the same day as that of the President of the United States, and when the day came round I received from him the following friendly note :

6 Grosvenor Square, W.,
28th December, 1913.

Dear Sir Archibald,

I have my country's flag flying on the roof to-day, for it is the President's birthday; and I hope you will not object to my associating you in my thought and in honour and respect with the President, since you also are entitled to great thanks for being born into a world that you have made wiser and merrier by your presence and labour.

May you remain in it, and keep young for many a happy Christmas.

Very heartily yours,

Walter H. Page.

# CHAPTER XIII

## RETIREMENT IN THE COUNTRY

## (1914–1924)

THE business of the Royal Society having been for ten continuous years my chief occupation, its complete cessation involved a rearrangement of the daily routine of life. There was now no need to reside in London. Some years before this time I had acquired a piece of land on the hill to the south of the village of Haslemere in Surrey, and had built on it a house that was planned by my wife and daughters, commanding a view of some twelve miles across a richly wooded landscape to the blue uplands of the Selborne country. The same group of artists laid out the ground with much taste and skill, converting the original bare meadow into a garden that now has its crops and its shade. To this pleasant country abode we eventually brought all our household effects, and made it our permanent home. The train service being excellent, it was always easy to go to London for the day and return in the evening, until for me the infirmities of age made railway travelling difficult.

It was at this quiet retreat on the last day of 1913 that a special messenger brought me the intimation that " the King had great pleasure in conferring upon me the Order of Merit, in recognition of the services rendered by me to my Country and to the world at large, in the Science of Geology." As in this most exclusive of the Orders the Sovereign retains in his own hands the bestowal of this honour, I deeply appreciated it as the King's personal gift.

It has become the kindly custom of the Fellows of the Royal Society to preserve visible memorials of their Presidents in the form of portraits in oil, which they add to the interesting historical canvasses that form so conspicuous a feature on the walls of Burlington House. In my case, the commission was given to Mr. R. G. Eves, one of the ablest among the younger artists of the day. The portrait when completed was reproduced in photogravure, of which each subscriber received a copy. About the same time another personal memorial, connecting me with the Geological Survey, was set on foot by my old colleague, then Director of that Service, Sir Aubrey Strahan. A subscription, started at home and abroad, enabled him to place in the hands of Professor Lantéri, head of the Royal College of Art at South Kensington, a commission to execute my bust in marble, to be added to the gallery of geologists at the Jermyn Street Museum. At the inauguration of this fine work of art, a duplicate by the same sculptor was presented to me, which is now in the Library of my alma mater, the University of Edinburgh, where it was welcomed as "a speaking likeness."

In the spring of 1914 I lost no time in returning to Italy, with the hope of being able to repeat the visit in succeeding years, making pilgrimages to the homes of a few Latin poets. I began my explorations by a renewed and more thorough study of the valley of the Licenza among the Sabine Hills. This visit convinced me that there cannot be any reasonable doubt that the little picturesque glen of that stream is Horace's *vallis reducta*, and that the site of the villa, which has been unearthed, is the only likely spot on which his dwelling could be placed. It is quite conceivable, however, as Professor Sellar pointed out, that the tessellated floors and other evidence of a large mansion may have been the work of a later occupant, who, though he may have been proud to possess the home of the Sabine bard, required an ampler and more luxurious abode than that which sufficed for Horace.

While in Rome I attended a meeting of the Accademia

dei Lincei; went over the new building of the British School at Rome with the Director and Mrs. Strong, as the Committee in London, of which I am a member, had asked me to do; lunched at the Embassy and had a pleasant talk with Sir Rennel Rodd, our Ambassador, about the arrangements of the School; spent an hour or two with my old friend Count Balzani; had the architectural structure of the Baths of Diocletian expounded on the spot by Commendatore Rivoira; accompanied Signore Boni in an underground excursion to see his *Mundus* on the Palatine; visited the recently disclosed *Mithraeum* under the foundations of the Church of San Clemente; motored with Mrs. Mond and party to Porta Prima, and after a busy fortnight left for the Lago di Garda, in order to make a renewed and more detailed study of the home of Catullus at Sirmio. This journey to the north was broken by a halt of two nights, "to see fair Padua, nursery of arts." My pilgrim feet naturally first led me to the University, which, although Gibbon spoke of it as, in his time, "a dying taper," is still, after some seven centuries, a centre of educational life. I saw the frescos of Andrea Mantegna and Giotto, meditated for a while by the shrine of St. Anthony of Padua, and perambulated the streets, not without thoughts of the delightful Paduan society with which Shakespeare has filled "The Taming of the Shrew." Since my visit, Mr. Horatio F. Brown, at the request of the University of Padua, has edited a list of the English and Scottish pilgrims who, during the seventeenth and eighteenth centuries, inscribed their names in the Academic Register. It appears that between the years 1618 and 1765 no fewer than 2038 left their signatures in the volume. It does not appear how many were actual students, working at the University, probably a large proportion were only doing the "grand tour," and visited the renowned University of Padua as a matter of course.

I had arranged with Mr. Brown that we should meet at Desenzano, and make the visit to Sirmione together. We had taken our seats in one of the carriages of the short railway which connects Desenzano with the main line, and in answer to my query about the progress of his researches among the Venetian archives, he launched into a vituperation of one of the mediaeval secretaries who wrote so illegibly as to give him a great deal of trouble which would doubtless continue until replaced by the handwriting of his successor. "I wish he were dead!" exclaimed Mr. Brown, as he thought of it all. There happened to have come into the carriage an English lady who had evidently been listening to our conversation, and when that wish was loudly ejaculated, she broke into the conversation by asking "Don't you think that is a most uncharitable wish of yours?" The Venetian annalist, with his keen sense of humour, was taken aback by this emphatic reproof, and had evidently much difficulty in restraining his laughter, when I intervened and assured the kindly questioner that the individual referred to must have been in his grave already for several centuries.

At Sirmione, where my companion was received as a welcome guest in the little hotel, we spent three days on the islet; walking over the whole of its surface, and boating all round it to see the details of the coast-line, together with the little caves excavated by the waves of the lake. I found that the level of the water was several feet higher than it had been in the previous year, for the beach or platform on which I then walked round it was now under water. The force of the waves generated by a violent northerly gale, blowing along the whole length of the lake, must be a powerful agent in the waste of the rocky shores and islets. Before starting homeward I sailed up to the northern fjord-like end of the lake, and was greatly interested in the many proofs of active erosion that meet the eye. Garda is probably one of the best lakes in Europe in which to study the effects of lacustrine waves. But I did not let geological problems prevent me from trying to group in my mind the various natural features of the landscape most likely to appeal to the poetic eye of Catullus.

The outbreak of the Great War in August 1914 put an abrupt end to my dream of further Italian journeys. I had gathered, however, material enough to form the subject of an essay on each of two Latin poets. As long as the reminiscences had all their freshness I wrote them down. The paper on "Catullus at Sirmio" formed an article in the *Quarterly Review* in December 1914 and that on "Horace at his Sabine Farm" appeared in the same journal for April 1916.

As the summer of the year 1914 advanced and the political sky grew more and more ominous, the portents recalled in my home the anxious days in 1870 when France took up the sword with alacrity and, as we then thought, not without provocation from Germany. But now, while France and Great Britain were striving earnestly to prevent any rupture of the peace, vast armies were mustering on the borders of France and Belgium, and we knew that some of our cousins in France would be in the fight, should war be declared. It was therefore with added personal interest and anxiety that we followed the early stages of the commotion. The stemming of the German torrent by the Battle of the Marne was nowhere welcomed with more earnest thankfulness than in our little France-loving household. By my daughter Elsie, who had been for some months in sadly failing health, the daily bulletins were awaited and scanned with such distressing anxiety that the malady from which she suffered seemed to be given fresh hold on her, cancelling all the progress that the summer had brought. The end came very suddenly; after only a few days in bed she died on the 23rd February, 1915. A more loving, gentle and sympathetic soul could not be, and this sweetness of disposition was combined with a love of literature, more especially in the domain of history, which made her a charming companion. Her death was a grievous loss to our little family circle, and most of all to her invalid mother, who in her increasing feebleness clung to Elsie as her constant help and joy.

A few days later another unlooked-for family bereavement befell us. My brother James died in Edinburgh on 1st March after a brief illness. He succeeded me in the

Chair of Geology at the Edinburgh University, and held it with success for thirty-two years, only retiring from it in the previous year. He had well maintained the renown of the Scottish Geological School, and the loss of his kindly nature and genial companionship was mourned by a wide circle of friends.

My wife, who had never recovered from the shock of her son's death, was still further broken down in spirit by the loss of Elsie. Though wasted to a shadow and too weak to leave her bed, she continued to take a keen interest in the progress of the war. Otherwise her thoughts were mostly in another world with her hopes of meeting again the dear ones she had lost. She lingered in life until 21st January, 1916. The inevitable end had long been in sight, but it could not come at last without a pang of poignant sorrow to us all, softened only by the thought that to her it brought a release from her prolonged sufferings. She had been the light of my life, the gentle and warmly sympathetic sharer of all my joys and sorrows; and it was with feelings, not only of the keenest grief, but of the deepest gratitude and affection that, in accordance with her oft-expressed desire, I laid her ashes in Brookwood Cemetery, in the same grave with those of her beloved son.

To the beginning of 1916 belongs the marriage of my youngest daughter, Gabrielle, to Harold J. Behrens, an officer of a Territorial Battalion of the West Yorkshire Regiment. When he went to the front, she returned to our Haslemere home, and her son, Derick, was born there in February 1917. After the war they settled at Ilkley, Yorkshire, where, as long as I was able to travel, I appeared on an annual visit, and Derick and I became fast friends. Now, as often as possible, he spends a week or two with me at Haslemere; and in the intervals between his visits, he sends me letters full of affection and humour, expressed in a phonetic spelling of his own device. It is one of my greatest joys to watch the expansion of this young mind, and to picture the future of which he gives such promise. (See Plate II.)

In the course of the war a peaceful countryside like that of Haslemere underwent a rapid and widespread change. The familiar idlers disappeared. Work of some kind was found for everybody. Committees were formed for different helpful and benevolent purposes. In the end every available man in the parish was discovered and marched off to the military or other service which he could do either at home or abroad. Not a few of them went to the front, where some sixty-two lost their lives. Large military camps and hospitals were created a few miles to the east and west of the village, which were mainly filled with Canadian soldiers in training for active service. The variety of uniforms to be met with everywhere on the country roads was a never-ending source of wonderment to the native rustic. Especially noteworthy were the kilts, bare legs and variety of tartans of the Colonial Scottish regiments. The Surrey peasantry had never seen or imagined such a uniform. Even a travelled Englishmen could not but start, if, in threading one of the narrow lanes between high hedgerows, he came unexpectedly upon a dozen of these kilted warriors from the other side of the globe.

Haslemere possesses an admirable "Educational Museum," founded and maintained by the eminent surgeon, Sir Jonathan Hutchinson, for the purpose of interesting the population of the district in the common objects of the countryside, and teaching them the elements of natural history. This institution during the War was a great centre of attraction to the soldiers in the surrounding camps. On a Sunday afternoon two hundred of them might be seen intently poring over the contents of the cases, and reading the descriptive labels. Sir Jonathan died in 1913, but without having endowed the Museum. In February 1914, at a public meeting specially called for the purpose, a Committee was formed with the view of endeavouring to collect funds for a permanent endowment. The outbreak of the War six months later put an abrupt end to this endeavour. All that the Committee could then do was to gather funds

Fellow-Students, Haslemere, October 1923

enough each year for the annual upkeep of the Museum, and this arrangement has been followed up to the present time. Having been elected Chairman of the Museum Committee at the start, I have remained at that post, and with my colleagues have so far succeeded in raising, with more or less difficulty, each year the £400 or £500 needed for the efficient upkeep of the institution.

Besides the Museum, Haslemere has a local Natural History Society which was founded some five and thirty years ago and continues in active life. The members annually elect me their President, and I have given them an occasional address. At one of their anniversaries I discoursed on English science and its literary caricaturists in the seventeenth and eighteenth centuries, contrasting the treatment of the men of science in those times with their position to-day, and bringing home to the present generation what the " Virtuosi " had then to endure at the hands of the more noted literary men, illustrated with quotations from the writings of these critics. On another occasion, when the Shakespeare tercentenary was approaching, I gave an address on " The Birds of Shakespeare," which a few months later was published as an illustrated little volume from the admirable press of Messrs. MacLehose of Glasgow.

From the outside world there still came to me in my rural retirement requests to undertake literary work of different kinds. Thus at the beginning of 1916 my associates of the Royal Society Club appealed to me to prepare a new and fuller history of this Club than was contained in an earlier pamphlet which was nearly out of print. The whole archives of the Club were put into my hands, and in looking through them I soon saw that they contained ample material for what was desired. Having now no official position that demanded my time, and realising that a volume of great interest in connection with the history of the Royal Society might be written from the Club's journals and papers, I accepted the proposal of the Club. It came to me at a time of deep domestic sorrow, and offered the kind of employment that could best distract one's thoughts. Seldom has it been my good fortune to find a piece of literary work so congenial as this. From beginning to end it never lost its attraction; it occupied most of my time throughout the year and brought me, as it were, into the social life of successive generations of the Fellows of the Royal Society and their friends during two centuries. The result appeared in the summer of 1917 as an octavo volume of 500 pages, with upwards of three dozen of portraits of men described or referred to in the text.[1]

Having long been impressed with the value of the writings of John Michell, Woodwardian Professor of Geology at Cambridge in 1762, to which, as it seemed to me, but scant justice had been done, and having had occasion to make enquiries about him in connection with his visits to the Royal Society Club, I was led to pursue these enquiries, and to collect all the information about his career which could now be obtained. These materials, together with such illustrative text as I could supply, made a little volume which the Cambridge University Press printed and published. Sir Joseph Larmor was good enough to contribute an appreciation of the value of Michell's papers on astronomical subjects. On the appearance of the book in June 1918 the present occupant of the Woodwardian Chair, Professor Marr, sent me a friendly letter in which he wrote: "It is with great satisfaction that I now find such full justice done to the work of one of my predecessors in the Woodwardian Chair. Though I had long admired Michell's geological work, I had no notion how big a man he really was until I read your book."

Having now withdrawn from the busy world to spend my last years in the quiet of the country, it was with no little surprise that on 1st December, 1919, I received a letter from Mr. Ian Macpherson, head of the Irish Office, informing me that "the Government proposes to appoint

[1] Its full title is *Annals of the Royal Society Club: the Record of a London Dining Club in the Eighteenth and Nineteenth Centuries*. The portraits were obtained from a large collection of prints in the possession of the Royal Society.

a Royal Commission of Inquiry into Dublin University, similar to that which has been established for Oxford and Cambridge, and would feel greatly indebted to you if you would kindly consent to act as Chairman of the Commission." I have already expressed in the foregoing chapters my admiration and affection for Trinity College, Dublin, which for some nineteen years, during my frequent visits to Ireland, was to me always a haven of quiet rest and kindly friendship. If it were possible for me to render any service to that venerable and admirable institution, I could have no greater pleasure than to give it. Yet I could not but doubt whether a man about to enter on his eighty-fifth year ought to accept so onerous an engagement. Mr. Macpherson was good enough to say at the end of his letter that he would be happy to see me at the Irish Office to discuss the proposal. Having to be in London two days after the receipt of his letter, I had the opportunity of talking the matter over with him; Lord French, the Lord Lieutenant of Ireland, being also present. The outcome of the interview was that I agreed to accept the duty.

This Royal Commission met in April 1920 in Trinity College, Dublin, where in the course of a few days we heard statements by the Provost, Bursar, Fellows, Professors and others; made a personal examination of class-rooms, and of the accommodation provided for the different schools, and also visited the Botanic Gardens and the Observatory of Dunsink. We likewise met for a dozen of times at the Irish Office in London. As the result of our enquiries and deliberations we came unanimously to the conviction that the existing resources of the University of Dublin, and of its single College, Trinity, were no longer adequate for its needs. We recommended that they should be increased from public funds by an immediate non-recurrent grant of £113,000, and an annual subsidy of £49,000. Much has happened in Ireland since we finally signed our Report on 12th November, 1920. It remains to be seen how far the Irish Free State will be disposed or able to carry out our recommendations. In an impressive Conclusion the Report affirmed that " this venerable seminary, generation after generation, has continued to train men of all creeds and classes for the business of life, sending forth also distinguished men of affairs, brilliant scholars and eminent men of science. Up till now this beneficent and uninterrupted labour in the interests of the whole community has been accomplished entirely on its own resources. But the great and manifold changes consequent upon the War have made it no longer possible for the College to maintain the same high standard of University training without appealing to the State for material assistance."

When the journey to London was becoming increasingly difficult owing to the advance of old age, I felt it to be my duty, as representative of the Royal Society on the Governing Body of Harrow School, to resign with regret the post which I had filled with pleasure to myself for thirty years. In his reply Lord George Hamilton, Chairman of the Board, was so good as to say : " All the Governors will greatly regret the severance of your connection with them. You have been for so many years their guide and mainstay in all questions relating to the teaching of Science that we shall feel quite adrift without you. Still I think you may leave the Governing Body with the knowledge that it is mainly due to your influence that the improvement in Science teaching during your tenure of office has placed Harrow quite in the forefront of our great Public Schools." Had it not been, however, for a succession of excellent science masters at Harrow, beginning with Mr. Ashford, now head of the Naval College at Dartmouth, the acknowledged eminence of Harrow in the teaching of science would never have been attained. From time to time I visited the science department, had long conferences with the senior science master as to the needs of his department, and was always delighted with the enthusiasm which he and his colleagues showed in their work.

Pleasant it is to continue the correspondence which for

many years I have enjoyed with scientific friends all over the world, though year by year the number of these friends decreases. Pleasant too to be remembered by the Societies and Academies who long years ago enrolled my name among those whom they wished to honour, and who still send me the records of their doings. The recognition amply shown to me in this way by my friends in France has always been specially appreciated. Towards the end of 1917 they conferred on me the highest distinction which they have to bestow—the "Associé Étranger de l'Institut de France," for it is a national compliment, being approved by a decree with the sign manual of the President of the French Republic. It was accompanied with a sheaf of telegrams and letters of congratulation from my kind-hearted French friends. The vacancy which I thus filled had arisen from the death of my old friend and correspondent E. Suess : and the honour of the election was increased in my estimation by my being deemed worthy to succeed him.

Two years later yet another link was added to the chain of kindly relations that has long bound me to France. The Council of the University of Strasbourg, once more restored, with Alsace and Lorraine, to France, arranged to hold a festival at the celebration of this restoration, and to bestow the degree of Honorary Doctor of the University upon a number of foreign men of science. Though I was unable to attend the festival in person, the diploma was duly sent to me.

I close this last page of the Record of a long life, grateful to God that having been privileged to retain a measure of bodily health, with mind and memory still undimmed, I can pass my last days in a home of quiet and restful beauty, tended with all the care which a daughter's love can devise.

# APPENDIX

## IN MEMORY OF RODERICK GEIKIE

Born at Boroughfield, the home of his parents at Edinburgh, on 11th April, 1874, Roderick spent the first eight years of his life, in a happy infancy and childhood, with his sister Lucy. Under the care of an intelligent and affectionate governess they were well grounded in elementary education and learnt to enjoy country walks. After the family removed to London in 1883 he was eventually placed in Rev. Morgan Brown's preparatory school at Hunstanton in Norfolk. His sojourn for several years on that coast was beneficial to him both in mind and body.

At Easter 1887 he entered Harrow School as a home boarder. In the belief that the earlier years of a boy's career at a large public school are best spent by him living under the parental roof, we had provisionally removed to Harrow, where we had a house and large garden, at which his friends among his class-fellows were frequent visitors. Eventually, however, as we realised that the full advantages of the school can hardly be enjoyed by a home boarder, we placed him in the house of F. E. Marshall, of which he advanced to be head in 1891. He became the Head of Harrow School in September 1892, and retained that position till he left for Cambridge at the end of the summer term in the following year. His school prizes were numerous. They included the Gregory Medal for Latin prose, the Pember prize for Greek and Latin Grammar, the Prior prize for Divinity, and the blue ribbon of Harrow, the Gregory Scholarship on leaving school. Of his influence as Head of the School, the headmaster, Dr. Welldon, spoke in the warmest terms of commendation. He gained the respect and esteem of the boys not only from his position in the School, but from his versatile dramatic gifts. On each Speech-day he took a foremost part in the various scenes that were acted in Greek, Latin, French, or English. His familiarity with the French language and French modes of speech enabled him to represent a French inn-keeper or peasant with much humour, and often with appropriate "gag," which added much to the force and the fun of his part.

He entered King's College, Cambridge, in 1893. His original intention had been to take the profession of Engineering, and during his first term he attended the lectures on Mechanism and Applied Mechanics by Professor Ewing, who thought that he showed much promise. But it was soon obvious to the boy, as it had for some time been to his parents, that his tastes and aptitudes lay rather in literary and historical pursuits. He never showed much zeal for Natural Science, though fond of the objects of the countryside, and always interested in his rural rambles. He accordingly read for the History Tripos, and passed in the 1st Class with the B.A. in 1896, and also gained the Winchester Reading Prize. Both at Harrow Chapel and in the King's College Chapel his reading of the lessons was unusually impressive. His talent as an actor found larger scope at the Cambridge Dramatic Union. His rendering of the part of the heroine in Euripides' "Iphigenia in Aulis," acted in Greek, was much praised. In various English and French plays, also, he gained a reputation which extended beyond the university.

He inherited his mother's love of music. While still a small boy in Edinburgh he was taught the violin, and both at Harrow and in London he had further training on the instrument. His violin was a great resource to him, and gave pleasure to those around him. He had also a remarkably fine voice, and sang with taste and feeling.

On leaving College he passed a Civil Service examination with a view to entering the Public Service. There happened in 1897 to be no vacancies in one or other of the Departments which he would have wished to join, so that he had meanwhile to be contented with a place in the Local Government Board. In that office the excellence of his digests of evidence attracted the attention of the head of the Medical Department, Sir Richard Thorne-Thorne, who was confident that he would rise high in the service. But it was desirable that he should have a wider scope for his abilities than the Local Government Board afforded. Through the influence of one of the Governing Body of Harrow School, Sir Matthew White Ridley, then Home Secretary, he was transferred in 1899 to the Colonial Office, where he soon found himself in deeply interesting and congenial work.

Outside his official duties, his leisure hours for more than a year were spent in making researches, partly among the papers preserved by the Duke of Marlborough at Blenheim, on the history of part of Marlborough's wars, with the object of writing an essay on the Barrier Treaty of 1713, as a Thesis for a Fellowship at King's College. He took his M.A. degree, and was elected to the Fellowship in 1900.

His work at the Colonial Office grew increasingly absorbing as years passed, and he threw himself into it with zeal, gaining the approbation of his seniors. The years of Cambridge life had given him many friends, with not a few of whom he was able to keep in touch in London. Endowed with a charm of manner which made him everywhere a favourite, he seemed to be certainly assured of a happy future. Besides the promotion to be expected in the Colonial Office there was always in his mind the possibility, or even, as he would sometimes say, the probability of his being sent on a mission of enquiry to some Crown Colony or other Colonial centre. Meanwhile, his income was more than sufficient for his wants. From the rooms which he rented in Queen's Road, Bayswater, he had the pleasant Hyde Park on his daily walk to Whitehall. In the evenings his books and his violin were companions in his solitude; though at times, when the office work was heavy, he would bring it back with him to his quiet quarters and advance it there.

This active and promising life was brought to a premature close by an accident on the Underground Railway on the morning of December 6, 1910. The shock to the family fell with the severest blow upon his mother, whose health had been failing for some years. Her son was the greatest joy of her life, and he was devoted to her. From this crushing bereavement she never recovered, but lingered in increasing feebleness for five years longer. As she earnestly desired, her ashes were laid at the Brookwood Cemetery in the same grave with those of her beloved son.

Out of the many letters of sympathy and appreciation which came to the family, a few words may be quoted here from three. Dr. Welldon, headmaster of Harrow School, wrote: "The memory of his bright and pure and honourable life at Harrow comes home to me now with a vividness which makes his untimely fate seem the more distressing and the more bewildering. For you and yours I feel more than I dare try to say." Dr. James, Provost of King's College, said: "We were all so fond of Roderick, and so proud of him as a Scholar and Fellow. I cannot attempt to measure what this blow must mean to you all; but whatever comfort is derivable from the memory of his life of kindliness and honour and distinguished achievement, but most of all, of truth and goodness—that is yours, and so too is the assurance of sympathy and sorrow from all here." The head of the Colonial Office, Mr. (afterwards Viscount) Harcourt, in a letter of kindly sympathy wrote: "I have only been Colonial Secretary for a short time; but it was sufficient for me to discover the brilliant abilities of your son, whose work I had especially noticed in the last few days. It may be some consolation to you to know how much he was beloved by all his colleagues."

# INDEX

PRINTED IN GREAT BRITAIN BY ROBERT MACLEHOSE AND CO. LTD.
THE UNIVERSITY PRESS, GLASGOW

Printed in the United States
By Bookmasters